SUPERCONDUCTIVITY

ELEMENTARY TOPICS

SUPERCONDUCTIVITY

ELEMENTARY TOPICS

Keshav N. Shrivastava
School of Physics, University of Hyderabad, India

World Scientific
Singapore • New Jersey • London • Hong Kong

Published by

World Scientific Publishing Co. Pte. Ltd.

P O Box 128, Farrer Road, Singapore 912805

USA office: Suite 1B, 1060 Main Street, River Edge, NJ 07661

UK office: 57 Shelton Street, Covent Garden, London WC2H 9HE

British Library Cataloguing-in-Publication Data
A catalogue record for this book is available from the British Library.

SUPERCONDUCTIVITY
Elementary Topics

ISBN 981-02-4451-7

Printed in Singapore by World Scientific Printers

Preface

Discovered in 1911 by Kammerlingh Onnes, superconductivity is still an important area of research. The research includes fundamentally important theories of pairing of quasiparticles by Bardeen, Cooper and Schrieffer and that of tunneling of electron pairs by Josephson. The single particle tunneling of electrons determines the gap in the dispersion relation as measured by Giaever and Esaki. The scattering of electrons by magnetic atoms which reduces the transition temperature was calculated by Abrikosov and Gorkov. The Gorkov equations of two particle Green functions provided the completeness of the pairing theory. It has been studied extensively both experimentally as well as theoretically and important discoveries are available in the literature, so what new discovery can be made and why study superconductivity? Most of the experimental work required the use of liquid helium so that it was confined to a few countries only. The transition temperatures were so low that only a small range of temperatures was available.

The largest transition temperature known for many years was only 23.2 K but Bednorz and Müller found in 1986 that there should exist a compound of La, Ba, Cu and O with transition temperature of atleast 35 K. During 1987, superconductors with transition temperatures of the order of 100 K were discovered which opened the large temperature range in which the properties of superconductors could be measured. Since the experiments could now be done by use of liquid nitrogen many countries became interested in superconductivity. Large transition temperatures resulted in the debate on well established BCS theory. There are very good books by Schrieffer, Tinkham, Kresin and Wolf and others. The fundamentally important papers have been collected by Bogoliubov and by Pines.

During 1987 a new microwave absorption was discovered which shows that there are oscillations in the imaginary part of the susceptibility. The Meissner effect, which is the expulsion of magnetic lines of force from a superconductor with negative (diamagnetic) susceptibility was discovered in 1933 but in recent past it has been found that in field cooled samples, at small fields, the susceptibility becomes positive. This new phenomenon was called the paramagnetic Meissner effect (PME).The theories of levitation of a superconductor and a magnet became clear only in recent years when this phenomenon was demonstrated easily by use of liquid nitrogen.

The recoilless fraction which measures the mean-square value of the fluctuation in the distance between atoms is found to be changed while going from the normal to the superconducting state. Similarly, the quantities measured from the Mössbauer spectra, such as the second-order Doppler shift and the charge density are affected by pairing of electrons in the superconducting state.

While the d.c. resistivity is zero, it is found that there is a finite a.c. resistivity. Since the superfluid density is proportional to the inverse square of the London penetration depth, the a.c. resistivity has been used to measure the London penetration depth. It has been found that in the mixed state, application of magnetic field leads to a change in the penetration depth. The flux is known to be quantized in units of $\phi_o = hc/2e$ but it has recently been found that the minimum value should be $\phi_o/4$. Fractional quantization of ϕ_o has also been reported. The Knight shift of the nuclear magnetic resonance and relaxation times are related to the coherence length and hence give useful information about superconductors. It is thus quite clear that many new discoveries have occured in recent years.

Sometime ago, after the publication of my article in *Physics Reports* (1991), I decided to reorient the teaching of a course such that a new book on superconductivity could be written, the emphasis being on learning superconductors rather than quantum field theory on which the books are already available. Some of the topics mentioned above have not been discussed by Schrieffer or by Tinkham. Therefore this book has been written. It can be read by scientists interested in superconductivity and can be used as a supplementary reading material by research scholars. It may also be used to teach a course.

Acknowledgements

I am grateful to Prof. C. N. Yang (Fig. 2.3), Prof. David Thouless (Fig. 2.4), Prof. A. Abrikosov (Fig. 3.2), Prof. A. David Caplin (Fig. 1.3), Prof. Michael Tinkham (Fig. 2.5), and Dr. David J. Bishop (Fig. 9.2) for their kind permission to use the figures which they had used. I would also like to thank the publishers of *Physical Review Letters*, *Physical Review*, *Soviet Physics JETP* and *Nature* for the same. The remaining figures have been generated from my own work.

Keshav N. Shrivastava

April 2000

Suggested References

0.1 H. Kammerlingh Onnes, Akad Watenchappen. **14**, 113, 818 (1911).

0.2 J. Bardeen, L.N. Cooper and J.R. Schrieffer, Phys. Rev. **108**, 1175 (1957).

0.3 B.D. Josephson, Phys. Lett. **1**, 251 (1962).

0.4 I. Giaever, Phys. Rev. Lett. **5**, 147, 464 (1960).

0.5 L. Esaki, Phys. Rev. **109**, 603 (1958).

0.6 A.A. Abrikosov and L.P. Gorkov, Sov. Phys. JETP **12**, 243 (1961).

0.7 L.P. Gorkov, Sov. Phys. JETP **34**, 505 (1958).

0.8 L.R. Testardi, J.H. Wernick and W.A. Royer, Solid State Commun. **15**, 1 (1974).

0.9 J.G. Bednorz and K.A. Müller, Z. Phys. B**64**, 189 (1986).

0.10 J.R. Schrieffer, Theory of superconductivity, W.A. Benjamin Inc., Reading, Ma. 1964.

0.11 M. Tinkham, Introduction to superconductivity, McGraw-Hill Inc., Intl. Edition 1996.

0.12 V.Z. Kresin and S.A. Wolf, Fundamentals of superconductivity, Plenum Press, New York, 1990.

0.13 N.N. Bogoliubov, Gordon and Breach, New York, 1968.

0.14 D. Pines, The many-body problem, W.A. Benjamin Inc., Reading, Ma. 1962.

0.15 K.N. Shrivastava, J. Phys. C**20**, L789 (1987); Superconductor Science and Technol. **4**, S 430 (1991); Oscillations in χ''.

0.16 W. Meissner and R. Ochsenfeld, Naturwiss **21**, 787 (1933).

0.17 K.N. Shrivastava, Solid State Commun. **90**, 589 (1994).

0.18 K.N. Shrivastava, Phys. Lett. A**188**, 182 (1994); Discovery of paramagnetic Meissner effect.

0.19 M.W. Coffey, Phys. Rev. B**51**, 424 (1995).

0.20 T.E. Cranshaw, B.W. Dale, G.O. Longworth and C.E. Johnson, Mössbauer spectroscopy and its applications, Cambridge University Press, 1985.

0.21 V.I. Goldanskii and R.H. Herber, Chemical applications of Mössbauer spectroscopy, Acad. Press, New York, 1968.

0.22 L.S. Lingam and K.N. Shrivastava, J. Phys.: Condens. Matter **7**, L 231 (1995).

0.23 P.K. Ghosh, N.M. Krishna and K.N. Shrivastava, J. Phys.: Condens. Matter. **9**, L 663 (1997).

0.24 N.M. Krishna, L.S. Lingam, P.K. Ghosh and K.N. Shrivastava, Physica C**294**, 243 (1998); Minimum flux is $\phi_o/4$.

0.25 H.R. Shea and M. Tinkham, Phys. Rev. Lett. **79**, 2324 (1997).

0.26 P.S. Deo, et al., Phys. Rev. Lett. **79**, 4653 (1997).

0.27 A.J. Millis, H. Monien and D. Pines, Phys. Rev. B**42**, 167 (1990); The use of coherence length to describe the line width is given in an earlier dated paper, K.N. Shrivastava, J. Phys. (Paris) **48**,C8-2239 (1988).

0.28 K.N. Shrivastava, Phys. Rep.**200**, 51 (1991).

0.29 K.N. Shrivastava, Phys. Today **51**[No.9],15 (1998).

Contents

Chapter 1

Introduction

1.1 General properties

In 1911 it was found by Kammerlingh Onnes that the d.c. resistivity of mercury linearly reduces upon cooling until a transition temperature is reached at which it drops to zero as shown in Fig.1.1. The conductivity, which is the inverse of resistivity, tends to infinity below this transition temperature. Therefore, it was called superconductivity. Many pure elements show this phenomenon. The transition temperatures of pure elements are given in Table 1.1 and some of the superconducting compounds are given in Appendix A.

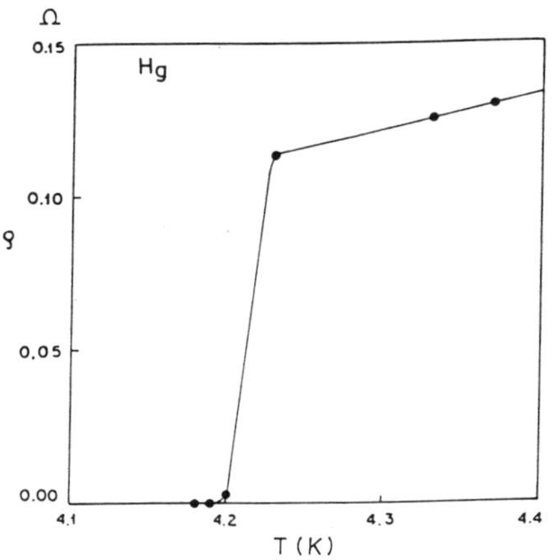

Fig.1.1: The d.c. resistivity of mercury drops to zero at T_c. Below the transition temperature the resistivity is zero. Above the transition temperature, the resistivity increases linearly with increasing temperature.

Table 1.1. The Superconducting transition temperatures of normal elements.

Element	$T_c(K)$	Element	$T_c(K)$
Al	1.14	Sn	3.70
Ti	0.39	La	6.00
V	0.38	Hf	0.12
Zn	0.87	Ta	4.48
Ga	1.09	Re	1.40
Zr	0.54	Os	0.65
Nb	9.50	Ir	0.14
Mo	0.92	Hg	4.15
Tc	7.80	Tl	2.39
Ru	0.51	Pb	7.19
Gd	0.56	Th	1.37
In	3.40	Pa	1.40

We write the current in terms of vector potential as,

$$\vec{j} = -\frac{c}{4\pi\lambda_L^2}\vec{A} \ . \tag{1.1}$$

Taking curl on both sides

$$curl\vec{j} = -\frac{c}{4\pi\lambda_L^2}curl\vec{A} \tag{1.2}$$

but $culr\vec{A} = \vec{B}$, the magnetic induction, so that

$$curl\vec{j} = -\frac{c}{4\pi\lambda_L^2}\vec{B} \ . \tag{1.3}$$

According to one of the Maxwell equations,

$$curl\vec{B} = \frac{4\pi}{c}\vec{j} \ . \tag{1.4}$$

Again taking curl on both sides,

$$curl\ curl\vec{B} = \frac{4\pi}{c}curl\vec{j} \tag{1.5}$$

so that upon using (1.3) we find

$$\nabla^2 B = -\frac{B}{\lambda_L^2} \tag{1.6}$$

the solution of which is,

$$B = B_o e^{-x/\lambda_L} \tag{1.7}$$

which shows that the magnetic field exponentially vanishes. Therefore, in most of the material the magnetic induction is zero except in a small distance λ_L which is called the London penetration depth, introduced in (1.1).

The diamagnetism implies that the rate of change of magnetic induction, B, is zero,

$$\frac{dB}{dt} = 0 \ , \tag{1.8}$$

which means that according to,

$$\begin{aligned} \nabla \times E &= -\frac{\partial B}{\partial t} \ , \\ \text{and} \qquad curl\ E &= 0 \end{aligned} \tag{1.9}$$

$E = 0$ means that $j = \frac{1}{\rho}E$ suggests that $\rho = 0$. Thus perfect diamagnetism (1.8) is consistent with zero resistivity. The field does not vary with time. If a time independent field pattern is formed, then also $dB/dt = 0$. Therefore, a field pattern which is independent of time, is consistent with diamagnetism. Therefore zero resistivity or superconductivity is consistent with exponentially decaying field which leads to zero field in most of the sample. The magnetic induction is,

$$B_{in} = B_a + 4\pi M \ . \tag{1.10}$$

If the field inside is zero, then

$$B_a = -4\pi M \tag{1.11}$$

or

$$\chi = \frac{M}{B_a} = -\frac{1}{4\pi} \tag{1.12}$$

which means that the susceptibility for zero induction is negative or diamagnetic. Thus zero resistivity is consistent with diamagnetism, $dB/dt = 0$, and then the susceptibility is negative, approaching the value of $-1/4\pi$ approximately as shown in Fig. 1.2. Since the London penetration depth is ignored in writing (1.11), the expression (1.12) is subject to small modifications. Since the field is almost zero inside the sample, except that there is penetration depth, any effort to apply magnetic field leads to pushing the lines of force outside the sample. This kind of pushing out of lines of force is called the Meissner effect. That means that field can be pushed inside the sample upto a distance of London penetration depth. Efforts to push the field inside the sample leads to flux creeping out of the sample. The continuous field in the interior of the sample is zero. What happens if we make a hole in the superconductor and watch as to how the field goes inside the sample. We can make a ring of the superconductor and see if the field goes in or not. In such a case the field goes through in the form of quantized fluxons but not as a continuous field.

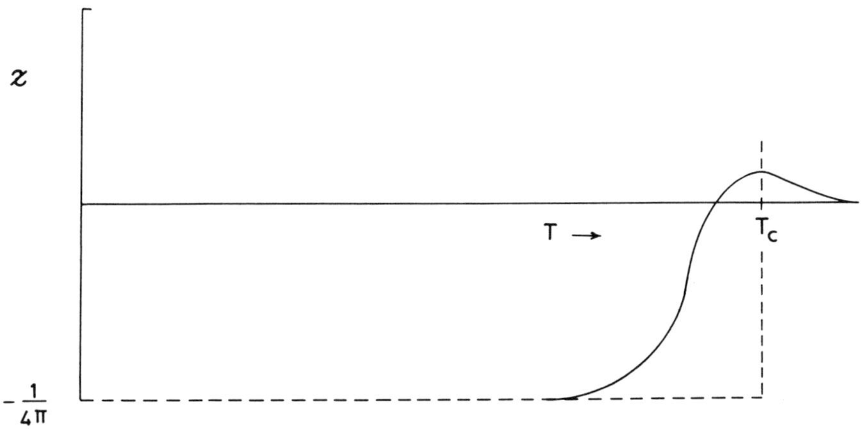

Fig.1.2: The susceptibility above T_c is showing paramagnetic, $\chi = c/T$, behaviour. Below T_c it becomes negative approaching the value of $-1/4\pi$.

In a superconducting ring, threaded by magnetic flux ϕ, the phase of the wave function that describes the superconducting quasiparticles advances steadily around the ring, and accumulates an increment of $2\pi\phi q/h$ for each circuit where h is the Planck's constant and q is the charge of the carriers. The wave function at any point on the ring has to be single valued so one circuit of the ring corresponds to an exact integral number of phase cycles, $\theta = 2n\pi$, forcing the contained flux to be quantized in units of $\phi_o = hc/q$ which has been measured by a pickup coil as shown in Fig.1.3.

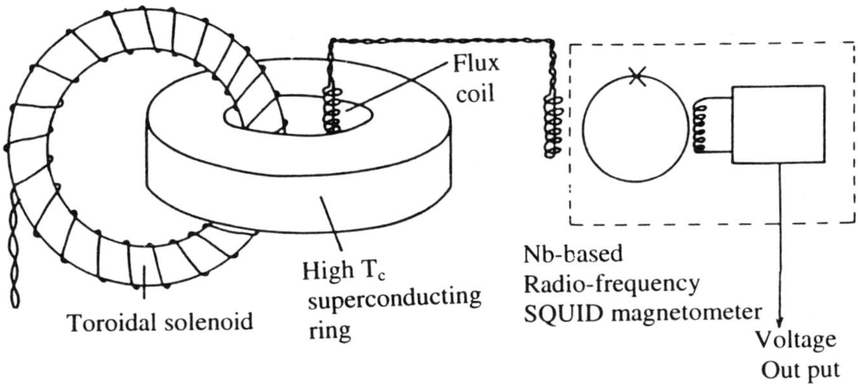

Fig.1.3: A coil brought near a superconducting ring can pick up quantized flux. Reprinted by permission from Nature 326:827(1987) ©(1987) Macmillan Magazines Ltd.

Consider a junction of a layer of a superconductor on which a layer of an oxide is deposited and on the oxide another layer of the superconductor is deposited. Such a junction is called a Josephson junction. A d.c. current flows across the junction in the absence of any electric or magnetic field. A d.c. voltage applied across the junction causes r.f. current oscillations across the junction. An r.f. voltage applied with the d.c. voltage can cause a d.c. current across the junction. The current-voltage characteristics of a junction placed inside a microwave cavity resonant at $\nu = 0.3$ GHz with a few tens of microwatt power become noisy and gradually vanish resulting in regions of zero slop in which the current lies at fixed voltage across the sample. The voltage at which the zero-slope regions occur is equal to $\pm h\nu/2e$, which shows that the charge of the quasiparticles in the superconductor is $q = 2e$. As the power is increased further zero-slope regions appear at higher voltages, and the origin alternates being stable and unstable. The current amplitude in a given zero-slope region varies periodically with increasing power. The interval in voltage from one-slope region to the next is not always $h\nu/2e$; some times a step is missing so that the voltage interval is $h\nu/e$. The system remains biased, e.g., at $+h\nu/2e$ not only when positive current is flowing but even when the current is zero and even when current is reversed and made negative. At some value of negative current, the voltage switches to $-h\nu/2e$. Thus the zero slope regions which occur with voltage $+h\nu/2e$ across the sample are connected by what appears to be a negative resistance at the origin. Negative resistance like regions occur between virtually every pair of zero-slope regions. At power levels of about ten milliwatts, these negative-resistance-like regions gradually vanish and the "zero-slope" regions exhibit a finite slope. These slopes are called "Shapiro steps" because they prove that the charge of a quasiparticle is $2e$ in a superconductor. The slopes occur at $\hbar\omega = 2eV$ so that the value of e/h can be measured by accurate measurement of the microwave frequency and the voltage across the junction.

Application of magnetic field beyond a critical value destroys the superconductivity. Just as there is a critical temperature, there is a critical field. The free energy of a normal material is F_n and that of the superconducting state is F_s. The applied magnetic field is represented by $H^2/8\pi$ so that the critical field, H_c is determined by

$$F_n - F_s = H^2/8\pi \ . \tag{1.13}$$

The critical field varies with the temperature as,

$$H_c = H_c^o \left(1 - \frac{T}{T_c}\right)^{1/2} \ . \tag{1.14}$$

The superconductors containing magnetic impurities or sublattices of magnetic atoms are called type-II superconductors. Their transition temperature is reduced due to scattering of electrons by the electrons belonging to magnetic atoms,

$$k_B T_c^{II} = k_B T_c - \frac{\hbar}{\tau_{AG}} \tag{1.15}$$

where T_c^{II} is the transition temperature of a superconductor with magnetic atoms whereas it would be T_c without magnetic atoms. Here τ_{AG} is the scattering time of electrons of the pure superconductor by the magnetic atoms. Such an idea was first given by Abrikosov and Gorkov (AG). The magnetic atoms work as pinning centres so that magnetic field enters the sample. The theory of Meissner effect creates the regions of zero flux so that the entered field forms localized vortices. Since the field enters, there appear terms in (1.13) which are linear in magnetic field so that instead of one critical field as in (1.14) there are two critical fields in type-II superconductors, which are called the lower critical field, H_{c1} and the upper critical field, H_{c2}. Due to flux quantization, the field can be written in terms of a distance. Therefore, the two critical fields can be described as distances, called the London penetration depth λ_L and the coherence length, ξ,

$$H_{c1} = \frac{\phi_o}{\pi\lambda_L^2} , \tag{1.16a}$$

$$H_{c2} = \frac{\phi_o}{\pi\xi^2} . \tag{1.16b}$$

These fields may be understood in terms of magnetization as shown in Fig. (1.4). In the type-I superconductors, there is only one critical field which is the solution of (1.13). For field larger than H_c the system becomes normal. Below H_c, the magnetization is negative (diamagnetic). In the presence of mangetic impurities, there are two critical fields. Below H_{c1}, the magnetization is negative but the vortices are not formed. For $H_{c1} < H < H_{c2}$ the vortices are formed and the magnetization is negative. Above H_{c2} the normal magnetization exists.

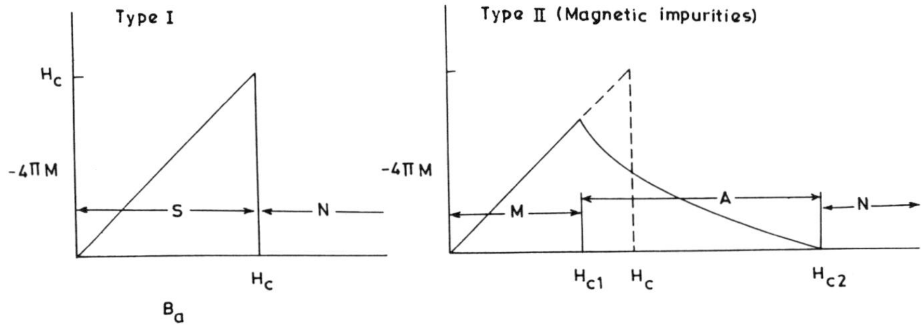

Fig.1.4: Magnetization of a type-I superconductor and that of a type-II superconductor. The region S is superconducting. In type-II superconductors, S as well as A are superconducting but vortices are formed in the A region. Above H_{c2} the system acquires normal resistivity, N.

1.2 Bosons

When the force is linearly proportional to the displacement, $F = kx$. The gradient of the potential energy gives the force, $-\partial V/\partial x = F$. Therefore, $V = \frac{1}{2}kx^2$ for a linear harmonic oscillator. The Hamiltonian of the harmonic oscillator is thus,

$$\mathcal{H} = \frac{p^2}{2m} + \frac{1}{2}kx^2 \tag{1.17}$$

where p is the linear momentum, $p = \hbar k$. The variables p and x are conjugate variables. Therefore, their commutator is $i\hbar$,

$$xp - px = i\hbar \ . \tag{1.18}$$

In terms of matrix elements (1.17) may be written as

$$< k|\mathcal{H}|l >= E_k < k|l >= \frac{1}{2m} < k|p|j >< j|p|l > +\frac{1}{2}k < k|x|j >< j|x|l > \tag{1.19}$$

We represent a complex conjugate by a star superscript and a hermitian conjugate by a dagger. The hermitian conjugate is obtained by replacing i by $-i$ and by interchanging rows into columns and columns into rows. In the case of a linear product, the order of operators is interchanged. Therefore, the hermitian conjugate differs from the complex conjugate. The matrix elements are thus

$$< j|p|l >=< l|p^\dagger|j >^*=< l|p|j >^* \tag{1.20}$$

Both of the terms of (1.17) are positive. The eigenvalues can be zero only when matrix elements $< k|p|j >$ and $< k|x|j >$ are zero for all j. However, the uncertainity principle also comes in the picture and makes sure that the minimum value of the energy is $\frac{1}{2}\hbar\omega$ where ω is the classical frequency, $\omega = (k/m)^{1/2}$, determined by the square root of the ratio of the force constant to mass of the oscillator. We calculate the commutator of x and that of p with $\mathcal{H}$:

$$x\mathcal{H} - \mathcal{H}x = \frac{i\hbar}{m}p \tag{1.21}$$

$$p\mathcal{H} - \mathcal{H}p = -i\hbar kx. \tag{1.22}$$

We choose a representation so that one of the factors in the commutator is diagonal in which case the difference between two eigenvalues appears when a matrix element is taken. The eq.(1.21) becomes,

$$\begin{aligned} < k|x|j >< j|\mathcal{H}|l > - < k|\mathcal{H}|j >< j|x|l > &= (E_l - E_k) < k|x|l > \\ &= \frac{i\hbar}{m} < k|p|l > \ . \end{aligned} \tag{1.23}$$

In this fashion, (1.22) becomes

$$(E_l - E_k) < k|p|l >= -i\hbar k < k|x|l > . \tag{1.24}$$

We eliminate $< k|x|l >$ from (1.23) and (1.24) to find that either,

$$< k|x|l >=< k|p|l >= 0 \tag{1.25}$$

or

$$E_l - E_k = \mp\hbar\left(\frac{k}{m}\right)^{1/2} = \mp\hbar\omega \tag{1.26}$$

where $\omega = (k/m)^{1/2}$ is the frequency of the classical harmonic oscillator. The eigen values differ from each other by integer multiples when the matrix elements of both p and x are not zero. When p and x are replaced by Δp and Δx, respectively, and the uncertainity product $\Delta p\Delta x \geq \hbar/2$ is used along with (1.18), by minimizing the energy we find that the minimum energy is $\frac{1}{2}\hbar\omega$ so that the full eigenvalues are determined by $(n+\frac{1}{2})\hbar\omega$, where n is a positive integer or zero. The equation (1.25) is also not consistent with the commutator (1.18). Therefore (1.26) is a solution and not (1.25). Thus we conclude that the eigen values are equally spaced and there is a minimum value of $\frac{1}{2}\hbar\omega$. We multiply (1.23), first by $-im\omega$ and add it to (1.24) so that

$$(E_l - E_k - \hbar\omega) < k|p - im\omega x|l >= 0 \tag{1.27}$$

in which the second factor is not zero when $E_k = E_l - \hbar\omega$ which means that the result of operating with $p - im\omega x$ on $|l>$ is some multiple of $|k>$, the energy of which is lower by $\hbar\omega$. Similarly, the hermitian conjugate of that is $p + im\omega x$ and raises energy of any state by $\hbar\omega$. Repeated application of the energy lowering operator to any wave function $|j>$ ultimately leads to a difficulty since the energy must be positive, unless there is a lowest energy eigenstate $|0>$ such that

$$(p - im\omega x)|0 >= 0. \tag{1.28}$$

The lowest energy eigenvalue can be found from (1.28) by operating on it from the left by $p + im\omega x$. Then,

$$\begin{aligned}(p+im\omega x)(p-im\omega x)|0> &= [p^2 + m^2\omega^2x^2 + im\omega(xp - px)]|0> \\ &= (p^2 + m^2\omega^2x^2 - m\hbar\omega|0> \\ &= 2m(\mathcal{H} - \frac{1}{2}\hbar\omega)|0>. \end{aligned} \tag{1.29}$$

Thus $|0>$ is an eigenstate of $\mathcal{H}$ with an eigenvalue $\frac{1}{2}\hbar\omega$. Application of energy-raising operator $p + im\omega x$ to $|0>$ gives a multiple of the wave function $|1>$ that has energy $\frac{3}{2}\hbar\omega$. Further application generates an infinite sequence of eigen states that are conveniently labeled $|n>$ with energy eigen values.

$$E_n = \left(n + \frac{1}{2}\right)\hbar\omega \qquad (n = 0, 1, 2, \cdots\cdots). \tag{1.30}$$

We arrange the energy raising and lowering operators in dimensionless form by multiplying them with $-i(2m\hbar\omega)^{-1/2}$ and $i(2m\hbar\omega)^{-1/2}$, respectively. The raising and lowering operators obtained in this way are called $a^\dagger$ and a, respectively, as shown below:

$$\begin{aligned} aa^\dagger &= \frac{\mathcal{H}}{\hbar\omega} + \frac{1}{2} \\ a^\dagger a &= \frac{\mathcal{H}}{\hbar\omega} - \frac{1}{2}. \end{aligned} \tag{1.31}$$

Thus a and $a^\dagger$ satisfy the commutation relations,

$$aa^\dagger - a^\dagger a = 1 \tag{1.32}$$

and

$$\mathcal{H} = (a^\dagger a + \frac{1}{2})\hbar\omega. \tag{1.33}$$

It follows from (1.30) and (1.33) that the eigenvalues of $a^\dagger a$ are the positive integers and zero. Therefore $< a^\dagger a >$ is called the number operator. This result is a direct consequence of the commutation relation (1.32). The only non-zero matrix elements of a are $< n-1|a|n >$, which we call λ_n. Similarly, the only nonvanishing matrix elements of $a^\dagger$ are $< n|a^\dagger|n-1 >$ equal to λ_n^*. Then the diagonal matrix element of $a^\dagger a$ is,

$$< n|a^\dagger a|n > = < n|a^\dagger|n' >< n'|a|n > = |\lambda_n|^2 \tag{1.34}$$

where the summation over the complete set $|n' >$ is implied. Since this diagonal element is equal to n, λ_n is equal to $n^{1/2}$ except that there may be a phase factor of unit magnitude. Thus the matrices for a and $a^\dagger$ are,

$$a = \begin{pmatrix} 0 & 1 & 0 & 0 & 0 & \\ 0 & 0 & \sqrt{2} & 0 & 0 & \\ 0 & 0 & 0 & \sqrt{3} & 0 & \\ 0 & 0 & 0 & 0 & \sqrt{4} & \\ & & & & & \cdots \\ & & & & & \cdots \\ & & & & & \cdots \end{pmatrix}, \quad a^{\dagger} = \begin{pmatrix} 0 & 0 & 0 & 0 & 0 & \\ 1 & 0 & 0 & 0 & 0 & \\ 0 & \sqrt{2} & 0 & 0 & 0 & \\ 0 & 0 & \sqrt{3} & 0 & 0 & \\ 0 & 0 & 0 & \sqrt{4} & 0 & \\ & & & & & \cdots \\ & & & & & \cdots \\ & & & & & \cdots \end{pmatrix}. \tag{1.35}$$

The product $a^{\dagger}a$ is diagonal with eigenvalues $0, 1, 2, \cdots\cdots$. The expressions for a and $a^{\dagger}$ in terms of x and p are

$$x = \left(\frac{\hbar}{2m\omega}\right)^{1/2} (a + a^{\dagger}) \tag{1.36}$$

and

$$p = i\left(\frac{m\hbar\omega}{2}\right)^{1/2} (a^{\dagger} - a). \tag{1.37}$$

In a many body system of many wave vectors, the Hamiltonian is replaced by

$$\mathcal{H} = \sum_k \hbar\omega_k ({a^{\dagger}}_k a_k + \frac{1}{2}) \tag{1.38}$$

where the summation is over all of the wave vectors from zero to a cut off value. In a lattice with interatomic distance a, the wave vector of the reciprocal lattice is $2\pi/a$ so that the maximum value of the wave vector or the cut off value becomes, $k_{max} = 2\pi/a$. The number of particles of a given wave vector k is $n_k =< {a^{\dagger}}_k a_k >$. The eigenvalues of the operators ${a^{\dagger}}_k$ and a_k are given by,

$$\begin{aligned} {a^{\dagger}}_k|n_k> &= (n_k+1)^{1/2}|n_k+1> \\ a_k|n_k> &= n_k^{1/2}|n_k-1> \; . \end{aligned} \tag{1.39}$$

Actually, there are product wave functions,

$$\psi = |n_{k_1}>|n_{k_2}>|n_{k_3}>\cdots|n_{k_D}> \tag{1.40}$$

in which a particular operator operates only on that product to which is associated its wave vector,

$$\begin{aligned} a^{\dagger}{}_{k_2}|\psi> &= (n_{k_2}+1)^{1/2}|n_{k_1}>|n_{k_2}+1>|n_{k_3}>\cdots|n_{k_D}> \\ a_{k_3}|\psi> &= n_{k_3}^{1/2}|n_{k_1}>|n_{k_2}>|n_{k_3}-1>\cdots|n_{k_D}> \end{aligned} \tag{1.41}$$

The commutation relations for the bosons are,

$$\begin{aligned} a_k a^{\dagger}{}_{k'} - a^{\dagger}{}_{k'} a_k &= \delta_{k,k'} \\ a^{\dagger}{}_k a^{\dagger}{}_{k'} - a^{\dagger}{}_{k'} a^{\dagger}{}_k &= 0 \\ a_k a_{k'} - a_{k'} a_k &= 0 \end{aligned} \tag{1.42}$$

instead of (1.32). These relations are sufficient to find the eigenvalues and the quasiparticle life time of interacting bosons.

1.3 Fermions

The Dirac equation is based on symmetry of space and time. Therefore, the kinetic energy is linear in momentum to be symmetric with single time derivative in the energy operator, $i\hbar\partial/\partial t$. An immediate consequence of having the hamiltonian linear in p, is the generation of γ_i Dirac matrices which anticommute. Since the momentum is not conserved, another angular momentum, called the spin has to be added. Thus the Dirac equation with anticommuting matrices is believed to be a good representation of the motion of the electron. The commutators for bosons become anticommutators for fermions,

$$[A,B] = AB + BA \tag{1.43}$$

Instead of (1.42) for bosons, we have the following anticommutators for fermions,

$$\begin{aligned} [c_k, c_i]_+ &= c_k c_i + c_i c_k = 0 \\ [c^{\dagger}{}_k, c^{\dagger}{}_i]_+ &= c^{\dagger}{}_k c^{\dagger}{}_i + c^{\dagger}{}_i c^{\dagger}{}_k = 0 \\ [c_k, c^{\dagger}{}_i]_+ &= c_k c^{\dagger}{}_i + c^{\dagger}{}_i c_k = \delta_{i,k}. \end{aligned} \tag{1.44}$$

We define $N_k = c^{\dagger}{}_k c_k$ as before and notice that each N_k commutes with all the others so that they can be diagonalized simultaneously. The eigenvalues of N_k can be obtained from the matrix equation,

$$N_k^2 = c^{\dagger}{}_k c_k c^{\dagger}{}_k c_k = c^{\dagger}{}_k(1 - c^{\dagger}{}_k c_k)c_k = c^{\dagger}{}_k c_k = N_k \tag{1.45}$$

where third of (1.44) has been used. If N_k is in diagonal form and has the eigenvalues $n'_k, n''_k, \cdots$, it is apparent that N_k^2 is also in diagonal form and has the eigenvalues $N_k'^2, n''^2, \cdots$. Thus the matrix equation (1.45) is equivalent to the algebraic equations

$$n_k'^2 = n'_k \quad ; \quad n_k''^2 = n''_k \tag{1.46}$$

for the eigenvalues. These are quadratic equations that have two roots; 0 and 1. Thus the eigenvalues of each N_k are 0 and 1 and the particles obey the exclusion principle. The eigenvalues of $N = \sum N_k$ are positive integers and zero as before. The expression for the hamiltonian is still $\mathcal{H} = \sum_k \epsilon_k c^{\dagger}{}_k c_k$ and the energy eigenvalues are $\sum_k f_k \epsilon_k$, where $f_k =< c^{\dagger}{}_k c_k >$. These are the spinless fermions. We include the spin, σ such that (1.43) remains unchanged but (1.44) becomes,

$$\begin{aligned}
[c_{k\sigma}, c_{i\sigma}]_+ &= c_{k\sigma}c_{i\sigma} + c_{i\sigma}c_{k\sigma} = 0 \\
[c^{\dagger}{}_{k\sigma}, c^{\dagger}{}_{i\sigma}]_+ &= c^{\dagger}{}_{k\sigma}c^{\dagger}{}_{i\sigma} + c^{\dagger}{}_{i\sigma}c^{\dagger}{}_{k\sigma} = 0 \\
[c_{k\sigma}, c^{\dagger}{}_{i\sigma'}]_+ &= c_{k\sigma}c^{\dagger}{}_{i\sigma'} + c^{\dagger}{}_{i\sigma'}c_{k\sigma} = \delta_{k,i}\delta_{\sigma,\sigma'}
\end{aligned} \tag{1.47}$$

The number density now becomes $f_{k\sigma} =< c^{\dagger}{}_{k\sigma}c_{k\sigma} >$, and the number of particles which can be accomodated in a level can be 0,1 or 2 out of which in case 2 particles are accomodated, their spins will be reversed, one up and the other down, etc.

1.4 Perturbation theory

The hamiltonian of a system is written in two parts. A large part containing only diagonal part of the interaction along with the kinetic energy, called the unperturbed hamiltonian and a small part of the interaction containing largely off diagonal elements called the interaction,

$$\mathcal{H} = \mathcal{H}_o + \mathcal{H}' \ . \tag{1.48}$$

The eigenfunctions are obtained from the unperturbed hamiltonian to which n-th order correction can be added. The eigenvalues of the above occur as unperturbed eigenvalue which is the expectation value of the unperturbed hamiltonian and the n-the order correction. Thus the wave function is written as,

$$\begin{aligned}
\psi_k &= |n_k> \\
\psi_o &= |\psi_o> \\
\psi &= |n_k> + \sum_{k'} \frac{<n_{k'}|\mathcal{H}'|n_k>}{E^o_{k'} - E^o_k} |n_{k'}>
\end{aligned} \tag{1.49}$$

which shows the first order correction to $|n_k>$. The energy denominators are to be obtained from the unperturbed hamiltonian only,

$$E^o_k = <n_k|\mathcal{H}_o|n_k> \ . \tag{1.50}$$

The second-order correction to the eigenfunctions is given by

$$\begin{aligned}
\psi &= |n_k> + \sum_{k'} \frac{<n_{k'}|\mathcal{H}'|n_k>}{E^o_{k'} - E^o_k} |n_{k'}> \\
&+ \sum_{k'} \sum_{k''} \frac{<n_{k''}|\mathcal{H}'|n_{k'}><n_{k'}|\mathcal{H}'|n_k>}{(E_{k'} - E^o_k)(E_{k''} - E^o_k)} |n_{k''}>
\end{aligned} \tag{1.51}$$

which can be written to infinite orders. However, in most of the problems, $<\mathcal{H}'> << <\mathcal{H}_o>$, so that corrections upto the second order are quite sufficient. If not, the hamiltonian may be transformed or higher-order corrections may be obtained by trivial extension of the above. The eigenvalues are given by

$$E_o = <n_k|\mathcal{H}_o|n_k> \tag{1.52}$$

$$\delta E_1 = <n_k|\mathcal{H}'|n_k> \tag{1.53}$$

$$E_1 = E_o + \delta E_1 \tag{1.54}$$

upto the first order. The second-order correction is given by

$$\delta E_2 = \sum_{k'} \frac{< n_k|\mathcal{H}'|n_{k'} >< n_{k'}|\mathcal{H}'|n_k >}{E_{k'} - E_k} \tag{1.55}$$

$$E_2 = E_o + \delta E_1 + \delta E_2 \tag{1.56}$$

The unperturbed hamiltonian gives infinite life time to the quasiparticles. The energy corrections (1.55) belongs to the system as a whole so that (1.56) gives the energy of the system of particles. The system energy includes all wave vectors. Hence the energy required to create one particle of wave vector k is,

$$\sum\nolimits_k = \frac{\partial}{\partial n_k}(\delta E_2) \tag{1.57}$$

which is called the self energy of the particle. The real part of this energy gives the energy shift of the particle due to the interaction correct upto the second order in the perturbation theory and the imaginary part gives the life time of that particle,

$$\frac{1}{\tau} = -\frac{2}{\hbar} Im \sum\nolimits_k \tag{1.58}$$

The real and imaginary parts are separated by using Dirac's identity,

$$\frac{1}{x-a} = \frac{P}{x-a} + i\pi\delta(x-a) \tag{1.59}$$

and by changing the summation into integration by use of suitable density of states. In this way the life time of the particles is obtained from the second quantized hamiltonians.

1.5 B.C.S. theory

We construct the interaction in such a way that the CPT is conserved. The charge is conserved by ensuring that the number of electrons created is equal to the number of electrons annihilated. The parity is conserved by taking the amplitudes of the interaction and that of the hermitian conjugate as equal and time reversal invariance is obtained by adding the hermitian conjugate. The hamiltonian of the unperturbed phonons is,

$$\mathcal{H}_{op} = \sum_q \hbar\omega_q {a^\dagger}_q a_q \tag{1.60}$$

where $a^{\dagger}{}_q$ and a_q are the creation and annihilation operators of the phonons, $\hbar\omega_q$ is the single quasiparticle frequency of the phonons and q is the wave vector summed over the full Brillouin zone from zero to a cutoff value $2\pi/a$ where a is the lattice constant. The electrons are filled below the Fermi energy, ϵ_F so that the unperturbed hamiltonian for the electrons is,

$$\mathcal{H} = \sum_{k\sigma} (\epsilon_{k\sigma} - e_{F\sigma}) c^{\dagger}{}_{k\sigma} c_{k\sigma} \tag{1.61}$$

where $\epsilon_{k\sigma}$ is the single particle energy for electrons and $c^{\dagger}{}_{k\sigma}(c_{k\sigma})$ are the creation (annihilation) operators for electrons. The crystal potential is modulated by phonons. Therefore, generalizing (1.36) to a many-body system, the electron-phonon interaction may be written as

$$\mathcal{H}' = \sum_{k,k',q} \left(\frac{\partial V}{\partial R}\right)_o \left(\frac{\hbar}{2M\omega_k}\right)^{1/2} c^{\dagger}{}_{k\sigma} c_{k'\sigma}(a_q + a^{\dagger}{}_{-q})\delta(k'+q-k) + h.c. \tag{1.62}$$

where the derivative of the potential is evalauted at the equilibrium. The displacement of an atom depends on the square root of the inverse mass which when integrated becomes the density of the solid. The phonon frequency may be assumed to depend on the inverse square root of the mass per atom. The phonon wave vector is within the first Brillouin zone. The Dirac δ function ensures that momentum is conserved. We define two states,

$$\begin{aligned} \psi_1 &= |f_k, f_{k'}, n_q, \cdots > \\ \psi_2 &= |f_k + 1, f_{k'} - 1, n_q - 1, \cdots > \end{aligned} \tag{1.63}$$

and the unperturbed energy separation between them as,

$$\begin{aligned} E_2^o - E_1^o &= \epsilon_k(f_k + 1) + \epsilon_{k'}(f_{k'} - 1) + \hbar\omega_q(n_q - 1) \\ &\quad -[\epsilon_k f_k + \epsilon_{k'} f_{k'}) + \hbar\omega_q n_q] \\ &= \epsilon_k - \epsilon_{k'} - \hbar\omega_q \ . \end{aligned} \tag{1.64}$$

We calculated the second-order perturbation energy from (1.62) for one phonon leaving the electron operators unoperated. The matrix element of the first term with $k' = k - q$ is

$$E^{(2)} = \sum_{k,q} D^2 \frac{c^\dagger{}_{k-q} c_k c^\dagger{}_k c_{k-q}}{\epsilon_k - \epsilon_{k-q} - \hbar\omega_q} \tag{1.65}$$

where $D = (\partial V/\partial R)_o(\hbar/2M\omega_k)^{1/2}$. Since there are several terms in the interaction, combinations in the energy denominators occur which are of the form $(\epsilon_k - \epsilon_{k'})^2 \pm (\hbar\omega)^2$, so that the effective interaction between electrons can be written as,

$$\mathcal{H}^{(2)} = D^2 \sum_q \sum_{k,k'} \frac{\omega_q}{(\epsilon_k - \epsilon_{k-q})^2 - (\hbar\omega_q)^2} c^\dagger{}_{k'+q} c_{k'} c^\dagger{}_{k-q} c_k \tag{1.66}$$

when the electronic energies are small, the sign of the above can be negative (attractive) compared with the positive kinetic energy. Thus electrons become attractive and form pairs of zero spin and zero moemntum, $k\uparrow, -k\downarrow$. We look for the two particle bound state equation including the attractive interaction from (1.66). The hamiltonian of the two electron system is given by

$$\begin{aligned} \mathcal{H} &= \frac{1}{2m}(p_1^2 + p_2^2) + \mathcal{H}'' \\ &= \frac{p^2}{m} + \mathcal{H}'' \end{aligned} \tag{1.67}$$

for $p_1^2 \simeq p_2^2 \simeq p^2$. We look for an eigenfunction of the form,

$$\chi(x) = \sum_k g_k e^{ik.x} = \sum_k g_k e^{ik.x_1} e^{-ik.x_2} \tag{1.68}$$

where

$$\begin{aligned} x &= x_1 - x_2 \\ X &= \frac{1}{2}(x_1 + x_2) \\ k &= \frac{1}{2}(k_1 - k_2) \\ K &= k_1 + k_2 \end{aligned} \tag{1.69}$$

in the centre of mass coordinates. If λ is the eigenvalue of $\chi(x)$, $(\mathcal{H} - \lambda)\chi(x) = 0$. Taking a matrix element of this equation,

$$\int dx e^{-ik(x_1-x_2)}(\mathcal{H} - \lambda) \sum_{k'} g_{k'} e^{ik'.(x_1-x_2)} = 0 \tag{1.70}$$

using $\epsilon_k = k^2/m$, the above becomes,

$$(\epsilon_k - \lambda)g_k + \sum_{k'} g_{k'} < k, -k|\mathcal{H}''|k', -k' >= 0 \tag{1.71}$$

where $k = k' + q$ and $-k = -k' - q$. Changing the summation into integration using the density of states, $\rho(\epsilon)$, the secular equation becomes,

$$(\epsilon - \lambda)g(\epsilon) + \int d\epsilon' \rho(\epsilon')g(\epsilon') < \epsilon|\mathcal{H}''|\epsilon' >= 0 \tag{1.72}$$

where

$$< \epsilon|\mathcal{H}''|\epsilon' >= -V \tag{1.73}$$

is the attractive potential for the one electron states above the Fermi energy from ϵ_F to $\epsilon_F + \hbar\omega_D$ or from k_F to k_m,

$$\frac{1}{2m}(k_m^2 - k_F^2) = \epsilon_m - \epsilon_F = \hbar\omega_D \tag{1.74}$$

which we use to set the limits of integration in (1.72) so that

$$(\epsilon - \lambda)g(\epsilon) = V \int_{2\epsilon_F}^{2\epsilon_m} d\epsilon' \rho(\epsilon')g(\epsilon') = C. \tag{1.75}$$

The middle term is independent of ϵ so that

$$g(\epsilon) = \frac{C}{\epsilon - \lambda}. \tag{1.76}$$

Substituting in (1.75)

$$1 - V \int_{2\epsilon_F}^{2\epsilon_m} d\epsilon' \frac{\rho(\epsilon')}{\epsilon' - \lambda} = 0 \tag{1.77}$$

within the limits of integration, the electrons are attractive near the Fermi level only so that $\rho(\epsilon') \simeq \rho_F$ or

$$\frac{1}{\rho_F V} = \int_{2\epsilon_F}^{2\epsilon_m} \frac{d\epsilon'}{\epsilon' - \lambda} = \log\frac{2\epsilon_m - \lambda}{2\epsilon_F - \lambda} = \log\frac{2\epsilon_m - 2\epsilon_F + \Delta}{\Delta} \tag{1.78}$$

where the lowest eigenvalue is

$$\lambda_o = \epsilon_1 + \epsilon_2 - \Delta = 2\epsilon_F - \Delta. \tag{1.79}$$

From (1.78),

$$\frac{\Delta}{2\epsilon_m - 2\epsilon_F + \Delta} = e^{-1/\rho_F V} \tag{1.80}$$

or

$$\Delta = \frac{2\hbar\omega_D}{e^{1/\rho_F V} - 1}. \tag{1.81}$$

The binding energy of the pair is thus related to the density of states, the attractive potential and the Debye frequency. Define a pair operator,

$$\begin{aligned} b_k &= <0|c_{-k\downarrow}c_{k\uparrow}|0>= -b_{-k} \\ b^{\dagger}{}_k &= <0|c^{\dagger}{}_{k\uparrow}c^{\dagger}{}_{-k\downarrow}|0>= b^{\dagger}{}_k \end{aligned} \tag{1.82}$$

so that the gap becomes,

$$\begin{aligned} \Delta_k &= V\sum_{k'}{}' b_{k'} \\ \Delta_k^* &= V\sum_{k'}{}' b^{\dagger}{}_{k'} \end{aligned} \tag{1.83}$$

within the attractive region. The reduced hamiltonian then becomes,

$$\mathcal{H} = \sum_{k\sigma} \epsilon_{k\sigma} c^{\dagger}{}_{k\sigma} c_{k\sigma} - \sum_{k} \Delta_k c^{\dagger}{}_{k\uparrow} c^{\dagger}{}_{-k\downarrow} + \Delta_k^* c_{-k\downarrow} c_{k\uparrow} - \Delta_k^* b_k^o \,.$$

We transform $c^{\dagger}{}_{k\sigma}$ and $c_{k\sigma}$ to another variables,

$$\begin{aligned} c_{k\uparrow} &= u_k\gamma_{ko} + v_k^*\gamma_{k1}^* \\ c_{-k\downarrow} &= -v_k\gamma_{ko} + u_k^*\gamma_{k1}^* \end{aligned} \tag{1.84}$$

where u_k and v_k are chosen to make the hamiltonian diagonal. One of the two relations is obtained by demanding that $c^{\dagger}{}_{k\sigma}(c_{k\sigma})$ anticommute. Another relation is obtained by setting the coefficient of non-diagonal terms such as $\gamma_{ko}^*\gamma_{k1}^*$ and $\gamma_{ko}\gamma_{k1}$, equal to zero. These two relations,

$$\begin{aligned} 2\epsilon_k u_k v_k + \Delta_k v_k^2 - \Delta_k^* u_k^2 &= 0 \\ u_k^2 + v_k^2 &= 1 \end{aligned} \tag{1.85}$$

lead to

$$\begin{aligned} u_k^2 &= \frac{1}{2}\left(1 + \frac{\epsilon_k}{E_k}\right) \\ v_k^2 &= \frac{1}{2}\left(1 - \frac{\epsilon_k}{E_k}\right) \end{aligned} \tag{1.86}$$

where the solution of the hamiltonian is,

$$E_k = (\epsilon_k^2 + \Delta^2)^{1/2} \ . \tag{1.87}$$

Substituting (1.84) in the hamiltonian,

$$\mathcal{H} = \sum_k E_k(\gamma_{ko}^* \gamma_{ko} + \gamma_{k1}^* \gamma_{k1}) + \sum_k (\epsilon_k - E_k + \Delta_k^* b_k^o) \tag{1.88}$$

The thermodynamic average is

$$b_k^o = Tr\left[\exp(-\beta\mathcal{H}) c_{-k\downarrow} c_{k\uparrow}\right] / Tr\ \exp(-\beta\mathcal{H}) \tag{1.89}$$

where $\beta = 1/k_B T$, or

$$b_k^o = u_k v_k^* [1 - 2f_k]. \tag{1.90}$$

Substituting this result in (1.83)

$$\Delta_k = -\sum_{k'} V_{k,k'} \Delta_{k'} \frac{1 - 2f_k}{2\epsilon_k} \tag{1.91}$$

$\Delta_k = 0$ at $T = T_c$,

$$1 = V \sum_{k, \epsilon_k < \hbar\omega_D} \frac{1 - 2f_k}{2\epsilon_k}. \tag{1.92}$$

Making the change of variables, and replacing the summation into integration,

$$\frac{1}{NV} = \ln \frac{\hbar\omega}{k_B T_c}(1 - 2f) + 2\int_o^{\hbar\omega_D/k_B T} dx \ln x \frac{d}{dx}\left(\frac{1}{e^x + 1}\right) \tag{1.93}$$

which can be written as,

$$\frac{1}{NV} = \ln(A\hbar\omega/k_B T_c) \tag{1.94}$$

where

$$\ln A = 2\int_o^\infty dx \ln x \frac{d}{dx}\left(\frac{1}{e^x + 1}\right) \tag{1.95}$$

or $A = 2\gamma/\pi = 1.13$. It follows that

$$k_B T_c = 1.13\hbar\omega e^{-1/N(o)V} \tag{1.96}$$

which means that the transition temperatures are small and proportional to the phonon frequency, $\hbar\omega$ which for a harmonic oscillator varies with the inverse square root of mass of an atom, so that

$$T_c \propto \frac{1}{\sqrt{M}}. \tag{1.97}$$

This is called the isotope effect in B.C.S. theory and provides the justification for the use of the electron-phonon interaction. The superconductivity is thus caused by the electron pairs of spin singlet and zero momentum which become attractive due to the electron-phonon interaction below a transition temperature.

1.6 Hybridization with a localized band

The coupling constant of the electron-phonon interaction V, the density of states at the Fermi level $N(o)$ and the phonon frequencies, $\hbar\omega$, are usually so adjusted that the transition temperature is quite small, $T_c < 15 - 20\text{K}$. These days, very large transition temperatures are found which can not be easily accomodated within the frame work of the BCS theory. The charge of $2e$ for the Cooper pairs gives superconducting ground state and this value of charge is also confirmed by experimental measurements. However, at low temperatures other values of the quasiparticle charge are also reported. Therefore, alternative mechanisms of superconductivity are being searched. We consider the phonon-induced intersite hopping along with a hybrid interaction in which a cannonical transformation leads to an interaction which permits transfer of electrons from the impurity valence band to the conduction band by virtual phonon transitions. This interaction is enhanced by reducing the energy gap between the conduction band and the impurity valence band. The energy region in which the electrons are attractive now varies from $-\epsilon_i - \hbar\omega_D$ to $\epsilon_i + \hbar\omega_D$ which is larger than in the B.C.S. theory. Because of the hopping in the hamiltonian, ions can occur with different valencies at different sites. We consider the hamiltonian of our system as,

$$\mathcal{H} = \sum_{i\sigma} \epsilon_{i\sigma} c^{\dagger}{}_{i\sigma} c_{i\sigma} + \sum_{k\sigma} \epsilon_{k\sigma} c^{\dagger}{}_{k\sigma} c_{k\sigma}$$

$$+\sum_q \hbar\omega_q \beta_q^\dagger \beta_q + \sum_{ik\sigma} V({c^\dagger}_{i\sigma} c_{k\sigma} + {c^\dagger}_{k\sigma} c_{i\sigma})$$
$$+\sum_{ijq\sigma} \lambda_{ijq\sigma}[{c^\dagger}_{i\sigma} c_{j\sigma}(\beta_q + \beta_{-q}^\dagger) + {c^\dagger}_{j\sigma} c_{i\sigma}(\beta_q^\dagger - \beta_{-q})] \quad (1.98)$$

where the first term represents the kinetic energy of electrons localized at i-th sites and $\epsilon_{i\sigma}$ is the single electron energy of spin $\sigma, {c^\dagger}_{i\sigma}(c_{i\sigma})$ are the creation (annihilation) operators for electrons at the i-th sites. The second term describes a conduction band with single particle energy given by $\epsilon_{k\sigma}$ for the wave vector k and spin σ which does not overlap with i-th localized band. Due to the fluctuating valence, electrons are created at random sites which form the localized band of single particle energy, $\epsilon_i(i \neq k)$. The wave vector k of the electrons in the conduction band is limited by the Fermi wave vector, k_F. The third term describes the phonon continuum with $\hbar\omega_q$ as the single phonon energy and $\beta_q^\dagger(\beta_q)$ are the phonon creation (annihilation) operators with wave vector q. The next term is a hybrid interaction containing creation and annihilation operators for electrons in the fluctuating impurity valence band as well as in the conduction band so that when a suitable transformation is carried out, the electrons can be promoted from the localized sites to the conduction band and vice versa. The next term in (1.98) is a phonon-induced intersite coupling which permits the hopping of electrons by phonons. The phonon mediated hopping between impurities may be energetically unfavourable but second-order virtual hopping from site to site is permitted. In particular, if an orbit on one of the sites is vacant, while occupied on another site, a phonon-induced virtual transition is feasible. We apply a cannonical transformation on the total hamiltonian (1.98) where the transformed hamiltonian $\tilde{\mathcal{H}} = e^{-s_1}\mathcal{H}e^{s_1}$ is obtained by using a generating function s_1 obtained from the solution of,

$$\sum_{ik\sigma} V({c^\dagger}_{i\sigma} c_{k\sigma} + {c^\dagger}_{k\sigma} c_{i\sigma}) + [\mathcal{H}_o, s_1] = 0 \ . \quad (1.99)$$

The generating operator is found to be,

$$s_1 = \sum_{ik\sigma} V(\epsilon_{k\sigma} - \epsilon_{i\sigma})^{-1}({c^\dagger}_{i\sigma} c_{k\sigma} - {c^\dagger}_{k\sigma} c_{i\sigma}) \ . \quad (1.100)$$

The single-particle frequencies in the transformed hamiltonian $\tilde{\mathcal{H}}$, are slightly modified and the transformed hamiltonian appears as,

$$\begin{aligned}\tilde{\mathcal{H}} &= \tilde{\mathcal{H}} + \sum_{k_1,k_2,i,j,\sigma} \lambda_{ijk_2\sigma}(\epsilon_{k_1\sigma} - \epsilon_{i\sigma})^{-1} \\ &\times \left[c^{\dagger}{}_{j\sigma}c_{k_1\sigma}(\beta_{k_2} + \beta^{\dagger}_{-k_2}) + c^{\dagger}{}_{k_1\sigma}c_{j\sigma}(\beta^{\dagger}_{k_2} + \beta_{-k_2})\right] \qquad (1.101)\end{aligned}$$

Thus we have generated an interaction by which electrons can be promoted from the impurity valence band to the conduction band by absorption of a phonon. The coupling constant of this interaction has the energy difference between the two bands in the denominator of (1.100) so that as the conduction band is brought closer to the impurity valence band, the strength of the interaction increases. We now apply a second transformation to (1.101) as $\tilde{\tilde{\mathcal{H}}} = e^{-s_2}\tilde{\mathcal{H}}e^{s_2}$ such that the generating function is,

$$\begin{aligned}s_2 &= \sum_{k_1,k_2,j\sigma} D_{jk_2\sigma}\left\{(\epsilon_{k_1\sigma} - \epsilon_{j\sigma} + \hbar\omega_{k_2})^{-1}c^{\dagger}{}_{j\sigma}c_{k_1\sigma}\beta_{k_2}\right. \\ &+(\epsilon_{k_1\sigma} - \epsilon_{j\sigma} - \hbar\omega_{-k_2})^{-1}c^{\dagger}{}_{j\sigma}c_{k_1\sigma}\beta^{\dagger}_{-k_2} \\ &+(\epsilon_{j\sigma} - \epsilon_{k_1\sigma} - \hbar\omega_{k})^{-1}c^{\dagger}{}_{k_1\sigma}c_{j\sigma}\beta^{\dagger}_{k_2} \\ &\left.+(\epsilon_{j\sigma} - \epsilon_{k_1\sigma} - \hbar\omega_{k})^{-1}c^{\dagger}{}_{k_1\sigma}c_{j\sigma}\beta_{-k}\right\} \qquad (1.102)\end{aligned}$$

where

$$D_{jk\sigma} = \sum_i \lambda_{ijk\sigma}V(\epsilon_{k\sigma} - \epsilon_{i\sigma})^{-1} . \qquad (1.103)$$

Then the new interaction becomes,

$$\mathcal{H}'' = \sum_{k_1,k_2,j,\sigma}\ \sum_{k'_1,k'_2,j',\sigma'} V_A c^{\dagger}{}_{k_s\sigma}c^{\dagger}{}_{k'_1\sigma'}c_{j\sigma}c_{j'\sigma'}\delta_{k'_2,-k_2} \qquad (1.104)$$

where

$$V_A = \frac{DD'\hbar\omega_{k_2}}{(\epsilon_{j'\sigma'} - \epsilon_{k'_1,\sigma'})^2 - (\hbar\omega_{k'_2})^2} \qquad (1.105)$$

which is capable of forming pairs either in the localized j band or in the conduction k band. There is complete flexibility about the singlet as well as the triplet spin pairing. Thus there are three components of the triplet and one singlet. The interaction is attractive as long as

$$-\epsilon_j - \hbar\omega_D < \epsilon_k < \epsilon_j + \hbar\omega_D \qquad (1.106)$$

where ϵ_j is the impurity valence band width and $\hbar\omega_D$ is the Debye cutoff for phonons. There are two types of gaps,

$$\begin{aligned} \Delta_1 &= \sum_j{}' V_A < c_{j\sigma} c_{j'\sigma'} > \\ \Delta_2 &= \sum_k{}' V_A < c_{k\sigma} c_{-k\sigma'} > \end{aligned} \quad (1.107)$$

for pairs in the insulating band Δ_1 and those in the conduction band Δ_2, respectively. In the later case, the hamiltonian

$$\begin{aligned} \mathcal{H} &= \sum_{k\sigma} \tilde{\epsilon}_{k\sigma} c^{\dagger}{}_{k\sigma} c_{k\sigma} + \sum_{j\sigma} \tilde{\epsilon}_{j\sigma} c^{\dagger}{}_{j\sigma} c_{j\sigma} \\ &+ \sum_{jj'\sigma'} \Delta_2 (c_{j,-\sigma'} c_{j',\sigma'} + c^{\dagger}{}_{j',\sigma'} c^{\dagger}{}_{j,-\sigma'}) \end{aligned} \quad (1.108)$$

with a similar hamiltonian in the former case. Thus pairs can be found in the insulating state as well as in the conducting state where they superconduct. The T_c is enhanced with respect to the B.C.S. value as $\hbar\omega_D$ is replaced by $\epsilon_j + \hbar\omega_D$ and the coupling constant of the pairing interaction is also enhanced because of the enhancement of D. Thus our model meets the requirements of a high T_c theory. The pairs in the insulating band create a fluid of suspended pairs while those in the conduction band create the superconducting liquid. In a system of random sites occupied by electrons, the attraction between electrons occurs by absorption or emission of phonons which have been accelerated by the phonon scattering.

The ordinary B.C.S. type pairing arises from the electron-phonon interaction in a momentum conserving periodic lattice,

$$\mathcal{H}' = D \sum_{k,k',\sigma} c^{\dagger}{}_{k\sigma} c_{k'\sigma} (a_q + a^{\dagger}{}_{-q}) \delta(k' + q - k) \quad (1.109)$$

In Fig. 1.5 we show the Bethe-Salpeter equation for pairing of electrons obtained in the present case from (1.98) and also the one obtained from the B.C.S. interaction. The comparison of the two equations shows that the nature of pairing in the two cases differs because of localized sites in the present theory.

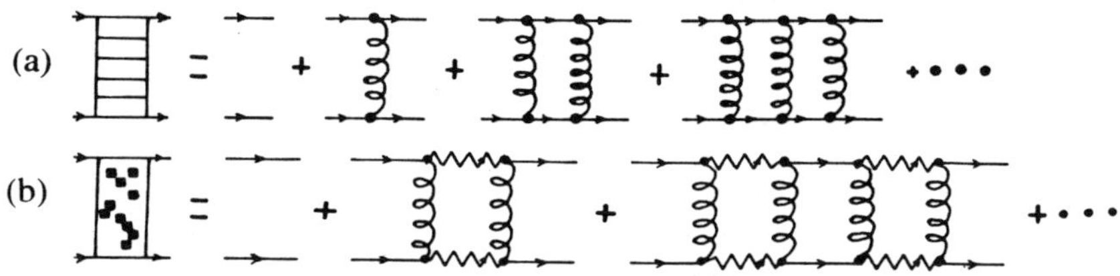

Fig.1.5: The Bethe-Salpeter equation for a (a) pair of electrons in the conduction band due to electron-phonon interaction and (b) pair of electrons with one localized band and one conduction band.

Suggested References

1.1. C. Kittel, Introduction to Solid State Physics, John Wiley, New York 1976.

1.2. C. Kittel, Quantum theory of solids, John Wiley, New York, 1963.

1.3. D. Caplin, Nature **326**, 827 (1987).

1.4. C.E. Gough, M.S. Colclough, E.M. Forgan, R.G. Jordan, M. Keene, C.M. Muirhead, A.I.M. Rae, N. Thomas, J.S. Abell and S. Sutton, Nature **326**, 855 (1987).

1.5. S. Shapiro, Phys. Rev. Lett. **11**, 80 (1963).

1.6. K.N. Shrivastava, Solid State Commun. **69**, 1173 (1989).

Chapter 2

Flux Quantization

2.1 Josephson's discovery

In a junction of two superconductors separated by a thin layer of an insulator, the pairs of electrons can tunnel through the insulating barrier causing a supercurrent. Josephson described the electron operators by use of Bagoliubov transformation and defined pair operators as,

$$
\begin{aligned}
a^{\dagger}{}_{k\uparrow} &= u_k \alpha^{\dagger}_{eko} + v_k \alpha_{hk1} && (2.1)\\
a^{\dagger}{}_{ek} \alpha_{hk} &= S_k^{+} n_k && (2.2)\\
\alpha^{\dagger}_{ek} \alpha_{ek} &= n_k && (2.3)
\end{aligned}
$$

with the number density N_l and N_r in the left and right side superconductors as shown in Fig.2.1. The unperturbed hamiltonian is

$$\mathcal{H}_o = \sum_k n_k E_k + \lambda_l N_l + \lambda_r N_r \qquad (2.4)$$

where E_k is the quasi-particle energy in the Bogoliubov theory. The commutators are

$$
\begin{aligned}
[\mathcal{H}_o, \alpha^{\dagger}_{ek}] &= [E_k + \lambda_k] \alpha^{\dagger}_{ek}\\
[\mathcal{H}_o, \alpha^{\dagger}_{hk}] &= [E_k - \lambda_k] \alpha^{\dagger}_{hk}\\
[\mathcal{H}_o, S_k^{+}] &= 2\lambda_k S_k^{+} \ . && (2.5)
\end{aligned}
$$

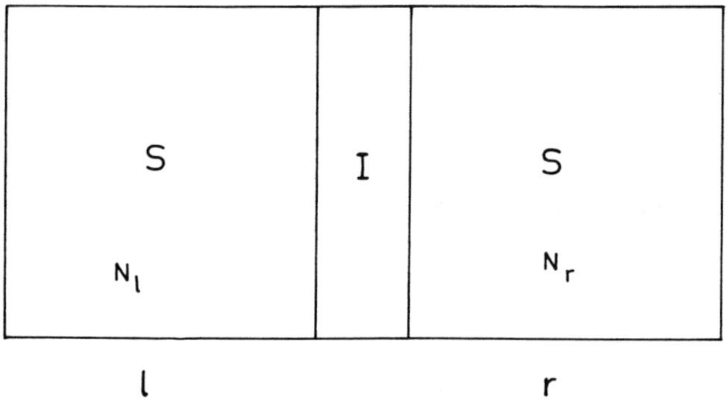

Fig.2.1: A junction of two superconductors separated by an insulator, I. For convenience l is marked for left and r for right-hand-side superconductors.

The tunneling hamiltonian is added to the unperturbed hamiltonian as in the perturbation theory,

$$\begin{aligned} \mathcal{H} &= \mathcal{H}_o + \mathcal{H}_T \\ \mathcal{H}_T &= \sum_{l,r}(T_{l,r}{a^\dagger}_k a_r + T_{rl}{a^\dagger}_r a_l) \ . \end{aligned} \tag{2.6}$$

All of the creation and annihilation operators are asymptotically independent of time. The particles are annihilated from the right and created on the left and the reverse process occurs through the hermitian conjugate of the interaction. The current can be defined by

$$J_{int}(t) = (ie/\hbar)[\mathcal{H}_T, N_r] \tag{2.7}$$

and it can be expanded in the form of a series with a constant term.

Retaining only first two terms

$$J = J_o + \frac{1}{2} J_1 S_l^+ S_r^- + \frac{1}{2} J_1^* S_r^+ S_l^- \tag{2.8}$$

which has pair creation and annihilation operators on the left- and right-hand side superconductors. The first term may be a constant current and the second and the third terms transfer pairs from the left to the right and from the right to the left. There is no frequency restriction on the various terms so they may oscillate with frequency,

$$\nu = 2eV/h \tag{2.9}$$

where $V = \lambda_l - \lambda_r$ is the applied voltage. As we have seen in the previous chapter, a large step surely occurs when the junction is subjected to microwaves of frequency given above. We can obtain a unitary transformation

$$J_H(t) = U^{-1}(t) J_{int}(t) U(t) \tag{2.10}$$

where

$$U(t) = \lim_{\epsilon \to 0^+} \left\{ T \exp \left[-\frac{i}{\hbar} \int_{-\infty}^{t} e^{\epsilon t'} \mathcal{H}_T(t') dt' \right] \right\} \tag{2.11}$$

or we use second-order perturbation theory to describe the current and separate the real and imaginary parts using Dirac's identity,

$$\begin{aligned} J_{int} \propto\ & 2ieu_l v_l u_r v_r T_{lr} T_{-l,-r} \left[(1 - f_{lo} - f_{r1}) \right. \\ & \times \left\{ P \frac{1}{eV - E_r - E_l} + i\pi\delta(eV - E_l - E_r) \right\} \\ & \left. -(f_{lo} - f_{ro}) \left\{ \frac{P}{eV + E_l - E_r} + i\pi\delta(eV + E_l - E_r) \right\} \right] \end{aligned} \tag{2.12}$$

There may be a d.c. current or an a.c. current with amplitude J_1 and frequency $2eV/\hbar$. At zero voltage, J_o is zero but a d.c. current upto a maximum of $|J_1|$ can occur. The oscillations in the voltage modulate the frequency of the supercurrent. From (2.9) the value of e/h can be measured and Josephson junctions can be used as particle or microwave detectors.

2.2 Flux quantum

The flux through any area with a boundary lying entirely in a superconductor is zero. However, if we make a ring of the superconductor or make a cup of the superconductor, then the flux through the superconducting ring or through the rim of the cup is quantized. Consider a superconducting body P with a tunnel O. The superonductor is larger than the penetration depth. According to the Meissner effect, the magnetic field inside a superconductor is zero and hence current in a loop normal to the surface of the superconductor is also zero. Due to finite penetration depth, surface currents exist and persist on the surfaces s_1 and s_2. The currents on the surfaces do not produce field in the interior but produce a net magnetic flux through O denoted by ϕ as shown in Fig. 2.2.

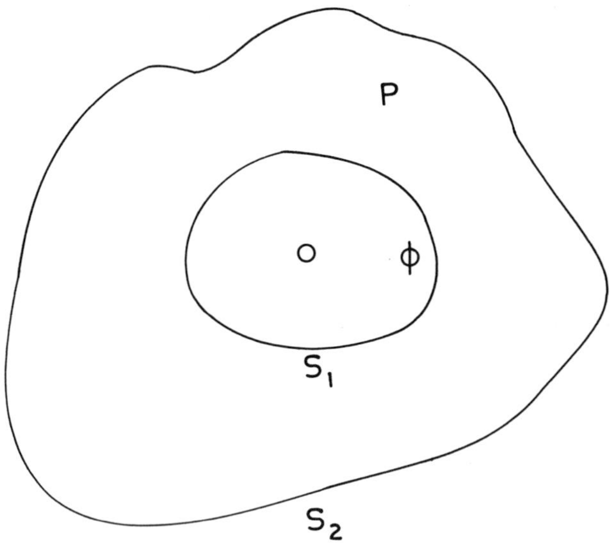

Fig.2.2: The superconductor with open space O and two boundaries, s_1 and s_2.

The eigenfunctions satisfy the Schrödinger equation,

$$\sum_j \frac{1}{2m}\left[-i\hbar\nabla_j + \frac{e}{c}A(r_j)\right]^2 \psi + V\psi = E\psi \tag{2.13}$$

where $\vec{A}$ is the vector potential due to surface currents. Inside the region P, we have $curl\, A = B$ so that $B = 0$ requires

$$\vec{\nabla} \times \vec{A} = 0 \ . \tag{2.14}$$

Hence

$$\vec{A} = \vec{\nabla}\chi \tag{2.15}$$

where χ is not single valued in P. Integrating the above,

$$\delta\chi = \oint \vec{A}.d\vec{l} = \int\int \vec{H}.d\vec{\sigma} = \phi \tag{2.16}$$

in a closed loop around the tunnel O once. We try the wave function,

$$\psi' = \psi \exp \sum_j i\frac{e}{c\hbar}\chi(\vec{r}_j) \tag{2.17}$$

which substituted in (2.13) gives

$$\sum_j \frac{1}{2m}(-i\hbar\nabla_j)^2\psi' + V\psi' = E\psi' \ . \tag{2.18}$$

The vector potential $\vec{A}$ is eliminated in (2.18). The boundary condition for ψ' is that when all electron coordinates are fixed, except for one, $\vec{r}_l$ and this vector, is taken around O once, ψ' changes by a constant factor,

$$\psi' \rightarrow \psi' \exp[i(e/c\hbar)\phi] \tag{2.19}$$

which is seen by using (2.16) and (2.17). The eigenvalues, E are determined by solving (2.18) with the use of boundary condition (2.19).

I. It is found that the energy levels are periodic in the magnetic flux ϕ with a period hc/e.

II. If the surfaces s_1 and s_2 are concentric cylinders on which $\psi = 0$, and $V = 0$, $\psi' \neq \psi$. If V is a real function of r_j by taking the complex conjugate of (2.18) and (2.19) it is found that the energy levels are even functions of ϕ.

III. From the above two results, it follows that, the partition function Q of the system is an even periodic function of ϕ with period hc/e. Accordingly at,

$$\phi = n\frac{hc}{2e} \tag{2.20}$$

where n is an integer,

$$\partial \ln Q/\partial\phi = 0 \tag{2.21}$$

and that $ln Q$ has the general form shown in Fig. 2.3a. The body current in the superconductor around O in Fig. 2.2 is

$$I = k_B T c \partial \ln Q/\partial\phi \tag{2.22}$$

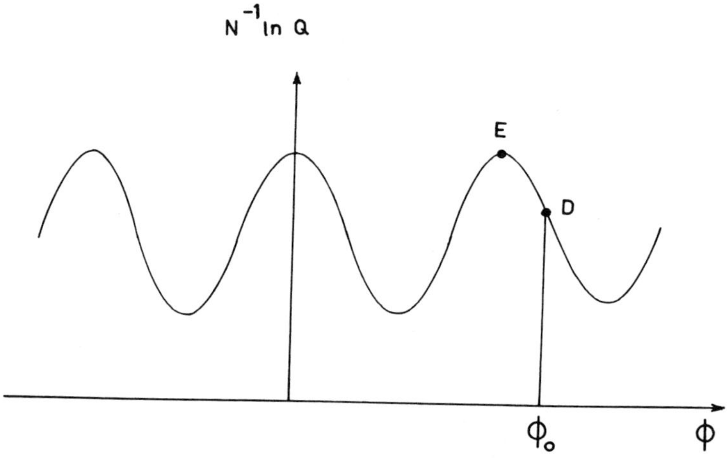

Fig.2.3a Maxima and minima in $N^{-1} lnQ$ as a function of flux.

where T is the temperature and c is the velocity of light. The current I is in a plane so that if the field normal to the plane is zero due to Meissner effect, then $I = 0$, which are the maxima and minima of the curve given in Fig. 2.3a.

IV. The point D in Fig. 2.3a is not an equilibrium state but the slope of the curve at that point indicates that if a flux ϕ_o is made to pass through

O, a body current would be induced opposite to the direction of I. The additional flux due to this body current causes the flux through O to decrease. The equilibrium state reached would therefore be E which is a maximum of the curve. Thus the superconducting state is given by the maxima of lnQ as a function of ϕ. If the external flux does not assume a value for which lnQ is a maximum, surface currents will flow on s_1 and s_2 to make up a total flux ϕ for which lnQ is a maximum. The flux through any surface whose boundary loop lies entirely in the superconductor is quantized in units of $hc/2e$. The requirement of this quantization in turn clearly implies that the flux through any small area in the interior of a superconductor is zero except in a small distance near the surface.

We will show that the maxima of lnQ occur at (2.20). We assume that there is a collection of noninteracting spinless electrons between two concentric cylinders s_1 and s_2. The electrons at a point at a distance r from the axis have momenta p_r, p_θ and p_z in the radial, azimuthal and z directions. Due to quantization rules,

$$p_\theta r = n\hbar \tag{2.23}$$

The kinetic energy of the electron is

$$\frac{1}{2m}\left[p_r^2 + p_z^2 + \frac{\hbar^2}{r^2}\left(n + \frac{e}{hc}\phi\right)^2\right] \tag{2.24}$$

The partition function Q and hence $N^{-1}lnQ$ to the lowest order does not depend on ϕ. Hence, for the electrons confined between s_1 and s_2 the flux is not quantized.

At $T = 0$, the one dimensional Fermi system has energy,

$$E = \frac{\hbar^2}{2mr^2}\sum\left(n + \frac{e}{hc}\phi\right)^2 \tag{2.25}$$

where $\sum$ is over all of the particles. The average energy per particle is then,

$$\frac{E}{N} = \text{const.} + \frac{\hbar^2}{2mr^2}\frac{e^2\phi^2}{\hbar^2c^2} \quad ; \quad \frac{e\phi}{hc} \leq \frac{1}{2} \tag{2.26}$$

for odd N and

$$\begin{aligned}\frac{E}{N} &= \text{const.} + \frac{\hbar^2}{2mr^2}\left(\frac{e\phi}{hc} - \frac{1}{2}\right)^2 \quad ; \quad \frac{1}{2} \geq \frac{e\phi}{hc} \geq 0 \\ &= \text{const.} + \frac{\hbar^2}{2mr^2}\left(\frac{e\phi}{ch} + \frac{1}{2}\right)^2 \quad ; \quad 0 \geq \frac{e\phi}{hc} \geq -\frac{1}{2} \qquad (2.27)\end{aligned}$$

Thus, the energy has a minimum at $e\phi/ch = \frac{1}{2}$ or 0. In the neighborhood of $\phi = 0^+$, the states with $n > 0$ have energies that increase with ϕ and those with $n < 0$ have energies that decreases with ϕ. The average energy for the two states n and $-n$, however, increases with ϕ like, const. + $(\hbar^2/2mr^2)(e\phi/ch)^2$, which are the parabolas in $N^{-1}lnQ$ versus $2e\phi/hc$ as shown in Fig. 2.3b. At $2e\phi/hc = 1$, pairing between,

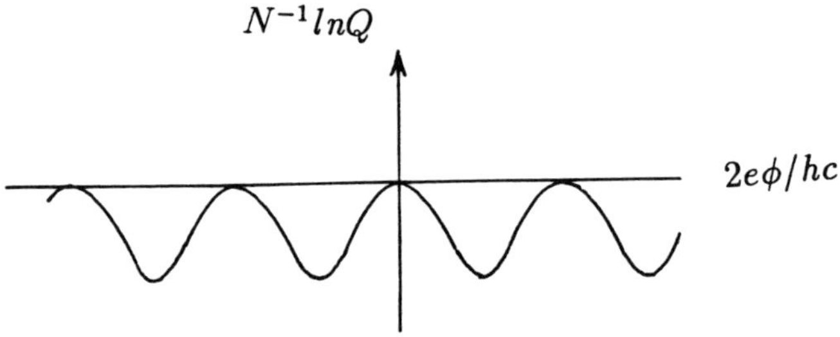

Fig.2.3b: $N^{-1}lnQ$ is plotted as a function of $2e\phi/hc$. The dependence near the maxima at $2e\phi/ch$ =integer is parabolic.

$$\begin{aligned}n + \frac{e}{hc}\phi &= n + \frac{1}{2} \\ \frac{1}{2} &\text{ and } -\frac{1}{2}\end{aligned}$$

$$\frac{3}{2} \text{ and } -\frac{3}{2} \tag{2.28}$$

occurs and the energy per particle remains the same as for the case of $\phi = 0$. In the neighbourhood of $2e\phi/ch = 1$, the additional energy for each of these pairs is again twice

$$\frac{\hbar^2}{2mr^2}\left[\frac{e\phi}{ch} - \frac{1}{2}\right]^2$$

which gives rise to the parabolas at $2e\phi/ch = \pm 1$ in Fig.2.3b. The parabolas are repeated at periods $\Delta(2e\phi/hc) = 1$, and the central parabola is given by

$$N^{-1} l\text{n}Q = -\frac{f}{k_B T}\frac{\hbar^2}{2m < r^2 >}\left(\frac{e\phi}{hc}\right)^2 + \text{constant} \tag{2.29}$$

where f is the fraction of electrons that are paired. For,

$$0 < \frac{2e\phi}{ch} < \frac{1}{2} \tag{2.30}$$

the body current is

$$I = -Nfc(e^2/mc^2)\phi/\{4\pi^2 < r^2 >\} \ . \tag{2.31}$$

The flux induced by this current calculated for a thin ring superconductor is given by,

$$\phi_{\text{induced}} = -ff_r\phi \tag{2.32}$$

where f_r is the number of electrons in a length e^2/mc^2 of the ring. Thus the trapped flux ϕ and the original flux ϕ_o are related in the following way:

$\phi_o/(hc/2e)$	ϕ
$-\frac{1}{2} \to \frac{1}{2}$	0
$\frac{1}{2} \to \frac{3}{2}$	1
$\frac{3}{2} \to \frac{5}{2}$	2

Thus the charge of $2e$ is consistent with the theory and also agrees with the experimental observations. However, it is not proved that other values of the charge can not occur in superconductors.

2.3 Fluxoid quantization in valence bonds

We consider an even number of bonded atoms between each pair of successive vacancies. The vacancies occur alternatively on odd and even site numbers. Any vacancy can hop to its next-nearest-neighbour site unless there is a vacancy on the intervening site. The hamiltonian of the vacancies can be written as,

$$\begin{aligned} \mathcal{H} &= \mathcal{H}_o + \mathcal{H}' \\ \mathcal{H}_o &= \sum_{\nu=1} \mu b^\dagger{}_\nu b_\nu \\ \mathcal{H}' &= -\sum_{\nu=1} V(b^\dagger{}_{\nu-1} b_{\nu+1} + b^\dagger{}_{\nu+1} b_{\nu-1})(1 - b^\dagger{}_\nu b_\nu) \end{aligned} \tag{2.33}$$

where $b^\dagger{}_\nu$ is a boson creation operator for a vacancy on the site ν, provided that it is understood that only states with not more than one vacancy per site are considered. It is also clear that νth site should be first occupied by an atom, then only a vacancy can be created by annihilating the vacancy. However, in the present model the atoms are not considered. The annihilated atom is equal to created vacancy and the created atom is equal to annihilated vacancy. We assume periodic boundary conditions for the system whose length L is an even multiple of the lattice constant a. The possible wave functions for N vacancies are of the form,

$$\det[e^{ik_j(x_n + na)}] \ , \tag{2.34}$$

in the region defined by $x_n < x_{n+1} + a$, $x_N < x_1 + L + a$, where the coordinates x_n are even or odd multiples of a for n even or odd. The condition for this wave function to vanish on the boundaries of the region is,

$$k_j = 2\pi(r_j + \delta)/(L + Na) \ , \tag{2.35}$$

where r_j is an integer and δ is a constant determined by the condition that the wave function is unchanged by the cyclic permutation $x_n \to x_{n+2}$. This condition gives,

$$-\sum_j 2ak_j + 2k_i(L + Na) = 2\pi n \tag{2.36}$$

where n is an integer. The solution of (2.35) and (2.36) gives,

$$\sum_j k_j = \frac{2\pi}{L}\sum_j (r_j + \frac{1}{2}) \tag{2.37a}$$

$$\text{or} \quad \sum_j k_j = \frac{2\pi}{L}\sum_j r_j \tag{2.37b}$$

Thus there are two inequivalent sets of states that cannot readily transform into one another. If we take the r_j to be the integers from $-N/2$ to $(N/2)-1$, then (2.37a) gives the ground state if the magnetic flux is zero but (2.37b) gives the ground state if the flux is $h/2e$. Which of these two families of states lies lowest is an oscillatory function of the enclosed flux with period h/e.

The mean square value of $\sum k_j$ is $8NVa^2/k_BT$ which is equivalent to $\delta = Nk_BT/32\pi^2V$, so thermal motion mixes the states when the temperature T is of the order of V/Nk_B and superconductivity is destroyed by thermal fluxtuations due to the occurrence of a transition temperature.

Consider a triangular lattice in which each row, transverse to the length of the strip, has an even number, $2K$, of atoms. The strip is connected to itself in the form of a ring of length, L. In the absence of vacancies each atom is bonded to one other atom, so that each atom has one bond ending at it. Each gap between two of the rows of $2K$ atoms is crossed by a number of bonds which is either even for each gap or odd, since an even number of bonds end on the row of atoms. If a row of atoms contains a single vacancy, then this number changes from odd on one side of the row to even on the other. The bonds can rearrange locally by electrons tunneling through doubly charged states of the atoms, but the parity of the number of bonds in a gap cannot change. Transport of a single vacancy right around the ring changes the parity in every gap as shown in Fig. 2.4. Therefore, the Bose statistics of the vacancies cannot lead to any superposition of states obtained from one another by any transport of the vacancies an odd number of times around the ring. The original configuration can be reached by transport of a pair of vacancies once around the ring or by a single vacancy twice around the ring. Therefore, the effective charge that determines the quantum of fluxoid quantization is $2e$ and not e.

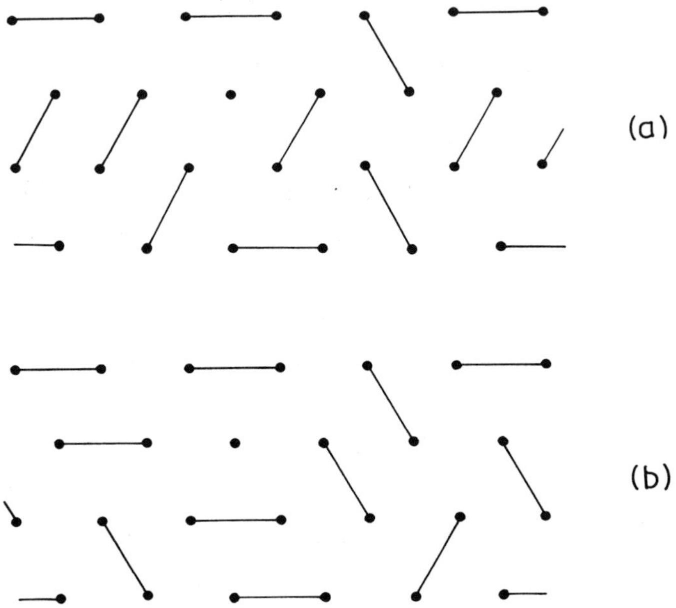

Fig.2.4: Triangular lattice with occupied and empty bonds. In both diagrams (a) and (b) there is one positively charged vacancy. In (a) the number of oblique bonds to the left of the vacancy is odd, 3, while those on the right hand side is even, 4. The configuration (b) is obtained from (a) by transporting the vacancy once around the ring.

2.4 Flux creep

It has been found that the product of the current and the magnetic critical field at a fixed temperature is a constant. If the current increases beyond this value, then the magnetic field creeps out of the sample of the superconductor. The constant product is a function of temperature,

$$\alpha(T) = J_{cr}(B_{cr} + B_o) \tag{2.38}$$

vanishing near the critical temperature, T_c. The temperature dependence of α can be explained by assuming that the mechanism of flux creep is the thermally activated motion of flux lines over free energy barriers coming from the pinning effect of inhomogenities, strains dislocations or other defects. The theory also explains the constant field B_o and predicts relaxation time effects. The magnetic field of a flux line extends over a distance of the order of the London penetration depth, λ_L. The length of a vortex of flux is of the order of d. The free energy barrier is the difference between the free energies of the normal state and that of the superconducting state,

$$(F_n - F_s) = (H_{CB}^2/8\pi)d^3 \quad , \tag{2.39}$$

where H_{CB} is the critical field in the bulk. The energy $H_c^2/8\pi$ is reduced by pinning by a factor of p. The Lorentz force on the flux vortex is,

$$f = \int \vec{J} \times \vec{B} \, d\tau \tag{2.40}$$

which contributes to the free energy approximately,

$$-JBxd^3 = -J\phi_B xd \tag{2.41}$$

since

$$Bd^2 = \phi_B \ . \tag{2.42}$$

The flux is quantized in the units of $\phi_o = hc/2e$. Thus the total free energy is

$$F(x) = F_i(x) - J\phi_B xd \ . \tag{2.43}$$

The free energy barrier which the vortex must climb to go from one stable minimum to another is of the order of,

$$\Delta F \simeq (pH_{cB}^2 d^3/8\pi) - J\phi_B d^2. \tag{2.44}$$

The rate of hopping for the flux vortex over the free energy barrier will be,

$$R = R_o \exp\left[-(1/k_B T)(pH_{CB}^2 d^3/8\pi - J\phi_B d^2)\right] \tag{2.45}$$

where R_o is an appropriate frequency. In the case of a tube of radius a and thickness w, with an interior field H' and an outer field H. If $H > H'$, the flux creep leads to the motion of flux vortices into the tube at a rate which may be estimated to be R times the number of vortices in the wall of the tube ($\simeq 2\pi a w H^*/\phi_B$) and divided by the number of barriers a vortex may see in traversing $w(= cw/d)$, where the factor c gives the affective number of barriers.

The equation determining the rate of creep is,

$$(H^*)^{-1}\left(\frac{d}{dt}\right)(H - H')$$
$$= -\left(\frac{2dR_o}{ca}\right)\exp\left[-\left(\frac{1}{k_B T}\right)\{(pH_{cB}^2 d^3/8\pi) - J\phi_B d^2\}\right]. \quad (2.46)$$

We define the rate,

$$K_1 = \frac{2dR_o}{ac}\exp\left[-\frac{pH_{CB}^2 d^3}{8\pi k_B T}\right] \quad (2.47)$$

and

$$\alpha(t) = J(t)[B(t) + B_o] \quad (2.48)$$

so that

$$\frac{d\alpha}{dt} = -H^*\left|\frac{d\alpha}{dH'}\right| K_1 \exp\left(\frac{\alpha d^4}{k_B T}\right) \quad (2.49)$$

because

$$\alpha = JB \quad (2.50)$$

for $x \simeq d$. Then the change in α is given by

$$\delta\alpha \simeq \text{const.} - \left(\frac{k_B T}{d^4}\right) lnt \quad (2.51)$$

$$\delta H' = \left(\frac{dH'}{d\alpha}\right)\left(\frac{k_B T}{d^4}\right) lnt \quad (2.52)$$

If dH'/dH and thus $dH'/d\alpha$ is steep, the rate becomes large and at a point instability would occur. The current and the magnetic field are related by the expression,

$$J(x) = J_1 \sin\left[(2\pi dHx/\phi_o) + \theta_o\right] \quad (2.53)$$

which for small argument of the sine function, $\sin\theta \simeq \theta$, becomes,

$$J(x) = J_1\left(\frac{2\pi dHx}{\phi_o} + \theta_o\right). \quad (2.54)$$

Thus the product $J_1H/J(x)$ is a constant and forms the basis of flux creep. Near the transition temperature, the resistivity is thermally activated,

$$\rho(T, H, \phi) = \rho_o \exp(-U_o/k_BT) \ . \tag{2.55}$$

For $H_\perp < 3T, \alpha = \frac{1}{2}$ otherwise α varies with temperature. Since both the directions of the hopping are permitted, in the direction of the Lorentz force, the rate of hopping is given by

$$R_f = \nu_o \exp\left[-(U_o - U_L)/k_BT\right] \tag{2.56}$$

and in opposite direction

$$R_b = \nu_o \exp\left[-(U_o + U_L)/k_BT\right]. \tag{2.57}$$

Therefore, the net hopping rate is

$$\nu_{eff} = \nu_o \exp\left[-U_o/k_BT\right] \sinh(U_L/k_BT) \tag{2.58}$$

The Lorentz force energy U_L is given by the Lorentz force density, $\vec{F}_L = \vec{J} \times \vec{B}$, the volume of the flux vortex V_c that moves independently of the other flux vortices and the rate of pinning potential r_p, as,

$$U_L = (J \times B)V_c r_p \ . \tag{2.59}$$

For current density J, for which $U_L \leq k_BT$ so that $\sinh(x) \simeq x$, the average flux velocity, $v_\phi = \nu_{eff}L$ with L the hopping distance, is given by,

$$v_\phi = 2\nu_o L \frac{JBV_c r_p}{k_BT} \exp(-U_o/k_BT) \ . \tag{2.60}$$

Thus $v_\phi(J)$ or $E(J)$ where E is the electric field, are linear in the current density. The current J is proportional to the electric field E only if $JBV_cr_p << k_BT$. For high fields, the flux vortex volume V_c can not be much larger than a_o^2d, which is the volume of one flux line, where a_o is the flux line separation and d is the sample thickness. If the correlation length, L_c is shorter than the samples thickness, d, the flux line lattice is said to be in the amorphous limit. The flux vortex volume is given by

$V_c = a_o^2 L_c$ where a_o is the hopping distance. The resistance caused by the flux motion becomes,

$$\rho = \frac{v_\phi B}{J} = \frac{2\nu_o \phi_o^2 L_c}{k_B T} \exp(-U_o/k_B T) , \tag{2.61}$$

where the preexponential factor ρ_o is related to the attempt frequency ν_o. For $L_c \simeq 0.1d$, the attempt frequency is about $\nu_o \simeq 10^{12}$ Hz. For large current densities within the flux creep regime, $J \geq k_B T/BV_c r_p$, the linear $E(J)$ turns into an exponential dependence.

2.5 Fractional flux

Consider an array of Josephson junctions with three parallel wires of aluminium in the superconducting state. These wires are slightly oxidized on the surface and two wires are put on them at an angle of 90°. The flux per unit cell is a fraction, f of ϕ_o and a is the lattice constant. The current density in the wire is given by,

$$J = \frac{e}{m}|\psi|^2 \left(\hbar \nabla \varphi - \frac{2e}{c} A\right) \tag{2.62}$$

where $|\psi|$ is the order parameter of the wires. We make the gauge as,

$$A = \frac{f x \phi_o}{a^2 \hat{y}}. \tag{2.63}$$

Setting J to zero implies that the phases φ^V of the vertical wires are position dependent with

$$\nabla \varphi^V = \frac{2\pi f x}{a^2 \hat{y}} \tag{2.64}$$

while the phases φ^H of the horizontal wires are constant. Above T_c the phases are uncorrelated. However, below the T_c, the phase coherence is expected. The Josephson energy at the transition temperature, $E_J(T = T_c)$ is related as follows,

$$2k_B T_c \simeq N E_J(T = T_c) \tag{2.65}$$

where

$$E_J(T) = \frac{\hbar i_c(T)}{2e} \tag{2.66}$$

and $i_c(T)$ is the critical current of a single junction at temperature T. The above expression can be written as,

$$E_J = \frac{i_c \phi_o}{2\pi} \tag{2.67}$$

which will be dicussed in Chapter V on the microwave absorption. The N dependence in (2.65) is unusual because it suggests that T_c depends on the size of the Josephson matrix. To keep the T_c of the array well below T_c^{wire}, we require

$$N\hbar i_c^o << 4ek_B T_c^{wire} \tag{2.68}$$

where i_c^o is $i_c(T=0)$. The phase drop in the wire is much less than that at the junction,

$$N^2 L_g i_c^o << \phi_o \tag{2.69}$$

where L_g is the geometric inductance of a cell in the array. When phase gradients along the wires due to current flow are much less than phase drops at the junction,

$$i_c^{\text{junction}} << i_c^{\text{wire}} \ . \tag{2.70}$$

The above conditions (2.67), (2.68) and (2.69) limit the value of i_c^b for a given value of N. The measurements of voltage and current for an array of Josephson junctions with long range interactions (ALRI), when current is in nA region, have been performed. The samples consist of 0.25 μm wide Al wires ($T_c^{\text{wire}} \simeq 1.7$ K) connected via oxides, to form Al-AlO-Al junctions cooled to 0.3 K. A small magnetic field is applied perpendicular to the array. The ratio of the maximum flux screened by the array to the flux quantum is much less than one, $N^2 Lgi_c^o/\phi_o \simeq 10^{-3} << 1$. The current voltage characteristics for single junctions do not show critical current at 0.3 K because the Josephson energy is smaller than $k_B T$ and hence a finite resistance is observed for all bias currents. However, array has a critical current. The jumps in the voltage correspond to,

$$eV = 2\Delta$$

and

$$eV = 4\Delta \ .$$

The differential resistance $R_d = dV/dI$ plotted as a function of field is found to be periodic in $f = \phi_{\text{cell}}/\phi_o$. As shown in Fig. 2.5 from the measurements carried out by Shea and Tinkham and interpreted by Shrivastava (see also Ch. 5). Shea and Tinkham used the same formalism as that found earlier in connection with microwave studies by Shrivastava.

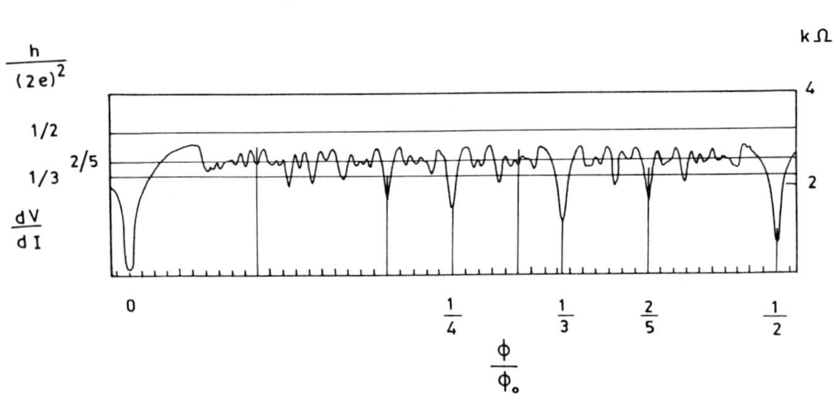

Fig.2.5: The resistivity showing dips at certain quantum of fluxes in a matrix of Al wires. When resistivity is measured in units of $h/(2e)^2$*, there is a plateau at 2/5.*

The minima are found at $f = 0, \frac{1}{2}, \frac{1}{3}, \frac{1}{4}, \frac{2}{5}$, etc., which shows the occurrence of fractional fluxes. The minimum observed flux is $\phi_o/4$ in a bulk superconductor as discussed in Ch.4 on magnetization.

Suggested References

2.1. M. Tinkham, Phys. Rev. B**6**, 1747 (1972).

2.2. B.S. Deaver, Jr., and W.M. Fairbank, Phys. Rev. Lett. **7**, 43 (1961).

2.3. N. Byers and C.N. Yang, Phys. Rev. Lett. **7**, 46 (1961).

2.4. C.N. Yang, Rev. Mod. Phys. **34**, 694 (1962).

2.5. D.J. Thouless, Phys. Rev. B**36**, 7187 (1987).

2.6. P.W. Anderson, Phys. Rev. Lett. **9**, 309 (1962).

2.7. Y.B. Kim, C.F. Hempstead and A.R. Strnad, Phys. Rev. Lett. **9**, 306 (1962).

2.8. T.T.M. Palstra, B. Batlogg, R.B. van Dover, L.F. Schneemeyer, J.V. Waszczak, Phys. Rev. B**41**, 6621 (1990).

2.9. H.R. Shea and M. Tinkham, Phys. Rev. Lett. **79**, 2324 (1997).

2.10. K.N. Shrivastava, Phys. Today **51**, [No.9], 15 (1998).

2.11. N. Murali Krishna, L. S. Lingam and K.N. Shrivastava, Mod. Phys. Lett. **11**, 849 (1997).

Chapter 3

Flux Lattices

3.1 Ginzburg-Landau model

We define an order parameter which has different values in different phases corresponding to the minima in the free energy. The minima of the free energy correspond to different phases of the system. The free energy may be written as,

$$F = \alpha|\psi|^2 + \frac{1}{2}\beta|\psi|^4 \ . \tag{3.1}$$

Minimizing the above free energy, $\partial F/\partial|\psi| = 0$, we find,

$$|\psi_o|[\alpha + \beta|\psi_o|^2] = 0 \tag{3.2}$$

which has two solution

$$|\psi_o| \ = \ 0 \tag{3.3a}$$

$$|\psi_o| \ = \ \pm\left(-\frac{\alpha}{\beta}\right)^{1/2} . \tag{3.3b}$$

There are two phases, one in which the order parameter is zero and the other in which there are two equal values except for the sign. This is analogous to the phase transition from a paramagnetic phase with zero spontaneous magnetization to a ferromagnetic phase with finite up or down magnetization. The parameter $\beta(T)$ is regular at the temperature $T = T_c$ whereas the parameter $\alpha(T)$ changes its sign from positive to negative at $T = T_c$. The Ginzburg-Landau model thus describes a phase transition by an order parameter. There can be many different terms in the free energy. Usually $|\nabla\psi|^2$ is used to provide the gradient from one value to another. In the presence of an electromagnetic field, $-i\hbar\vec{\nabla}$ is changed to $-i\hbar\vec{\nabla} - \frac{e}{c}\vec{A}$ where $\vec{A}$ is the vector potential of the electromagnetic field and the electromagnetic energy $H^2/8\pi$ is added so that the free energy becomes,

$$F = \alpha|\psi|^2 + \frac{1}{2}\beta|\psi|^4 + \frac{1}{2m}\left|\left(\frac{\hbar}{i}\nabla - \frac{e}{c}A\right)\psi\right|^2 + \frac{H^2}{8\pi} \tag{3.4}$$

where e and m are the charge and mass of the quasi-particles of the system. Due to the requirement of symmetry conservation at the critical temperature, the order parameter should be a complex quantity, $\psi(r) = |\psi(r)|\exp[i\theta(r)]$. An order parameter proportional to $\cos\theta$ can fix the phase to $\theta = 0$. The classical condensate wave function exhibits a spontaneously broken gauge symmetry implying that the average value of the order parameter is equal to the classical value. In the classical Ginzburg-Landau model, gauge symmetry can be restored by thermal fluctuations of the order parmeter. A plot of $F(\psi)$ in the complex ψ plane is given in Fig. 3.1. The minima of $F(\psi)$ consists of points on a circle of radius $|\psi_o|$. The fluctuations of ψ about the broken symmetry state, point A in Fig. 3.1 have two components, one along the tangent to the circle at $\theta = 0$, which corresponds to a pure phase fluctuation $\theta_1(r)$, and the other one which is amplitude fluctuation ψ_1.

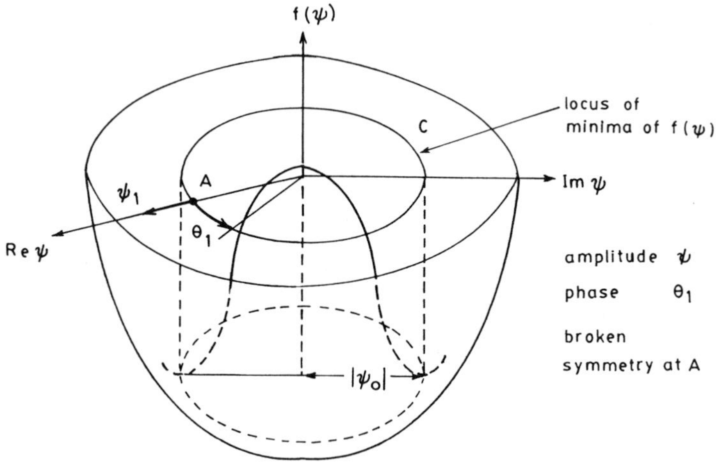

Fig. 3.1. Free energy as a function of order parameter.

3.2 Abrikosov vortices

In 1957 Abrikosov solved the Ginzburg-Landau model equations. It was found that in a plane perpendicular to the field direction, the magnetic field is nonuniformly distributed forming periodic patterns. There are points of high as well as low fields. There is a large field at the centre of the pattern called the vortex.

In the superconductors, there are two critical fields, $H_{c1} = \phi_o/\pi\lambda^2$ and $H_{c2} = \phi_o/\pi\xi^2$. For $H_{c1} < H < H_{c2}$, the superconductivity is maintained. For $H < H_{c1}$, Meissner effect is found but for $H > H_{c2}$, the superconductivity is destroyed. The two fields also define the two characteristic lengths, λ, the London penetration depth and ξ the coherence length.

The Ginzburg-Landau equations used by Abrikosov can be written as,

$$\left[i\frac{\nabla}{\kappa} + A\right]^2 \psi = \psi - \psi|\psi|^2 \tag{3.5}$$

$$-\text{curl curl}A = |\psi|^2 A + \frac{i}{2\kappa}(\psi^* \nabla \psi - \psi \nabla \psi^*) \tag{3.6}$$

The vector potential $\vec{A}$ is directed along the y-direction and the external field H_o is parallel to the z-direction. Near the transition $|\psi|^2 < 1$ so that the effect of ψ on H_o can be neglected. Therefore,

$$H = \text{constant} = H_o \tag{3.7}$$

$$A = H_o x \ . \tag{3.8}$$

Inserting (3.7) and (3.8) into (3.5), we obtain the oscillator type equation

$$\frac{d^2\psi}{dx^2} + \kappa^2(1 - H_o^2 x^2)\psi(x) = 0 \tag{3.9}$$

which has solutions when $H_o = \kappa/(2n+1)$. In particular, the lowest value $H_o = \kappa$ corresponds to,

$$\psi = \exp\{-(\kappa^2 x^2)/2\}. \tag{3.10}$$

In addition to such solutions, (3.5) is satisfied by

$$\psi = \exp\left\{iky - \frac{\kappa^2}{2}\left(x - \frac{k}{\kappa^2}\right)^2\right\} \tag{3.11}$$

so that the solution may be chosen as

$$\psi = \sum_{n=-\infty}^{\infty} C_n e^{ikny} \psi_n(x) \tag{3.12}$$

$$\psi_n = \exp\left\{-\frac{\kappa^2}{2}\left(x - \frac{kn}{\kappa^2}\right)^2\right\} \tag{3.13}$$

with arbitrary coefficients k and c_n. This solution is correct only when $H_o = \kappa$. For $H_o < \kappa$, we substitute (3.8) with $H_o = \kappa$ and ψ from (3.12) and (3.13) into (3.6) which gives $\partial^2 A/\partial x^2$ and $\partial^2 A/2x\partial y$. Then using $H = \partial A/\partial x$ we find that

$$\begin{aligned} H &= H_o - |\psi|^2/2\kappa && (3.14) \\ A &= H_o x - (2\kappa)^{-1}\int |\psi|^2 dx && (3.15) \end{aligned}$$

We now substitute (3.13) into (3.5) taking into account $\psi|\psi|^2$ term. Taking (3.12) and (3.13), adding $\psi_n^{(1)}(x)$ to all the $c_n\psi_n(x)$, we find,

$$\left(\frac{\kappa - H_o}{\kappa}\right)\overline{|\psi^2|} + \left(\frac{1}{2\kappa^2} - 1\right)\overline{|\psi|^4} = 0 \ . \tag{3.16}$$

The magnetic induction may be written into the form of

$$B = H_o - \overline{|\psi|^2}/2\kappa = H_o - (\kappa - H_o)/(2\kappa^2 - 1)\beta \tag{3.17}$$

where

$$\beta = \overline{|\psi|^4}\Big/(\overline{|\psi|^2})^2 \ . \tag{3.18}$$

The free energy per unit volume, F_1, using (3.5) and the Ginzburg-Landau formalism, is found to become,

$$F_1 = \frac{1}{2} + B^2 - \frac{(\kappa - B)^2}{[1 + (2\kappa^2 - 1)\beta]} \tag{3.19}$$

which is minimum when β is minimum for a constant B. The function β is minimum when

$$\frac{\partial\beta}{\partial c_n^*} = 0 \ . \tag{3.20}$$

The periodicity of ψ is determined from the recursion relation for c_n,

$$c_{n+N} = c_n \ . \tag{3.21}$$

The lowest energy corresponds to $N = 1$. In this case β is given by

$$\beta = \frac{k}{\kappa\sqrt{2\pi}}\left[\sum_n \exp\left\{-\frac{\kappa^2 n^2}{2\kappa^2}\right\}\right]^2 \ . \tag{3.22}$$

The minimum value of β is obtained when $k = \kappa(\sqrt{2}\,\pi)$ and is equal to

$$\beta = \theta_3^2(0,1) = \left[\sum_n e^{-\pi n^2}\right]^2 = 1.18 \tag{3.23}$$

where

$$\theta_3(z,q) = 1 + 2\sum_{n=1}^{\infty} q^{n^2} \cos 2nz \ . \tag{3.24}$$

Then ψ is,

$$\psi = ce^{-(\kappa^2 x^2)/2}\theta_3\left[\sqrt{2\pi}\,\kappa i(x+iy), 1\right] \ . \tag{3.25}$$

The rotation by $\pi/2$ multiplies ψ by the phase factor $e^{i\kappa^2 xy}$ dut to θ_3 but ψ remains unchanged otherwise. Then $|\psi|^2$ has the symmetry of a square lattice. When $x = y = (1/\kappa)(\sqrt{\pi/2})$, the function vanishes, $\psi = 0$. According to (3.17) H attains its maximum value H_o at three points and its minimum value is,

$$H_{min} = H_o - (H_{c2} - H_o)/(2\kappa^2 - 1)1.18 \tag{3.26}$$

The field and current distribution is shown in Fig. 3.2. The lines correspond to $|\psi|^2$ = constant, normalized to

$$|\psi|^2_{max} = 1 \ (H = \text{constant}) \tag{3.27}$$

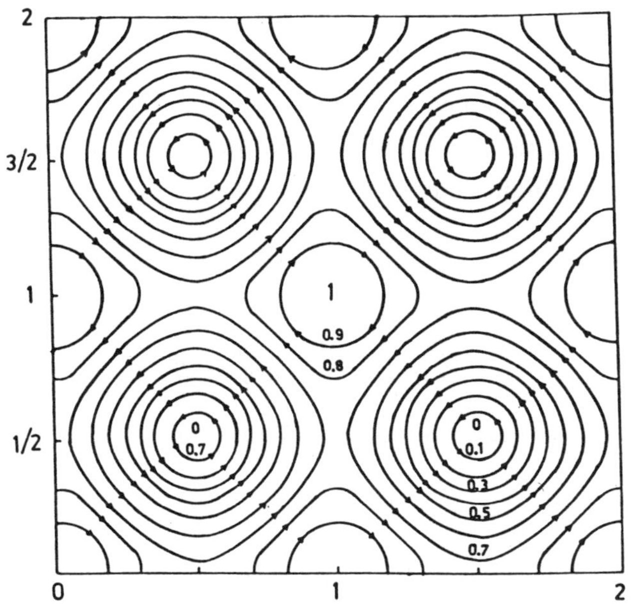

Fig. 3.2. Abrikosov's square lattice of vortices.

We introduce the phase and the modulus of the function such that,

$$\psi = f e^{i\phi} \ . \tag{3.28}$$

Then the free energy per unit length of a filament is

$$\epsilon = 2\pi \int_o^\infty \left[H^2 + \frac{1-f^4}{2} \right] r\, dr. \tag{3.29}$$

The scalar quantity Q is the absolute value of $(A - \nabla\phi/\kappa)$ so that

$$H = -\frac{1}{r}\frac{d}{dr}(rQ) \ . \tag{3.30}$$

The free energy (3.29) is then given

$$\epsilon = \pi \int_o^\infty \left[(1-f^4)r - f^2 \frac{d}{dr}(r^2Q^2) \right] dr. \tag{3.31}$$

The integrand decays exponentially when $r > 1$. For $ln\kappa >> 1$, the largest contribution to the integral occurs from the interval $1/\kappa << r << 1$ so that we can write the above, approximately, as,

$$\epsilon = \frac{2\pi}{\kappa^2}(ln\kappa + 0.081) \ . \tag{3.32}$$

For the formation of filaments, the quantity $F_1 - 2H_oB$ must be negative. The number of filaments per unit area, n, determines the free energy, F_1 and the magnetic induction,

$$F_1 = n\epsilon \tag{3.33}$$

and

$$B = n \int H ds = n \oint \vec{A}.d\vec{l} \ . \tag{3.34}$$

At large distances $Q = 0$, so that $A = \nabla\phi/\kappa$ and

$$B = 2\pi n/\kappa \ . \tag{3.35}$$

Therefore, the formation of filaments starts when the field strength H_o attains the value,

$$H_{c1} = \epsilon\kappa/4\pi \ . \tag{3.36}$$

When $\kappa >> 1$,

$$H_{c1} = (1/2\kappa)(ln\kappa + 0.08) \ . \tag{3.37}$$

The equation for the field is

$$\Delta H - H = -(2\pi/\kappa)\sum_m \delta(r - r_m) \tag{3.38}$$

when r_m are the coordinates of the centres of the filaments.
The solution of this equation is,

$$H = \frac{1}{\kappa}\sum_m K_o(|r - r_m|) \ . \tag{3.39}$$

The energy can be written in the form,

$$F_1 = \overline{H(H - \Delta H)} \tag{3.40}$$

$$= \frac{2\pi n}{\kappa^2}\sum_m K_o(|r_m|) \ . \tag{3.41}$$

The filament surface density, n and the periodicity of the lattice in (3.41) can be expressed in terms of the magnetic induction B by (3.35). For a square-lattice,

$$F_1 = \frac{\kappa B}{2\pi}\epsilon + \frac{B}{\kappa}\sum_{l^2+m^2\geq 1} K_o\left[\left\{\frac{4\pi(l^2+m^2)}{\kappa B}\right\}^{1/2}\right] \tag{3.42}$$

and for a triangular lattice

$$F_1 = \frac{\kappa B}{2\pi}\epsilon + \frac{B}{\kappa}\sum_{l^2+m^2+lm\geq 1} K_o\left[\left\{\frac{4\pi(l^2+m^2+lm)}{\sqrt{3}\,\kappa B}\right\}^{1/2}\right] \ . \tag{3.43}$$

For small fields, the triangular lattice is favoured over the square lattice. As the field is increased from H_{c1} the triangular lattice goes to the square lattice via a first-order transition. The triangular solution shown in Fig. 3.3 was discussed by Kleiner et al after Abrikosov's work.

Fig. 3.3. The triangular lattice was found to have lower energy than the square lattice. The triangular lattice is displayed (Kleiner et al). In real materials, both the square lattice as well as the triangular lattices have been found.

Each fluxoid, or vortex, is associated with a single quantum of flux represented by

$$\phi_o = \frac{hc}{2e} = 2.07 \times 10^{-7} Gcm^2 \tag{3.44}$$

and is surrounded by a circulating supercurrent, J, of spatial extent, λ. As the applied field increases, the fluxoids begin to interact and as a consequence assemble themselves into a lattice. A simple geometrical argument for the spacing, d of a triangular lattice then gives the flux quantization condition,

$$Bd^2 = \frac{2}{\sqrt{3}} \phi_o \tag{3.45}$$

where B is the induction, to be compared with (3.35). The flux quantization is known to occur according to (3.45). However, other values of the flux are not excluded.

3.3 Flux-lattice melting

The solution of the Ginzburg-Landau phenomenological free energy is useful for the understanding of the Abrikosov flux lattice. The coordinate-dependent order parameter $\phi(r|0)$ describes the flux vortices of the periodicity of a triangular lattice. Fluctuations from $\phi(r|0)$ change the state to ψ. the minimization of the free energy with respect to ψ gives the ground state, $\phi(r|0)$. The free energy is given by,

$$F = \alpha|\psi|^2 + \frac{1}{2}\beta|\psi|^4 + (2m)^{-1}|(-i\hbar\vec{\nabla} - 2eA/c)\psi|^2 + (B-H)^2/8\pi \quad (3.46)$$

where $\vec{B} = \text{curl}\ \vec{A}$ and the average magnetic induction is $\vec{B}(-y,0,0)$. The free energy has solutions of vortices of triangular form. The coordinates of the three vertices of a triangular vortex are given by $(0,0),(l,0)$ and $(\frac{1}{2},\frac{\sqrt{3}}{2})l$. The fluctuation from the ground state corresponding to that of the triangualr lattice is given by,

$$\delta\psi = ia\left[e^{iqz}\phi(r|r_o) + e^{-iqz}\phi(r|-r_o)\right] \quad (3.47)$$

so that for small fluctuations,

$$\delta\psi = \psi - a_1\phi(r|0)\ . \quad (3.48)$$

The free energy correct upto the order $O(a^2)$ is

$$F = F_{MF} + \frac{1}{2}\sum_{q,k}(\rho_s q^2 + c_{66}P^2k^4)a^2/a_1^2 \quad (3.49)$$

where F_{MF} is the mean field value of the free energy and the next term gives the fluctuations with ρ_s as the superfluid density and c_{66} the shear modulus.

The deviation of the free energy from the mean-field value $F - F_{MF}$, with respect to the thermal energy, k_BT, can be used to obtain the physical properties of the fluctuations which are useful for the understanding of the melted vortex lattices in superconductors. The deviation from the triangular Abrikosov lattice is defined by,

$$D = \ < |\psi - a_1\phi(r|0)|^2 > /a_1^2 \quad (3.50)$$

which uses the spatial and thermal averages calculated with a relative probability $\exp(-F/k_BT)$. Classically,

$$D = \frac{k_BT}{F - F_{MF}} \tag{3.51}$$

measures the fluctuation from the triangular vortex state. The wave vector along the z-direction is called q and that in the xy plane is called k so that the three dimensional integration of the deviation may be written as

$$D = \int \frac{dq}{2\pi} \int_{3z} \frac{d^2k}{(2\pi)^2} \frac{k_BT}{(\rho_s q^2 + c_{66}P^2k^4)} \tag{3.52}$$

where the shear modulus of the flux lattice is

$$c_{66} = \frac{0.48H_{c2}^2(2\kappa^2 - 1)(1 - \bar{B}/H_{c2})^2}{8\pi[1 + (2\kappa^2 - 1)\beta]^2} \tag{3.53}$$

as calculated by Labusch and the superfluid density is,

$$\rho_s = \frac{\phi_o H_{c2}(1 - \bar{B}/H_{c2})}{4\pi^2[1 + (2\kappa^2 - 1)\beta]} \tag{3.54}$$

where $\phi_o = hc/2e$ and $\kappa = \lambda/\xi$, is the Ginzburg-Landau parameter. Twice the area of the triangular flux vortex is called $(\sqrt{3}/2)l^2 = 2\pi P$. The q-space integral in the above may be solved by using the method of contour integration for infinite limits, so that

$$\oint \frac{dq}{q^2 + c_{66}P^2k^4/\rho_s} = \frac{\pi\rho_s^{1/2}}{Pk^2c_{66}^{1/2}} \tag{3.55}$$

the substitution of which in (3.52) gives,

$$D = \frac{k_BT}{4\pi P\rho_s^{1/2}c_{66}^{1/2}} \int \frac{dk}{k} \,. \tag{3.56}$$

The value $k = 0$ gives a divergence which is removed by assuming a finite size sample. We consider L as the linear dimension of the sample in a direction perpendicular to the applied field so that the minimum value of the wave vector k becomes $1/L$. The maximum value of k is limited by the size of the vortex to $(2/P)^{1/2}$. The value of the integral is then found to give,

$$D = \frac{k_BT}{4\pi P\rho_s^{1/2}c_{66}^{1/2}} ln\frac{2^{1/2}L}{P^{1/2}} \tag{3.57}$$

the accuracy of which is limited by use of infinite limits of q so that the number of flux lines is also infinite. In order to obtain finite number of flux lines and hence finite energy of the flux lines, the infinite limits have to be removed. Substituting (3.53) and (3.54) and the temperature dependent upper critical field,

$$H_{c2} = H_{c2}^{o}(1 - T/T_c) \tag{3.58}$$

in (3.57) we find that

$$D = \frac{2(2\pi)^{1/2}\kappa^2\beta^{3/2}k_B T \ln(2^{1/2}L/P^{1/2})}{(0.48)^{1/2}\phi_o^{1/2}H_{c2}^{3/2}[(1-\bar{B}/H_{c2})(1-T/T_c)]^{3/2}} \,. \tag{3.59}$$

We use the Ginzburg criterion parameter,

$$\epsilon = \frac{16\pi^3\kappa^4(k_B T_c)^2}{\phi_o^3 H_{c2}} \tag{3.60}$$

to eliminate κ so that the fluctuation from the Abrikosov triangular state becomes,

$$D = \frac{\epsilon^{1/2}\ln(L2^{1/2}/P^{1/2})(T/T_c)\beta^{3/2}\phi_o}{(0.48)^{1/2}2^{1/2}\pi H_{c2}[(1-\bar{B}/H_{c2})(1-T/T_c)]^{3/2}} \tag{3.61}$$

when $\bar{B} = H_{c2}$ the fluctuation given by this expression diverges. At $T = T_c$ also there is a similar divergence which shows that there is a phase transition at this point, and the corresponding exponent is 3/2.

In order to introduce finite limits on q we write the expression (3.52) as,

$$D = \frac{k_B T}{(2\pi)^2\rho_s}\int\frac{dq}{q^2 + (c_{66}P^2k^4/\rho_s)}\int k\,dk \,. \tag{3.62}$$

We introduce the cutoff value of q as $q_m = (2b)^{1/2}/\xi_z$ where $b = \bar{B}/B_{c2}(T)$ and ξ_z is the coherence length along the z-direction. This is a very important effect as it removes the unphysical divergence in the number of flux lines and introduces coherence along the z-direction. Similarly, coherence effects occur in the xy plane by introducing finite cutoff in the k space. The effect of finite boundary conditions is therefore to introduce wave propagation in the vortex lattice which is absent when infinite boundary conditions are used. Performing finite limit integration in the q space from $-q_m$ to $+q_m$ we find,

$$D = \frac{k_B T}{(2\pi)^2 P \rho_s^{1/2} c_{66}^{1/2}} \int \frac{1}{k} \left[\tan^{-1}\left(\frac{q_m \rho_s^{1/2}}{P k^2 c_{66}^{1/2}}\right) - \tan^{-1}\left(-\frac{q_m \rho_s^{1/2}}{P k^2 c_{66}^{1/2}}\right) \right] dk \tag{3.63}$$

in which the integrals are of the form

$$I_{\pm} = \int \frac{1}{k} \tan^{-1}\left(\pm\frac{A}{k^2}\right) dk \tag{3.64}$$

with the lower limit, $k_{min} = 1/L$, corresponding to the inverse length of the sample and $k_{max} = (2/\rho)^{1/2}$ corresponding to the inverse size of the triangular vortex. We make the change of variables,

$$x = \tan^{-1}\left(\frac{A}{k^2}\right) \tag{3.65}$$

in (3.64) and integrate by parts to obtain,

$$\begin{aligned} I_{\pm} &= (1/2)\tan^{-1}(AL^2)\log\tan\{\tan^{-1}(AL^2)\} \\ &\quad -(1/2)\tan^{-1}(AP/2\log\tan\{\tan^{-1}(AP/2)\} \end{aligned} \tag{3.66}$$

where we have ignored a small contribution of

$$\int_{\tan^{-1}(AL^2)}^{\tan^{-1}(AP/2)} \log\tan x \, dx \tag{3.67}$$

as

$$\int_o^{\pi/2} \log\tan x \, dx = 0 \ . \tag{3.68}$$

In (3.67),

$$A = \frac{q_{in} \rho_s^{1/2}}{P c_{66}^{1/2}} \tag{3.69}$$

The fluctuation (3.63) is then given by

$$D = k_B T (I_+ - I_-)/[(2\pi)^2 P \rho_s^{1/2} c_{66}^{1/2}] \tag{3.70}$$

which diverges at

$$\tan\{\tan^{-1}(AL^2)\} = 0 \tag{3.71}$$

which is equivalent to $AL^2 = \tan(2n\pi)$ and hence

$$\left[\frac{q_m \rho_s^{1/2}}{P c_{66}^{1/2}}\right] L^2 = 0 \tag{3.72}$$

This means that the fluctuation diverges when the superfluid density $\rho_s = 0$. From (3.54) this is the same point where the average induction is cqual to the upper critical field. This is the superconducting to the normal state transition. Since q_m has the coherence length along the z direction in the denominator, the zero of q_m corresponds to infinity of ξ_z. Then the fluctuation becomes zero and we obtain the triangular flux lattice. The second term of (3.66), also leads to similar conclusions. Thus we find a degenerate infinity at the transition temperature as well as at the upper critical field. Since log zero$=-\infty$, we find that $\pm\infty$, with both signs occur in the fluctuation at the same point.

The Lindemann criterion suggests that the lattice melts when the displacement becomes equal to one tenth of the cell size. In our fluctuation in the free energy from that of the triangular lattice, the cutoff wave vectors $q_m = (2b)^{1/2}/\xi_z$ and $k_m = (2/P)^{1/2}$ occur which represent inverse distances. Application of the melting criterion to the distance ξ_z or to the distance $P^{1/2}$ suggests that about ten times larger wave vector limit should be taken than has been used in the above calculation. Although, the phonon-induced forces between electrons at random sites are attractive, there are pinning forces which are repulsive. The modulation of pinning forces in some way may led to a mechanism of melting. However, it is quite clear that the vortices of flux lines superflow as in Bose-Einstein condensation except that the present phenomenon is found in the superconducting state. We are able to find the divergences in the fluctuations of the Abrikosov triangular lattices. By relaxing the limits on the wave vectors, it is possible to shift the divergences below the $T_c(H_{c2})$ line so that a description of the flux lattice melting can be obtained in terms of suitable cutoff values for the wave vectors.

The mean square fluctuation in the distance between vortices is given by,

$$d^2(T) =< (u_x^2 + u_y^2) > \quad . \tag{3.73}$$

As in the case of (3.62), we take cylindrically symmetric flux lines so that the wave vector in the xy plane is k and along z direction, it is q so that (3.73) can be written as an integral,

$$d^2(T) = P^2 \int \frac{dq}{2\pi} \int \frac{d^2k}{(2\pi)^2} \frac{k_B T k^2}{(\rho_s q^2 + c_{66} P^2 k^4)} \ . \tag{3.74}$$

The wave vector k is restricted within the Brillouin zone. The superfluid density ρ_s is given by (3.54) and the shear modulus of the flux lattice is given by (3.53). The double integral in the above may be written as

$$d^2(T) = \frac{P^2 k_B T}{(2\pi)^3 \rho_s} \int \frac{dq}{q^2 + (c_{66} P^2 k^4/\rho_s)} \int k^2 d^2k \ . \tag{3.75}$$

We consider the dq part of the integral which along the large semicircle in the upper-half plane is zero as long as $c_{66}P^2k^4/\rho_s << q$. There are two poles, $q = \pm i(c_{66}P^2k^4/\rho_s)^{1/2}$ but only one is inside the contour in the upper-half plane as shown in Fig. 3.4. Therefore,

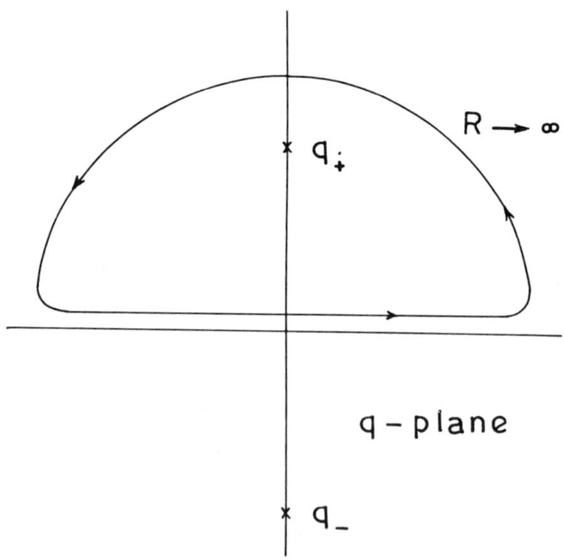

Fig. 3.4. The contour at $R \to \infty$ and the poles of the integral in (3.76) with $q_\pm = \pm i(c_{66}P^2k^4/\rho_s)$ in the complex q plane.

$$\oint \frac{dq}{q^2 + c_{66}P^2k^4/\rho_s} = \frac{\pi}{Pk^2}\left(\frac{\rho_s}{c_{66}}\right)^{1/2} \tag{3.76}$$

substituting (3.76) in (3.75) and taking dk integral over a circular Brillouin zone of radius Λ where $\Lambda^2 = 2/P$, we find,

$$d^2(T) = \frac{k_BTP}{4\pi(\rho_s c_{66})^{1/2}} \int_o^{(2/P)^{1/2}} k\, dk \tag{3.77}$$

the solution of which is given by

$$d^2(T) = \frac{k_BT}{4\pi(\rho_s c_{66})^{1/2}} \tag{3.78}$$

which is obtained by using infinite boundary conditions for q as in (3.76) and a circular zone for k. The infinite limit for q introduces unphysical divergence in the ground state energy as well as in the number of flux lines. Therefore, instead of using infinite limits as in (3.76) we obtain the q-space integral with a finite cutoff, q_m. The problem is slightly different from that of (3.77) as (3.75) becomes,

$$d^2(T) = \frac{Pk_BT}{(2\pi)^3(\rho_s c_{66})^{1/2}} \int d^2k \tan^{-1}\left[\frac{q_m}{Pk^2}\left(\frac{\rho_s}{c_{66}}\right)^{1/2}\right] . \tag{3.79}$$

Using the abbreviation,

$$a^2 = (q_m/P)(\rho_s/c_{66})^{1/2} \tag{3.80}$$

it is seen that the integral in (3.79) reduces into the form,

$$d^2(T) = \frac{Pk_BT\, I}{(2\pi)^2(\rho_s c_{66})^{1/2}} \tag{3.81}$$

where

$$I = \int_o^{(2/P)^{1/2}} k \tan^{-1}\left(\frac{a^2}{k^2}\right) dk . \tag{3.82}$$

Integrating by parts, the above integral gives

$$I = (1/2)k^2 \tan^{-1}\left(\frac{a^2}{k^2}\right)\Bigg|_o^{(2/P)^{1/2}} + a^2 \int_o^{(2/P)^{1/2}} \frac{k^3 dk}{(k^4 + a^4)} \tag{3.83}$$

the solution of which is

$$I = (1/P)\tan^{-1}\left[\frac{q_m}{2}\left(\frac{\rho_s}{c_{66}}\right)^{1/2}\right] + \frac{q_m}{4P}\left(\frac{\rho_s}{c_{66}}\right)^{1/2} \ln\left(\frac{4c_{66}+q_m^2\rho_s}{q_m^2\rho_s}\right) \tag{3.84}$$

so that the fluctuation in the distance between vortices becomes,

$$\begin{aligned} d^2(T) &= \frac{k_BT}{(2\pi)^2(\rho_s c_{66})^{1/2}}\left[\tan^{-1}\left\{\frac{q_m}{2}\left(\frac{\rho_s}{c_{66}}\right)^{1/2}\right\}\right. \\ &\left. + \frac{q_m}{4}\left(\frac{\rho_s}{c_{66}}\right)^{1/2} \ln\left(\frac{4c_{66}+q_m^2\rho_s}{q_m^2\rho_s}\right)\right] . \end{aligned} \tag{3.85}$$

Taking the value of the angle arising from the $\tan^{-1}$ term to be π, the first term of (3.85) exactly agrees with (3.78) but then the second term of (3.85) provides correction to (3.78). Thus the result (3.85) is more accurate than (3.78). The wave vector along the z direction varies from zero to q_m with the restriction $q_m = (2b)^{1/2}/\xi_z$ on the maximum value. The average value of the field and the upper critical field determine, $b = \bar{B}/B_{c2}(T)$ and ξ_z is the coherence length along the z direction.

The Lindemann's criterion is that when the fluctuation becomes equal to about one tenth of the lattice spacing, then the lattice melts. We use this rule on the fluctuation given by (3.85) along with the flux quantized lattice spacing. The flux-lattice melting temperature, T_{FM}, and hence the $H_{c2}(T_{FM})$ line depends upon the relative importance of the two terms of (3.85). Introduction of the critical behaviour opens up the possibility of a normal region above $T_c(H_{c2})$. Thus there is a normal region and two superconducting regions, one with a flux lattice and the other with a flux liquid.

Case I. We assume that $\tan^{-1}[(q_m/2)(\rho_s/c_{66})^{1/2}] \simeq \pi$ and the logarithmic term is small compared with the previous term. Then using (3.52) and (3.53), we find,

$$d^2(T) = \frac{(2\pi)^{1/2}k_BTH_{c2}^{-3/2}(1-\bar{B}/H_{c2})^{-3/2}f^{3/2}}{0.69\,\phi_o^{1/2}(2\kappa^2-1)^{1/2}} . \tag{3.86}$$

Using the Lindemann criterion, $d^2 = c^2l^2$, the critical behaviour of the upper critical field $H_{c2} = H_{c2}(1 - T/T_c)$, we find,

$$[(0.69)^{\frac{1}{3}}\phi_o^{\frac{1}{3}}\pi^{-\frac{1}{3}}2^{-\frac{1}{3}}(2\kappa^2 - 1)^{\frac{1}{3}}H_{c2}(1 - \frac{T_{FM}}{T_c})(1 - \frac{\bar{B}}{H_{c2}})(k_BT)^{-\frac{2}{3}}f^{-1}]$$
$$= c^{-4/3}l^{-4/3} \quad (3.87)$$

in which we apply the fluz-quantisation condition $l^2 = \phi_o/B$ so that (3.87) becomes

$$[(0.69)^{2/3}\phi_o^{1/3}2^{-1/3}\pi^{-1/3}(2\kappa^2 - 1)^{1/3}H_{c2}(1 - T_{FM}/T_c)(1 - \bar{B}/H_{c2})$$
$$\times(k_BT)^{-2/3}f^{-1}] = c^{-4/3}\phi_o^{-2/3}B^{2/3} \quad (3.88)$$

This result is equivalent to $(1 - T_{FM}/T_c) \simeq 10^{-5}c^{-4/3}B^{2/3}$ and its exponent is consistant with early experiments on the high temperature superconductors.

Case II. For a small angle, $\tan^{-1}x = x$ with logarithmic term smaller than $\tan^{-1}x$ term, the result is different from that found above. This approximation is equivalent to, $(1/2)q_m\rho_s^{1/2} << c_{66}^{1/2}$. In this case (3.85) becomes,

$$d^2(T) = k_BTq_m/2(2\pi)^2c_{66} \ . \quad (3.89)$$

Using (3.53), the Lindemann criterion we find,

$$\frac{8\pi k_BT_{FM}q_mf^2}{0.96(2\pi)^2H_{c2}^2(2\kappa^2 - 1)(1 - \bar{B}/H_{c2})^2} \simeq c^2l^2 \ . \quad (3.90)$$

In the length of the triangular vortex, we use the flux-quantization condition so that the above becomes,

$$\frac{8\pi k_BT_{FM}q_mf^2}{0.96(2\pi)^2(2\kappa^2 - 1)(1 - \bar{B}/H_{c2})^2H_{c2}^2(1 - T_{FM}/T_c)^2} = \frac{c^2\phi^2}{B^2} \quad (3.91)$$

which means that

$$1 - T_{FM}/T_c \simeq c^{-1}B \quad (3.92)$$

apart from the constants, so that the flux-melting temperature measured from the critical temperature becomes a linear function of the magnetic field. However, the cutoff wave vector along the z-direction introduces the average field so that (3.91) becomes,

$$\frac{0.96(2\pi)^2(2\kappa^2 - 1)(1 - \bar{B}/H_{c2})^2H_{c2}^2(1 - T_{FM}/T_c)^2\xi_zB_{c2}^{1/2}}{8\pi f^2k_BT_{FM}(2\bar{B})^{1/2}} = c^{-2}\phi_o^{-2}B^2$$
$$(3.93)$$

so that the critical phase boundary is predicted to be,

$$1 - (T_{FM}/T_c) \simeq c^{-1}B^{5/4} \tag{3.94}$$

which has much larger exponent than (3.88). However, (3.94) applies only when the maximum value of the wave vector along the z direction is small, i.e., $q_m << 2(c_{66}/\rho_s)^{1/2}$.

Case III. For large logarithmic term in (3.85) compared with the arctan term, the result is similar to (3.93) as the logarithmic term has only small variations for large changes in its argument. In high-temperature superconductors, there is a vortex transition below the transition temperature. The flux-lattice melting temperature as a function of the upper critical field is determined from the fluctuations of the flux-lattice spacing. We have assumed that there is a flux lattice so that there is wave propagation. Such a wave is quantized and obeys the Bose-Einstein statistics. Therefore, we investigate the effect of such waves on the melting curve. We define a dynamical mass from the wave vector such that the single particle energy of the fluxon is $\hbar v|k|$ which is anisotropic because of the anisotropy of the wave vector. The frequency wave vector dispersion relation then defines the velocity v of the fluxon. The correlation function is $2n_k + 1 = \coth(\hbar v|k|/2k_BT)$ becoming $2k_BT/\hbar v|k|$ in the high-temperature approximation. The correlated thermal energy is then $(1/2)\hbar v|k|\coth(\hbar v|k|/2k_BT)$ instead of the classical value k_BT. Since there is a Bose distribution, n, for fluxons of velocity v, we write for the square of the flux displacement in the xy plane as, instead of (3.74),

$$d^2(T) = P^2 \int \frac{dq}{2\pi} \int \frac{d^2k}{(2\pi)^2} \frac{\hbar v k^3(n+1/2)}{(\rho_s q^2 + c_{66}P^2k^4)} \tag{3.95}$$

in which we make the change of variables $\hbar kv/k_BT = x$, so that,

$$d^2(T) = P^2 \int \frac{dq}{(2\pi)^3} \int \left(\frac{k_BT}{\hbar v}\right)^5 \frac{\hbar v[(e^x-1)^{-1}+1/2]x^4dx}{[\rho_s q^2 + c_{66}P^2x^4(k_BT/\hbar v)^4]} \tag{3.96}$$

which has two terms. One with $n = 0$ gives the displacement due to zero-point fluxons in a rigid lattice and another, due to thermal excitation of fluxons so that,

$$d^2(T) = d^2(o) + d^2(n)\ . \tag{3.97}$$

The zero-point contribution is,

$$d^2(o) = \frac{P^2(k_BT)^5}{2(2\pi)^3\hbar^4v^4}\int dq\int \frac{x^4dx}{[\rho_s q^2 + c_{66}P^2x^4(k_BT/\hbar v)^4]} \ . \tag{3.98}$$

We integrate (3.96) without the approximation of small or large superfluid density. The improved solution of (3.96) is given below.
Integrating q from $-\infty$ to $+\infty$,

$$\oint \frac{dq}{q^2 + \frac{c_{66}P^2x^4}{\rho_s}\left(\frac{k_BT}{\hbar v}\right)^4} = \frac{\pi\rho_s^{1/2}}{c_{66}^{1/2}Px^2(k_BT/\hbar v)^2} \ . \tag{3.99}$$

Substituting (3.99) in (3.96) we find that

$$d^2(T) = \frac{Pk_B^3T^3}{4\pi\rho_s^{1/2}c_{66}^{1/2}}\int [(e^x-1)^{-1}+1/2]x^2\,dx \ . \tag{3.100}$$

We break up (3.100) into two terms, one due to the zero-point quantum mechanical oscillations and the other due to finite number density so that,

$$d^2(o) = P\hbar^3v^3k_m^3/12\pi\rho_s^{1/2}c_{66}^{1/2} \tag{3.101}$$

and the next term as,

$$d^2(T) = \frac{Pk_B^3T^3}{4\pi\rho_s^{1/2}c_{66}^{1/2}}\int \frac{x^2dx}{e^x-1} \ . \tag{3.102}$$

At low temperatures the fluctuations vary as T^3. When infinite limits for q as in (3.99) are not applicable, we obtain this integral in an alternative manner. We assume that the maximum value of q is $q_m = (2b)^{1/2}/\xi_z$, with $b = \bar{B}/B_{c2}(T)$. Then (3.96) upon integration from $-q_m$ to $+q_m$ becomes,

$$\begin{aligned} d^2(T) &= \frac{P(k_BT)^3}{(2\pi)^2(\hbar v)^2\rho_s^{1/2}c_{66}^{1/2}}\int\left[\tan^{-1}\left\{\frac{q_m(\hbar v)^2\rho_s^{1/2}}{Px^2(k_BT)^2c_{66}^{1/2}}\right\}\right. \\ &\quad \left. -\tan^{-1}\left\{-\frac{q_m(\hbar v)^2\rho_s^{1/2}}{Px^2(k_BT)^2c_{66}^{1/2}}\right\}\right] \\ &\quad \times \left[(e^x-1)^{-1}+\frac{1}{2}\right]x^2dx \end{aligned} \tag{3.103}$$

which we write as

$$d^2(T) = \frac{P(k_BT)^3}{(2\pi)^2(\hbar v)^2\rho_s^{1/2}c_{66}^{1/2}}\int\left[\tan^{-1}\left(\frac{Q}{x^2}\right)-\tan^{-1}\left(-\frac{Q}{x^2}\right)\right]$$
$$\times\left[(e^x-1)^{-1}+\frac{1}{2}\right]x^2dx \qquad (3.104)$$

where

$$Q = \frac{q_m(\hbar v)^2\rho_s^{1/2}}{P(k_BT)^2c_{66}^{1/2}} \ . \qquad (3.105)$$

The fluctuation then becomes,

$$d^2(T) = \frac{P(k_BT)^3}{(2\pi)^2(\hbar v)^2\rho_s^{1/2}c_{66}^{1/2}}[I_1(Q)-I_1(-Q)] \qquad (3.106)$$

where

$$I_1(\pm Q) = \int\tan^{-1}\left(\pm\frac{Q}{x^2}\right)[(e^x-1)^{-1}+1/2]x^2\,dx \ . \qquad (3.107)$$

When

$$tan^{-1}\left(\pm\frac{Q}{x^2}\right) = \frac{\pi}{2}$$

then

$$\frac{Q}{x^2} = \frac{q_m(\hbar v)^2\rho_s^{1/2}}{x^2P(k_BT)^2c_{66}^{1/2}} = \infty \qquad (3.108)$$

which we write as

$$\frac{q_m(\hbar v)^2\rho_s^{1/2}}{(\hbar kv/k_BT)^2P(k_BT)^2c_{66}^{1/2}} = \infty \qquad (3.109)$$

which is possible if c_{66} goes to zero. According to the Lindemann criterion, the lattice is said to melt when $d^2 = c^2l^2$ for $c \simeq 0.1$. Substituting $T = T_M$ in (3.106) and equating d^2 to c^2l^2 we obtain the flux lattice melting temperature. In this calculation, the superfluid density term is comparable with the term containing c_{66}. The introduction of bosons thus leads to interesting effects.

The order parameter, ψ of the Ginzburg-Landau free energy has a phase factor of $e^{i\phi}$. The gauge-invariant phase has a finite average value without any divergence what so ever. A divergence in the phase, if found, will be of interest from the view point of phase transitions in the flux lattice. Therefore, we search for possible divergences in the phase, ϕ. We find that the phase does not diverge except at the upper critical field. We may write the Ginzburg-Landau phenomenological free-energy density functional

$$F = -|\psi|^2 + \frac{1}{2}|\psi|^4 + H^2 + \left|\left(\frac{\vec{\nabla}}{i\kappa} - \vec{A}\right)\psi\right|^2 \tag{3.110}$$

which shows several values of ψ, one in each phase for which F is minimized. Here $\kappa = \lambda/\xi$, with λ as the London penetration depth and ξ as the coherence length. $\vec{A}$ is the vector potential of the electromagnetic wave. The thermal average of ψ due to elastic fluctuations is defined in terms of a phase factor such that,

$$<\psi>= |\psi| < e^{i\phi} > \tag{3.111}$$

where

$$<\phi^2>= b^2\kappa^4 \int\int \frac{dk_z k_\perp^2 dk_\perp^2 k_B T}{8\pi^2 k^4 [c_{66}^2 k_\perp^2 + c_{44} k_z^2]} \tag{3.112}$$

with $b = \bar{B}/B_{c2}$ and the components of $\vec{k}$ as k_z and $k_\perp$ and the elastic constants are c_{44} and c_{66}. The solution of the above integral when k_z varies from zero to infinity and $k_\perp$ from zero to a finite value shows that there is a logarithmic divergence in (3.112) as a function of size of the system. However, the lower limit of k may be replaced by a finite value so that this trivial divergence disappears. It is then of interest to see whether any divergence occurs or not. In the limit $k^2 \geq < \omega_o >$ the elastic constant, $c_{44} \sim (B^2/4\pi) < \omega_o > /k^2$ where,

$$<\omega_o>= 2\kappa^2(1-b)/[(2\kappa^2-1)\beta+1] \tag{3.113}$$

which has a zero at $b = 1$ but the divergence from $c_{44}(k)$ at $k = 0$ is removed. In this process c_{44} has a value of zero at $b = 1$ for finite $|\vec{k}|$, $\beta = 1.16$ for the triangular flux lattice and (3.42) becomes,

$$<\phi^2>\approx \frac{b^2\kappa^4 k_B T}{8\pi^2} \int dk_z \int \frac{k_\perp^2 dk_\perp^2}{(k_\perp^2 + k_z^2)[c_{66}k_\perp^4 + k_z^2(c_{66}k_\perp^2 + <\omega_o> B^2/4\pi)]} \tag{3.114}$$

This integral can be solved in several special approximations. In the case of cylindrical symmetry $k_\perp$ is small compared with k_z. The modern high-temperature superconductors such as $YBa_2Cu_3O_7$ have planar symmetry with the crystallographic dimensions $a \simeq b$ and $c >> a$. The coherence length in these systems is also highly anisotropic $\xi_x \simeq \xi_y \simeq 34\mathring{A}$ and $\xi_z \simeq 7\mathring{A}$ so that the integral in (3.114) may be solved using the approximation $\xi_z << \xi_x$. Using finite limits of integration in (3.114), the trivial divergence depending on the size of the sample, is removed. It is then of interest to determine if any other divergence occurs in (3.114).

Case I. For $k_z >> k_\perp$, solving the integral along the z direction using finite limits on k_z with the minimum value given by k_z^o and the maximum value k_z^m we find,

$$< \phi^2 >= -\frac{b^2\kappa^4 k_B T}{4\pi^2}\left\{\int \frac{dk_\perp}{k_\perp c_{66}}\left(\frac{1}{k_z^m}-\frac{1}{k_z^o}\right)+I^{\max}-I^{\min}\right\} \tag{3.115}$$

where,

$$I^{\max} = \int dk_\perp \frac{k_\perp^2[c_{66}k_\perp^2+<\omega_o>B^2/4\pi]}{c_{66}k_\perp^4[-4c_{66}k_\perp^4(c_{66}k_\perp^2+<\omega_o>B^2/4\pi)]^{1/2}}$$
$$ln\frac{2(c_{66}k_\perp^2+<\omega_o>\frac{B^2}{4\pi})k_z^m-[-4c_{66}k_\perp^4(c_{66}k_\perp^2+<\omega_o>\frac{B^2}{4\pi})]^{\frac{1}{2}}}{2(c_{66}k_\perp^2+<\omega_o>\frac{B^2}{4\pi})k_z^m+[-4c_{66}k_\perp^4(c_{66}k_\perp^2+<\omega_o>\frac{B^2}{4\pi})]^{\frac{1}{2}}} \tag{3.116}$$

and $I^{\min}$ can be written from $I^{\max}$ by replacing k_z^m by k_z^o. Assuming finite limits for $k_\perp$ from $k_\perp^o$ to $k_\perp^m$, the first term in (3.115) is itegrated so that,

$$< \phi^2 >= -\frac{b^2\kappa^4 k_B T}{4\pi^2}\left\{\frac{1}{c_{66}}ln\left(\frac{k_\perp^m}{k_\perp^o}\right)\left(\frac{1}{k_z^m}-\frac{1}{k_z^o}\right)+I^{\max}-I^{\min}\right\} \tag{3.117}$$

thus for finite limits, there is no divergence in the first term. It appears that there is a logarithmic divergence when the argument of the ln term in (3.116) goes to zero but such a zero occurs only at,

$$<\omega_o>\frac{B^2}{4\pi} = -c_{66}k_\perp^2\left[1+\frac{k_\perp^2}{(k_z^m)^2}\right] \tag{3.118}$$

in which the left hand size can be negative only when $b > 1$. This means that the average induction is larger than the upper critical field. Thus the logarithmic divergence separates the normal and superconducting states but there is no divergence in $< \phi^2 >$ within the superconducting state. For $b < 1$, the condition (3.118) is not satisfied for any physical solution of $< \phi^2 >$.

Case II. There is an alternative method of solving the double integral in (3.114). We solve the $k_\perp$ integral first and k_z integral later. Since $k_z >> k_\perp$ we find that,

$$\begin{aligned} < \phi^2 > &= \frac{b^2 \kappa^4 k_B T}{4\pi} \int \frac{dk_z}{k_z^2} \left[\frac{k_\perp^3 c_{66}^{1/2} \{F_- - F_+\}}{k_z [c_{66} k_z^2 - < \omega_o > B^2/\pi]^{1/2}} \right. \\ & -3[k_z \{c_{66} k_z^2 - < \omega_o > B^2/\pi\}^{1/2} f_-^{1/2}]^{-1} \\ & \times \int k_\perp^2 \tan^{-1} \frac{k_\perp c_{66}^{1/2}}{f_-^{1/2}} dk_\perp \\ & +3[k_z \{c_{66} k_z^2 - < \omega_o > B^2/\pi\}^{1/2} f_+^{1/2}]^{-1} \\ & \times \left. \int k_\perp^2 \tan^{-1} \frac{k_\perp c_{66}^{1/2}}{f_+^{1/2}} dk_\perp \right|_{k_\perp^o}^{k_\perp^m} \end{aligned} \tag{3.119}$$

where

$$\begin{aligned} F_\mp &= \left[c_{66} \left\{ \frac{1}{2} c_{66} k_z^2 \mp \frac{1}{2} k_z^2 c_{66} \left(1 - \frac{< \omega_o > B^2/\pi}{k_z^2 c_{66}} \right)^{1/2} \right\} \right]^{-1/2} \\ & \times \left[\tan^{-1} \frac{k_\perp c_{66}^{1/2}}{f_\mp} \right] \end{aligned} \tag{3.120}$$

and

$$f_\mp = \frac{1}{2} c_{66} k_z^2 \mp \frac{1}{2} k_z^2 c_{66} \left[1 - \frac{< \omega_o > B^2/\pi}{k_z^2 c_{66}} \right]^{1/2} \tag{3.121}$$

Taking one more approximation, $k_z^2 c_{66} >> < \omega_o > B^2/4\pi$, we remove the square root from (3.121) by using the binomial theorem expansion and retaining only the first two terms. Then F_+ is small compared with F_-. Integrating $\int x^2 \tan^{-1}(ax) dx$ type integrals we find that,

$$< \phi^2 > = \frac{b^2 \kappa^4 k_B T}{4\pi^2} \int \frac{dk_z}{k_z^2} \left[\frac{k_\perp^3 c_{66}^{1/2} \{F_- + F_+\}}{k_z [c_{66} k_z^2 - < \omega_o > B^2/\pi]^{1/2}} \right.$$

$$- \left[k_z \left(c_{66}k_z^2 - <\omega_o> B^2/\pi\right)^{1/2} f_-^{1/2}\right]^{-1} \left\{k_\perp^3 \tan^{-1} \frac{k_\perp c_{66}^{1/2}}{f_-^{1/2}}\right.$$

$$- \frac{k_\perp^2 f_-^{1/2}}{2c_{66}^{1/2}} ln \left(1 + \frac{k_\perp^2 c_{66}}{f_-}\right) + \frac{1}{2}\frac{f_-^{3/2}}{c_{66}^{3/2}}$$

$$\times \left[\left(1 + \frac{c_{66}k_\perp^2}{f_-}\right) ln \left(1 + \frac{c_{66}k_\perp^2}{f_-}\right)\right.$$

$$\left.\left.\left. - \left(1 + \frac{c_{66}k_\perp^2}{f_-}\right)\right]\right\} + \{f_- \to f_+\}\right|_{k_\perp^o}^{k_\perp^m}\Bigg] \tag{3.122}$$

where $f_- \to f_+$ means that this term is generated from the previous term by replacing f_- by f_+. The first and the second terms in (3.122) are of the order of,

$$\sim -\frac{b^2\kappa^4 k_B T}{4\sqrt{2} < \omega_o > B^2} \{(k_\perp^m)^3 - (k_\perp^o)^3\} \left[\frac{1}{(k_z^m)^2} - \frac{1}{(k_z^o)^2}\right] \tag{3.123}$$

which has a divergence only at $b = 1$ coming from $< \omega_o >= 0$, which occurs at the upper critical field due to (3.113). The third term in (3.122) is of the order of

$$\sim \frac{b^2\kappa^4 k_B T}{12\pi^2} k_\perp^3 \tan^{-1} \left(\frac{k_\perp c_{66}^{1/2}}{f_-^{1/2}}\right) k_z^{-3}\Bigg|_{k_z^o}^{k_z^m} \tag{3.124}$$

which does not diverge. The various other terms are also integrated to find that there are no divergences in the system except at $\bar{B} = B_{c2}$.

Case III. In the previous cases, the approximation used is that k_z is large compared with that in the xy plane as in the real high-temperature superconductors. However, we now see that such an approximation is not necessary to extract the divergences of the double integral. We solve the k_z integral in the limits of 0 to k_z^m in (3.114) without introducing any approximation to find,

$$\begin{aligned} < \phi^2 > &= \frac{b^2\kappa^4 k_B T}{\pi B^2 < \omega_o >} \left[\int_{1/L}^{\Lambda} \tan^{-1}\left(\frac{k_z^m}{k_\perp}\right) dk_\perp \right. \\ &\quad - \int_{1/L}^{\Lambda} dk_\perp \left\{1 + \frac{B^2 < \omega_o >}{4\pi c_{66} k_\perp^2}\right\}^{1/2} \\ &\quad \left. \times \tan^{-1}\left\{\frac{k_z^m}{k_\perp}\left(1 + \frac{B^2 < \omega_o >}{4\pi c_{66} k_\perp^2}\right)^{1/2}\right\}\right] . \end{aligned} \tag{3.125}$$

For the positive argument of the $\tan^{-1}$ function, $(x > 0)$, we use the relation $\tan^{-1} x = \cot^{-1}(1/x)$ and for the square root, we use the binomial theorem expansion in which we retain only the dominant term, so that

$$< \phi^2 >= \frac{b^2 \kappa^4 k_B T}{\pi B^2 < \omega_o >}[I_1 - I_z] \tag{3.126}$$

where

$$\begin{aligned} I_1 &= \int_{1/L}^{\Lambda} \cot^{-1}\left(\frac{k_\perp}{k_z^m}\right) dk_\perp \\ &= \Lambda \cot^{-1}\left(\frac{\Lambda}{k_z^m}\right) - L^{-1} \cot^{-1}\left(\frac{1}{k_z^m L}\right) + \frac{1}{2} k_z^m ln\left[\left\{(k_z^m)^2 + \Lambda^2\right\}(k_z^m)^{-2}\right] \\ &\quad -\frac{1}{2} k_z^m ln\left[\left\{(k_z^m)^2 + \frac{1}{L^2}\right\}(k_z^m)^{-2}\right] \end{aligned} \tag{3.127}$$

and

$$I_2 = \int_{1/L}^{\Lambda} \left(1 + \frac{B^2 < \omega_o >}{8\pi c_{66} k_\perp^2}\right) \tan^{-1}\left\{\frac{k_z^m}{k_\perp}\left(1 + \frac{B^2 < \omega_o >}{8\pi c_{66} k_\perp^2}\right)\right\} . \tag{3.128}$$

Using the expansion of $\tan^{-1} x$ and retaining only the most dominant term we find that,

$$I_2 = k_z^m \int_{1/L}^{\Lambda} \left[\frac{1}{k_\perp} + \frac{B^4 < \omega_o >^2}{64\pi^2 c_{66}^2 k_\perp^5} + \frac{B^2 < \omega_o >}{4\pi c_{66} k_\perp^3}\right] dk_\perp \tag{3.129}$$

which is integrated to find,

$$I_2 = k_z^m \left[ln(\Lambda L) + \frac{B^4 < \omega_o >^2}{256\pi^2 c_{66}^2}\left(L^4 - \frac{1}{\Lambda^4}\right) + \frac{B^2 < \omega_o >}{7\pi c_{66}}\left(L^2 - \frac{1}{\Lambda^2}\right)\right] \tag{3.130}$$

showing that there is no divergence except at $< \omega_o >= 0$ from (3.126) which is the point $b = 1$. We thus conclude that the only divergence that occurs in $< \phi^2 >$ is at $\bar{B} = B_{c2}$. In the case of Bose condensation, the distribution,

$$< n >= \frac{1}{\exp\{(E_k - \mu)/k_B T\} - 1} \tag{3.131}$$

diverges at $E_k \simeq \mu$, that is, when the single-particle frequency equals the chemical potential. In the present case the elastic propagator is found to be,

$$< s_i(k) s_j(k) >= \left(n + \frac{1}{2}\right) E_k \left[\frac{1}{c_{66} k_\perp^2 + c_{44} k_z^2} + \frac{1}{c_{11} k_\perp^2 + c_{44} k_z^2}\right] \tag{3.132}$$

which means that $< \phi^2 >$ diverges if $n \to \infty$. Then due to Bose condensation of fluxons, the mean square phase $< \phi^2 >$ diverges at the condensation temperature. Otherwise, there is no divergence in $< \phi^2 >$ except at B_{c2}. The gauge invariant phase of the wave function is given by

$$\tilde{\phi}(r) = \phi(r) - \kappa \int_o^r \vec{A}.d\vec{l} \tag{3.133}$$

The phenomenon of the flux-lattice melting requires that well below H_{c2} there should be a singularity. In such a case, we can replace $< \omega >$ by $< \tilde{\omega} >$ with a divergence at $\bar{B} = B_{FM}$ with $b_1 = \bar{B}/B_{FM}$ with a flux melting field, B_{FM}, such that

$$< \tilde{\omega} >= 2\kappa^2(1-b)(1-b_1)/[(2\kappa^2-1)\beta+1] \tag{3.134}$$

which also enters in c_{44} and has the advantage of showing a divergence at $k = 0$ and $b_1 = 1$. This means that our superconductor has a phase within the superconducting phase.

Usually the number of critical exponents is a constant. In the case of a phase transition from a paramagnetic state to a ferromagnetic state de Almedia and Thouless find that the phase boundary is given by,

$$H \geq (1 - T/T_c)^{3/2} \tag{3.135}$$

which, due to universality, is applicable to superconductors also. In one dimension, $d = 1$, the Ising model does not describe any phase transition, while in two dimensions, $d = 2$, a phase transition does occur. In the case of $d = 3$, numerical solutions show that there is a phase transition. Thus, due to universality, the lower critical dimension, $d_{l.c.}$ for which there is a phase transition should be $2 > d_{l.c} > 1$. In the case of the flux-lattice, the role of self-similarity becomes important and the effective dimensionality becomes less than three. The formation of domains is needed and the effective dimensionality is reduced. The lower critical dimensionality for a system of a flux lattice is not determined from the calculation of the phase factor of the order parameter of the Ginzburg-Landau free energy. We find that there is no divergence in the phase factor given by (3.111) except at H_{c2}. The long range order does not break down and hence there is no phase boundary inside the superconducting phase. However, a phase boundary can be introduced by considering a Bose condensation. It is also possible to modify the expressions of elastic constants such as that given by (3.134) which permits a phase below $H_{c2}(T_c)$.

References

3.1 V.L. Ginzburg and L.D. Landau, Zh. Eksp. Teor. Fiz. **20**, 1064 (1950).

3.2 E. Simanek, Inhomogeneous superconductors, Oxford University Press, New York.

3.3 A.A. Abrikosov, Sov. Phys. JETP **5**, 1174 (1957).

3.4 W.H. Kleiner, L.M. Roth and S.H. Autler, Phys. Rev. **133**, A1226 (1964).

3.5 G. Eilenberger, Phys. Rev. **164**, 628 (1967).

3.6 M.A. Moore, Phys. Rev. B**39**, 136 (1989).

3.7 R. Labusch, Phys. Stat. Solidi **32**, 439 (1969).

3.8 D.R. Nelson and H.S. Seung, Phys. Rev. B**39**, 9153 (1989).

3.9 E.H. Brandt, Phys. Rev. Lett. **63**, 1106 (1989).

3.10 P.L. Gammel, L.F. Schneemeyer, J.V. Waszczak and D.J. Bishop, Phys. Rev. Lett. **61**, 1666 (1989).

3.11 A. Houghton, R.A. Pelcovits and A. Sudbo, Phys. Rev. B40, 6763 (1989); **42**, 906 (1990).

3.12 K.A. Muller, M. Takashige and J.G. Bednorz, Phys. Rev. Lett. **58**, 1143 (1987).

3.13 K.N. Shrivastava, Phys. Rev. B**41**, 11168 (1990).

3.14 V.M. Vinokur, M.V. Feigel'man, V.B. Geshkenbein and A.I. Larkin, Phys. Rev. Lett. **65**, 259 (1990).

3.15 P. Esquinazi, Solid State Commun. **74**, 75 (1990).

3.16 S. Koka and K.N. Shrivastava, Solid State Commun. **75**, 911 (1990).

3.17 S. Koka and K.N. Shrivastava, Physica B**165**, 1097 (1990).

3.18 S. Koka and K.N. Shrivastava, Solid State Commun **77**, 225 (1991).

3.19 T.K. Worthington, W.J. Gallager and T.R. Dinger, Phys. Rev. Lett. **59**, 1160 (1987).

3.20 J.R.L. de Almeida and D.J. Thouless, J. Phys. A**11**, 983 (1978).

3.21 K.N. Shrivastava and C.K. Majumdar, Phys. Rev. B**21**, 3971 (1980).

3.22 A. Aharony, Y. Imry and S.K. Ma, Phys. Rev. Lett. **37**, 1364 (1976).

3.23 K.H. Fischer, Physica C**178**, 161 (1991).

3.24 S. Atkas, C.P. Poole and H.A. Farach, J. Phys. Condens. Mater **6**, 7373 (1994).

3.25 A. Aruna and K. N. Shrivastava, Mod. Phys. Lett. B **12**, 977 (1998).

3.26 S. Koka and K. N. Shrivastava, Solid State Commun. **80**, 933 (1991).

3.27 S. Koka and K. N. Shrivastava, J. Phys.: Condens. Matter **3**, 9249 (1991).

Chapter 4

Magnetization

4.1 Magnetization of a type-II superconductor

Usually the free energy is a function of magnetic field. The magnetization is the first derivative of the free energy with respect to the magnetic field and susceptibility is the second,

$$M = \frac{dF}{dH} \tag{4.1}$$

$$\chi = \frac{d^2F}{dH^2} = \frac{dM}{dH} \ . \tag{4.2}$$

For small x, the modified Bessel function of the second kind used by Abrikosov to make vortices, can be expanded as follows:

$$K_o(x) \simeq -ln\frac{x}{2} \ . \tag{4.3}$$

For large values of x,

$$K_o(x) = \sqrt{\frac{\pi}{2x}}\, e^{-x} \tag{4.4}$$

so that the function can be differentiated in both the limits. For the field in the range $H_{c1} < H < H_{c2}$, the average intervortex distance, L, may be smaller than the London penetration depth but larger than the coherence length, $\xi << L << \lambda$.

The free energy in this range is,

$$8\pi F = B^2 + (B\phi_o/2\pi\lambda^2)ln(L\beta/\xi) \tag{4.5}$$

where B is the magnetic induction, ϕ_o is the flux quantum and β is a constant which depends upon the flux-line lattice structure. The average spacing between lines is defined by $L^2 = \phi_o/\beta$. The logarithmic term in the above may be written as,

$$ln\frac{L\beta}{\xi} = ln\frac{\beta}{\xi} + \frac{1}{2}ln\frac{H_{c2}}{B} + \frac{1}{2}ln\frac{\phi_o}{H_{c2}} \tag{4.6}$$

in which only the second term depends on the magnetic field so that the relevant terms of (4.5) are found to be,

$$8\pi F = B^2 + \frac{B\phi_o}{4\pi\lambda^2}ln\frac{H_{c2}}{B}\ . \tag{4.7}$$

Differentiating the above with respect to magnetic field,

$$8\pi\frac{\partial F}{\partial B} = 2B + \frac{\phi_o}{4\pi\lambda^2}\left[ln\frac{H_{c2}}{B} - 1\right] \tag{4.8}$$

which gives the magnetization,

$$M = \frac{2B}{8\pi} + \frac{\phi_o}{8\pi.4\pi\lambda^2}\left[ln\frac{H_{c2}}{B} - 1\right]\ . \tag{4.9}$$

Since the logarithmic term is much bigger than the linear term,

$$-4\pi M \simeq \frac{\phi_o}{8\pi\lambda^2}ln\frac{H_{c2}}{B} \tag{4.10}$$

which means that from the measurement of magnetization as a function of magnetic field, the penetration depth can be measured.
From (4.9) the second derivative becomes,

$$\chi = \frac{dM}{dB} = \frac{1}{4\pi} - \frac{\phi_o}{8\pi.4\pi\lambda^2 B} \tag{4.11}$$

which shows that the susceptibility varies inversely with the magnetic induction.

For an ellipsoidal sample, the internal field H is not equal to the externally applied field because of the anisotropic demagnetization factor. Therefore, the results (4.10) and (4.11) hold only for isotropic samples. We introduce the anisotropy by replacing B by $(m_1B_x^2 + m_3B_z^2)^{1/2}$ in the xz plane. The z direction is parallel to the c axis and $\hat{x}$ is chosen as the intersection of the $-B$ plane with basal plane normal to $\hat{z}$ as shown in Fig. 4.1.

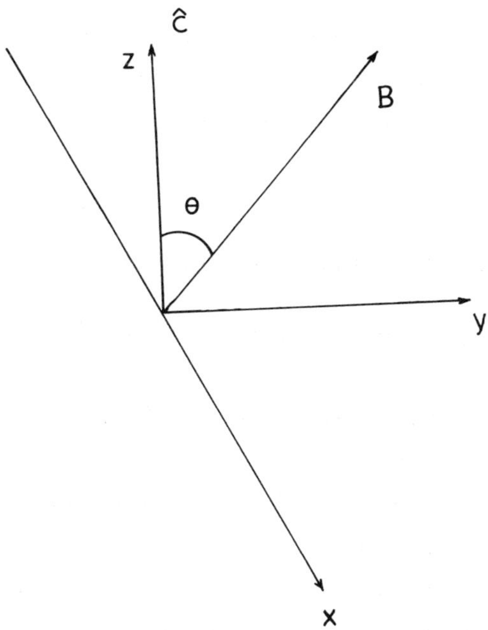

Fig. 4.1. Orientation of the magnetic field.

Therefore $B_z = B\cos\theta$ and $B_x = B\sin\theta$ with θ being the angle between $\hat{c}$ and B. The average penetration depth λ is defined so that λ^2 is proportional to the average mass, $M_{av} = (M_1^2 M_3)^{1/3}$ where the effective masses, M_1 and M_3 are components of M_{ik} along the principal crystal directions $\hat{x}$ and $\hat{y}$, respectively. We define the dimensionless masses $m_{ik} = M_{ik}/M_{av}$ and $m_1^2 m_3 = 1$. Differentiating F with respect to B one obtains the magnetization,

$$-M_z = \frac{H_z - B_z}{4\pi} = M_o \frac{m_3 \cos\theta}{\sqrt{m(\theta)}} \tag{4.12}$$

$$-M_x = M_o \frac{m_1 \sin\theta}{\sqrt{m(\theta)}} \tag{4.13}$$

where

$$M_o = \frac{\phi_o}{32\pi^2\lambda^2} \ln \frac{H_{c1}\beta}{H} \tag{4.14}$$

$$M(\theta) = m_1 \sin^2\theta + m_3 \cos^2\theta \ . \tag{4.15}$$

The induction may be assumed to be in the same direction as that of the applied field, so that the angle θ between B and $\hat{c}$ is equal to that between H_a and $\hat{c}$, as long as there is no hysteresis. The component of M parallel to the applied field, M_p is obtained from (4.12) and (4.15),

$$M_p = M_z \cos\theta + M_x \sin\theta = -M_o\sqrt{m(\theta)} \ . \tag{4.16}$$

The component, M_n normal to the applied field is,

$$M_n = M_x \cos\theta - M_z \sin\theta = M_o \frac{m_3 - m_1}{\sqrt{m(\theta)}} \sin\theta\cos\theta \ . \tag{4.17}$$

The component M_n vanishes when the field is directed along one of the principal crystal directions, $\theta = 0$ or $\pi/2$. However, the ratio $M_n/M_p = (m_3 - m_1)\sin\theta\cos\theta/m(\theta)$ reaches its maximum of $\sqrt{m_3/m_1}\,(m_3 - m_1)/2m_3$ at an angle θ_m such that $\tan^2\theta_m = m_3/m_1$.

For field applied along the $\hat{c}$ axis ($\theta = 0$) and perpendicular to it ($\theta = \pi/2$), from (4.14) and (4.16), we write,

$$-M(0) = \frac{\phi_o\sqrt{m_3}}{32\pi^2\lambda^2} ln\frac{(H_{c2}\beta)_0}{H} \tag{4.18}$$

$$-M(\frac{\pi}{2}) = \frac{\phi_o\sqrt{m_1}}{32\pi^2\lambda^2} ln\frac{(H_{c2}\beta)_{\pi/2}}{H} \tag{4.19}$$

In a polycrystal, the grains are randomly oriented so that the average value of the magnetization of each grain can be obtained by integration over all of the orientations as,

$$M_{\text{polycrys}} = N\int_o^{\pi/2} \mu_p(\theta)\sin\theta d\theta = \int_o^{\pi/2} M_p(\theta)\sin\theta\, d\theta \tag{4.20}$$

2where M_p is given by (4.16). The slope $dM_{\text{polycrys}}|d(ln H)$ can be evaluated as,

$$\frac{dM_{\text{polycrys}}}{d\ln(H/H_o)} = \frac{\phi_o}{32\pi^2\lambda^2}\int_o^{\pi/2} \sqrt{m(\theta)}\, \sin\theta\, d\theta$$

$$= \frac{\phi_o}{64\pi^2\lambda^2}\sqrt{m_1}\, f(r) \tag{4.21}$$

where H_o is an arbitrary scaling field and

$$r^2 = m_3/m_1 \tag{4.22}$$

$$f(r) = r + (r^2-1)^{-1/2} ln[(r^2-1)^{1/2} + r] \tag{4.23}$$

For isotropic material $r = m_1 = f/2 = 1$ and (4.21) agrees with (4.10). For $r^2 >> 1$, $f = r + O(1/r)$. The normalization $m_1^2 m_3 = 1$ combined with the definition of r yields $m_1 = r^{-2/3}$ and $m_3 = r^{4/3}$. Thus the slope in (4.21) and (4.23) can be used to measure the penetration depth.

4.2 Para-Meissner effect

In superconductors at very small magnetic fields, the susceptibility becomes positive. As the field is increased, the positive value becomes negative approaching the value of $-1/4\pi$, consistent with the Meissner effect. The temperature at which the susceptibility becomes zero is field dependent approaching lower values at higher fields. According to the Meissner effect, the susceptibility is negative. Hence the observation of positive susceptibility below the superconducting transition temperature has been named as the para-Meissner effect (PME). In real materials, it is quite common to find that the value of the susceptibility is larger than $-1/4\pi$ indicating the existence of a small positive term. In this state the field spreads in an area larger than $\pi\lambda_L^2$ so that the susceptibility becomes positive. A temperature at which the susceptibility becomes zero due to cancellation of positive and negative contributions occurs.

We consider that the spacing between flux lines is d so that the area in which the flux is measured is d^2. According to the flux quantization, $Bd^2 = \phi_o$ so that

$$d = (\phi_o/B)^{1/2} . \tag{4.24}$$

The areal density $n = B/\phi_o$, is the number of flux lines per unit area, $n = 1/d^2$. The ϵ_1 is the energy per unit length of single flux line. The free energy per unit volume is given by,

$$g(n) = g_o + \left(\epsilon_1 - \frac{H\phi_o}{8\pi}\right) n + n \left(\frac{3\phi_o^2}{8\pi^2\lambda^2}\right) K_o(d/\lambda) \tag{4.25}$$

where $K_o(d/\lambda)$ is the modified Bessel function which describes the distribution of field as a function of distance between flux lines in units of London penetration depth, λ. Two approximations of the modified Bessel functions are solvable.

Case I. For small d/λ, the separation between flux lines is smaller than the London penetration depth, the modified Bessel function can be written as,

$$K_o(d/\lambda) \simeq -ln(d/2\lambda) \ . \tag{4.26}$$

This approximation is valid for large magnetic fields, $B > \phi_o/\lambda^2$. Substituting (4.26) in (4.25), we find,

$$g(n) = g_o + \left(\epsilon_1 - \frac{B\phi_o}{8\pi}\right)\left(\frac{B}{\phi_o}\right) - \frac{3\phi_o B}{16\pi^2\lambda^2} ln(\phi_o/4\lambda^{1/2}) + \frac{3\phi_o B \ln B}{16\pi^2\lambda^2} \ . \tag{4.27}$$

Since the magnetization is defined as $\partial g/\partial B = M$, we differentiate the above expression to find,

$$M = \frac{\epsilon_1}{\phi_o} - \frac{B}{4\pi} - \frac{3\phi_o ln(\phi_o/4\lambda^{1/2})}{16\pi^2\lambda^2} + \frac{3\phi_o(1 + lnB)}{16\pi^2\lambda^2} \ . \tag{4.28}$$

The susceptibility is $\chi = \partial M/\partial B$ so that from above,

$$\chi = -\frac{1}{4\pi} + \frac{3\phi_o}{16\pi^2\lambda^2 B} \ . \tag{4.29}$$

As B increases, the positive term decreases and χ becomes more and more negative. We shall see later on, that this theoretical result is in accord with experimental measurements. The above susceptibility becomes zero for a field of,

$$B_o(\chi = 0) = 3\phi_o/4\pi\lambda^2 \ . \tag{4.30}$$

For $B > B_o$, χ becomes negative so that the Meissner effect is predicted. However for $B < B_o$, χ becomes positive. The field given above is of the order of $H_{c1} = \phi_o/\pi\lambda^2$. Therefore from (4.30), the field at which the susceptibility becomes zero becomes, $B_o \simeq (3/4)H_{c1}$. For B to be less than B_o, the only way it is possible is when the area in which the flux is quantized is larger than $\pi\lambda^2$. This requires that the field penetrate distances larger than λ to make the susceptibility positive. If the flux creeps large distances, then such a possibility occurs. The positive susceptibility is thus also anti-London in the sence that due to mesoscopic nature of the samples, the London equation with only one exponent is not strictly obeyed. If the decay of the field inside a superconductor is non-exponential, then such a possibility exists. In the case of Abrikosov phase, the flux penetrates the sample, but in that case the induction is larger than H_{c1} whereas in the present case, fields smaller than H_{c1} are envisaged. Hence this effect is different from that of the Meissner and we calle it para-Meissner effect. Very small field in (4.29) can make the positive term larger than the negative term and hence such a possibility may occur in a multigrain sample.

We now calculate the entropic contribution to the free energy and its effect on the susceptibility. The percolation parameter q which depends on the geometry of the system may be written as $1/c = 8\pi lnq$. Since the domain walls are randomly arranged, only an average distqance between walls which is l can be defined and the lattice constant is a. The appropriate reciprocal lattice vector is $q = l/a$. The collisions between flux lines make a contribution of $k_B lnq$ to the entropy of the system with $q > 1$. The entropic contribution to the free energy of the system per unit volume is,

$$g_{en} = (k_B T)^2 n^2 \ln q/\epsilon_1 \ . \tag{4.31}$$

Since the areal density is $B\phi_o$, the above may be written as

$$g_{en} = \ln q (k_B T)^2 B^2/(\epsilon_1 \phi_o^2) \ . \tag{4.32}$$

The field derivative of which gives the entropic contribution to the magnetization as,

$$M = \frac{\partial g_{en}}{\partial B} = [2 \ln q (k_B T)^2/(\epsilon_1 \phi_1^2)] B \tag{4.33}$$

which contributes to the susceptibility,

$$\chi_{en} = \left[2 \ln q (k_B T)^2/(\epsilon_1 \phi_1^2)\right] \ . \tag{4.34}$$

Adding this contribution to (4.29) we find the full susceptibility with entropic contribution taken into account,

$$\chi = -\frac{1}{4\pi} + \frac{3\phi_o}{16\pi^2\lambda^2 B} + \frac{2\ln q (k_B T)^2}{\epsilon_1 \phi^2} \ . \tag{4.35}$$

We look for the temperature at which the positive contributions to the susceptibility become equal to the negative contributions so that the susceptibility becomes zero. This happens at the temperature given by

$$T_{o+} = \frac{\phi_o}{k_B}\left[\frac{\epsilon_1}{2lnq}\left\{\frac{1}{4\pi} - \frac{2\phi_o}{16\pi^2\lambda^2 B}\right\}\right]^{1/2} \tag{4.36}$$

where we have left out a negative solution. A real solution is possible only for such B values for which $1/4\pi$ term is larger than the field dependent term. Therefore T_{o+} is approached only from the side where χ is negative and hence it is pro-Meissner.

Case II. We now obtain another expansion of the modified Bessel function. For large d/λ,

$$K_o(d/\lambda) \simeq \left(\frac{\pi\lambda}{2d}\right)^{1/2} e^{-d/\lambda} \ . \tag{4.37}$$

This expansion is valid for the distance between flux lines larger than the London penetration depth which happens at small magnetic fields,

$$\left(\frac{\phi_o}{B}\right)^{1/2} >> \lambda \ . \tag{4.38}$$

Small field, $B << \phi_o/\lambda^2$, requires that areas larger than λ^2 be involved, which is similar to the problem of flux creep. Substituting (4.37) in (4.25) we find that the free energy is given by,

$$g(n) = g_o + \left(\epsilon_1 - \frac{B\phi_o}{8\pi}\right)\frac{B}{\phi_o} + \frac{3\phi_o}{8\pi^2\lambda^2}\left(\frac{\pi\lambda}{2d}\right) e^{-d/\lambda} \ . \tag{4.39}$$

Since the distance between flux lines is quantized according to (4.24), the above becomes,

$$g(n) = g_o + \left(\epsilon_1 - \frac{B\phi_o}{8\pi}\right)\frac{B}{\phi_o} + \frac{3\phi_o^{3/4}B^{5/4}}{8\sqrt{2}\,\pi^{3/2}\lambda^{3/2}}\exp\left[-B^{-1/2}\phi_o^{1/2}/\lambda\right] \ . \tag{4.40}$$

The magnetization is determined by the field derivative of the above as $\partial g(n)/\partial B = M$ so that

$$\begin{aligned} M &= \frac{\epsilon_1}{\phi_o} - \frac{B}{4\pi} + \frac{15\phi_o^{3/4}}{16\sqrt{2}\,\phi^{3/2}\lambda^{3/2}}\exp\left(-\frac{B^{-1/2}\phi_o^{1/2}}{\lambda}\right) \\ &\times \left[B^{1/4} + \left(\frac{2\phi_o^{1/2}}{5\lambda}\right)B^{-1/4}\right] \ . \end{aligned} \tag{4.41}$$

The susceptibility from the above expression is found to become,

$$\chi = -\frac{1}{4\pi} + \frac{15\phi_o^{3/4}}{32\sqrt{2}\pi^{3/2}\lambda^{5/2}}\left[\frac{1}{2}\lambda B^{-3/4} + \phi_o^{1/2}\left(\frac{4}{5}B^{-5/4} + B^{-7/4}\right)\right]$$
$$\times \exp(-B^{-1/2}\phi_o^{1/2}\lambda^{-1}) \ . \quad (4.42)$$

The field dependence in the first positive term of (4.42) is $\sim [B^{3/4}\exp(\phi_o^{1/2}/\lambda B^{1/2})]^{-1}$ so that in conjunction with (4.38), it should be noted that the dependence of χ on B depends on the value of λ. At some value of the field the positive term increases with increasing field. Thus paramagnetic states become diamagnetic upon application of a small magnetic field, H_o. As the temperature is increased, the paramagnetic susceptibility p/T becomes negative at a temperature T_o because of the diamagnetic contribution of $-1/4\pi$. The temperature T_o is estimated by using (4.29) for the diamagnetic contribution and a simple paramagnetic term as,

$$\chi = -\frac{1}{4\pi} + \frac{3\phi_o}{16\pi^2\lambda^2 B} + \frac{p}{T} + \frac{M_s}{4\pi(H+H_o)} \quad (4.43)$$

so that a Curie type term exists in addition to the field dependent term with spontaneous magnetization, M_s. From (4.43), the $\chi = 0$ point occurs at a temperature of,

$$T_o = 4\pi p\left[1 - \frac{3\phi_o}{4\pi^2\lambda^2 B} - \frac{M_s}{H+H_o}\right]^{-1} \ . \quad (4.44)$$

As the field H increases, the spontaneous magnetization term reduces such that the denominator increases provided that

$$\frac{M_s}{(H+H_o)} << 1 \quad (4.45)$$

which means that T_o reduces. We shall see later on that the prediction of reducing T_o with increasing H is in agreement with the experimental measurements. There is a lower field limit below which the Meissner effect does not occur in the mesoscopic samples. For this purpose we write the magnetization as,

$$M = \frac{\epsilon_1}{\phi_o} - \frac{H}{4\pi} + \frac{3\phi_o}{8\pi^2\lambda^2}K_o(d/\lambda) + 2n(k_B T)^2/(8\pi c\epsilon_1\phi_o) \ . \quad (4.46)$$

For the Meissner effect to occur the magnetization is equal to $M = -H/4\pi$ so that from this condition and the above expression we find that the remaining terms of (4.46) should be zero. Actually all of this type of terms are positive and in general, much smaller than the negative term so that the resistivity is zero, consistent with perfect diamagnetism. Since we are interested in making the magnetization positive, we include a Curie type paramagnetic term in the magnetization such that $M/H = p/T$ with positive sign. Such a paramagnetic term may originate from the mesoscopic defect structure of the samples which contains unpaired electrons. However, a uniform paramagnetism is not possible because it will have finite resistivity. If vortices are formed within the paramagnetic regions, then resistivity may be maintained at zero value. For a Curie type paramagnetic magnetization, the contribution to the free energy is of the form, $\int_o^H pHT^{-1}dH + g_{op}$, with g_{op} as the constant of integration to be determined from the boundary conditions. The expression (4.46) can then be written as,

$$M = -\frac{H}{4\pi} + \frac{H_M}{4\pi} \tag{4.47}$$

where

$$\frac{H_M}{4\pi} = \frac{\epsilon_1}{\phi_o} + \frac{3\phi_o}{8\pi^2\lambda^2}K_o(d/\lambda) + 2n(k_BT)^2/8(\pi c\epsilon_1\phi_o) + pH/T \ . \tag{4.48}$$

For the flux lines bound in a lattice, such as those at the impurity sites due to pinning, the sign of ϵ_1/ϕ_o can be negative so that there exists a solution for T at which $H_M = 0$. Usually the flux lines are free so that the line energy per unit length is positive. In the case of pinning, we set the right hand side of (4.48) to be zero,

$$T\left[\frac{\epsilon_1}{\phi_1} + \frac{3\phi_oK_o(d/\lambda)}{8\pi^2\lambda^2}\right] + \frac{2nk_B^2T^3}{8\pi c\epsilon_1\phi_o} + pH = 0 \tag{4.49}$$

which has positive solutions for T. At these temperatures, the Meissner effect is correctly reproduced such that $\chi = -1/4\pi$. This means that the susceptibility is oscillating around three temperatures. For the susceptibility to be negative, the condition is that,

$$H_n < H \tag{4.50}$$

or the external field must be larger than H_M for the observation of perfect diamagnetism. Thus the Meissner effect is predicted for $H > H_M$ only. Usually, in the phase diagram of a type-II superconductor, the Meissner effect is shown to occur for all fields less than the local critical field, H_{c1}, without any stipulation of any lower limit. Thus introduction of the lower limit on the magnetic field produces a new phase in the phase diagram. The Meissner condition

$$0 < H < H_{c1} \tag{4.51}$$

thus requires to be modified. The zero value in (4.51) may thus be replaced by a finite field. The classical paramagnetic phase is avoided by the formation of vortices. In such a system, the Meissner condition (4.51) is replaced by

$$H_M < H < H_{c1} \tag{4.52}$$

which means that there is a region in the $H(T)$ phase diagram below a certain field called H_M. In this region the susceptibility is positive. The magnetization passes from negative values to positive values through a temperature at which $M = 0$. From (4.47) and (4.48) we see that three are there such temperatures for which $M = 0$. At low temperatures, M is determined by positive pH/T term. As the temperature is increased, pH/T term reduces so that M becomes negative.

We imagine that one grain of the sample resembles a Josephson junction. Due to mesoscopic nature of the sample, the grains are weakly linked through the Josephson supercurrent. The free energy of such a junction may be written as,

$$\epsilon = \frac{1}{2}LJ^2 - \frac{J_c\phi_o}{2\pi}\cos\left[(\phi_x + LJ)\right] \tag{4.53}$$

where L is the inductance, J the Josephson current, J_c the critical Josephson current, and $\phi_x = H.A$ is the flux threading the ring due to an external field, H with A as the area of cross section of the ring. Differentiating ϵ with respect to J and setting the differential to zero we find,

$$J + J_c \sin\left[\frac{2\pi}{\phi_o}(\phi_x + LJ)\right] = 0 \ . \tag{4.54}$$

For small phase factor, the current is given by

$$J = -\frac{2\pi J_c(\phi_x/\phi_o)}{1 + (2\pi L J_c/\phi_o)} \tag{4.55}$$

which is the stable current for minimum free energy. For very small phase factors, we can set from (4.53),

$$\frac{2\pi}{\phi_o}(\phi_x + LJ) = 0 \tag{4.56}$$

which gives,

$$L = -\frac{1}{J}\phi_x \ . \tag{4.57}$$

Substituting (4.56) and (4.57) in (4.53), we find,

$$\epsilon = -\frac{HAJ}{2} - \frac{J_c\phi_o}{2\pi} \tag{4.58}$$

where A is the area of cross section across the field H, so that for a general flux quantized system,

$$\epsilon_n = -\frac{HAJ}{2} - \frac{J_c n\phi_o}{2\pi} \ . \tag{4.59}$$

Thus

$$\epsilon_{n_1} - \epsilon_{n_2} = \frac{J(n_2 - n_1)\phi}{2\pi} \tag{4.60}$$

or leaving out a constant

$$\epsilon_n = \frac{nJ\phi}{2\pi} \tag{4.61}$$

which means that there are levels in the gap causing the microwave absorption. The magnetization from (4.53) is obtained by $\partial\epsilon/\partial H$,

$$M = AJ_c \sin\left[\frac{2\pi}{\phi_o}(\phi_x + LJ)\right] \tag{4.62}$$

so that the susceptibility, $\partial M/\partial H$ is,

$$\chi = \frac{2\pi A^2 J_c}{\phi_o}\cos\left[\frac{2\pi}{\phi_o}(\phi_x + LJ)\right] \ . \tag{4.63}$$

This value of χ is negative (Meissner) only if the phase angle is $\geq \pi$ which is equivalent to saying that Meissner effect is obtained for $H > H_o$ only, where

$$H_o = \left(\frac{\phi_o}{2} - LJ\right)\frac{1}{A} \ . \tag{4.64}$$

For $H < H_o$, χ is positive (para-Meissner). Thus the para-Meissner effect is predicted from the phase factors. We consider the energy of a system of N Josephson junctions with L as the inductance as,

$$E_{pn} = -Nn\phi_o J_c \cos\left\{\frac{2\pi}{N}\left(p - \frac{\phi}{\phi_o}\right)\right\} + \frac{1}{2}LJ_c^2\cos^2\left\{\frac{2\pi}{N}\left(p - \frac{\phi}{\phi_o}\right)\right\} \tag{4.65}$$

where J_c is the critical Josephson current, ϕ_o is the unit flux, $\phi_o = hc/2e$ and n and p are integers. We use $\phi_o S^{-1}$ for B so that the first term becomes derivable from the Lorentz force $(j \times B)/c$ when work is calculated with S as the surface area normal to the field and the second term is the heat loss. At $p = 1$, the cosine term becomes unity so that,

$$E_{1,n} = -Nn\phi_o J_c + \frac{1}{2}LJ_c^2 \tag{4.66}$$

and the energy differences become independent of the inductance,

$$E_n - E_{n+1} = N\phi_o J_c \ . \tag{4.67}$$

This is the same expression as deduced for the study of the microwave absorption. The magnetization is defined by the field derivative of the energy so that

$$\frac{\partial}{\partial\phi}E_{p,n} = -2\pi n J_c \sin\left\{\frac{2\pi}{N}\left(1 - \frac{\phi}{\phi_o}\right)\right\} + \frac{\pi LJ_c^2}{N\phi_o}\sin\left\{\frac{4\pi}{N}\left(p - \frac{\phi}{\phi_o}\right)\right\} \ . \tag{4.68}$$

We consider several special cases of this expression for the magnetization. For $L = 0$, the above magnetization is negative only as long as,

$$\frac{2\pi}{N}\left(1 - \frac{\phi}{\phi_o}\right) < \frac{\pi}{2} \tag{4.69}$$

from which we define a field,

$$H_{o1} = \frac{N\phi_o}{2A}\left(\frac{2}{N} - \frac{1}{2}\right) \tag{4.70}$$

so that the Meissner effect is predicted only for fields larger than H_{o1}, i.e.,

$$H > H_{o1} \ . \tag{4.71}$$

Here A is the area in which flux is quantized. In addition to (4.69) there is one more condition,

$$\frac{2\pi}{N}\left(1-\frac{\phi}{\phi_o}\right) > \frac{3\pi}{2} \tag{4.72}$$

for the magnetization to be negative. This condition is equivalent to

$$H < H_{o2} \tag{4.73}$$

where

$$H_{o2} = H_{o1} - \frac{N\phi_o}{2A} \tag{4.74}$$

for which negative magnetization occurs. Thus (4.71) and (4.73) both should be obeyed for the observation of negative magnetization. Thus combining the two conditions into one,

$$H_{o2} > H > H_{o1} \tag{4.75}$$

gives the region of the field in which negative magnetization is predicted. For large inductance, the magnetization is negative as long as

$$J_c\left[-2\pi n + \frac{\pi L J_c}{N\phi_o}\right] < 0 \tag{4.76}$$

and the in equalities (4.69) and (4.71) are simultaneously obeyed. From (4.76) we define $L_o = 2nN\phi_o/J_c$. In such samples, in which L exceeds L_o, the voltage drop across the superconductor becomes large and the result (4.75) is modified. For negative magnetization $\sin\frac{2\pi}{N}(1-\frac{\phi}{\phi_o})$ has to be positive while $\sin\frac{4\pi}{N}(1-\frac{\phi}{\phi_o})$ has to be negative, so that for negative magnetization,

$$H_{o4} > H > H_{o3} \tag{4.77}$$

where

$$H_{o3} = \frac{N\phi_o}{A}\left(\frac{1}{N}-\frac{1}{4}\right) \tag{4.78}$$

$$H_{o4} = \frac{N\phi_o}{A}\left(\frac{1}{N}-\frac{1}{8}\right) . \tag{4.79}$$

Another condition arises near the phase angle of $3\pi/2$. For negative magnetization,

$$H_{o5} > H > H_{o6} \tag{4.80}$$

with

$$H_{o5} = \frac{N\phi_o}{A}\left(\frac{1}{N}-\frac{3}{4}\right) \tag{4.81}$$

$$H_{o6} = \frac{N\phi_o}{A}\left(\frac{1}{N}-\frac{3}{8}\right) . \tag{4.82}$$

We calculate the susceptibility by finding the second derivative of $E_{p,m}$ from (4.68) with respect to the external magnetic field,

$$\chi = \frac{\partial^2 E_{p,n}}{\partial\phi^2} = \frac{(2\pi)^2 nJ_c}{N\phi_o}\cos\left\{\frac{2\pi}{N}\left(p-\frac{\phi}{\phi_o}\right)\right\} - \frac{4\pi^2 LJ_c^2}{(N\phi_o)^2}\cos\left\{\frac{4\pi}{N}\left(p-\frac{\phi}{\phi_o}\right)\right\} \tag{4.83}$$

which also shows oscillations in signs at $n\pi$.

The magnetization of several samples of $Bi_2Sr_2CaCu_2O_{8+\delta}$ has been measured by using a magnetometer with antimagnetic Mumetal shield. The magnetic field of the earth is almost completely eliminated. The measured values are shown in Figs. 4.2a and 4.2b.

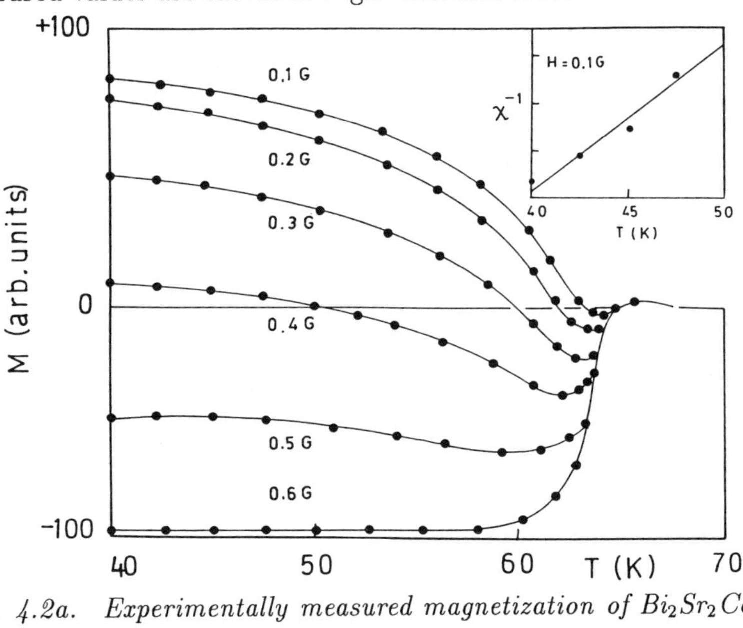

Fig. 4.2a. Experimentally measured magnetization of $Bi_2Sr_2CaCu_2O_{8+\delta}$ showing positive values and temperatures at which negative values turn towards positive values. The inset shows an example of inverse susceptibility obeying Curie type law, $\chi = 17.2/T$.

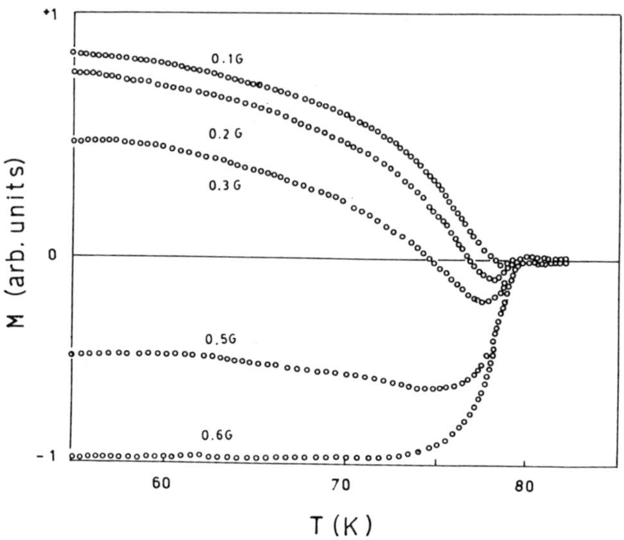

Fig. 4.2b. Experimentally measured magnetization of $Bi_2Sr_2CaCu_2O_{8+\delta}$ showing positive magnetization in low field cooled samples. Usual negative (Meissner) values are found at slightly larger fields.

Similar measurements have also bee carried out by using $Tl_2CaBa_2Cu_2O_8$ which are displayed in Fig.4.2c.

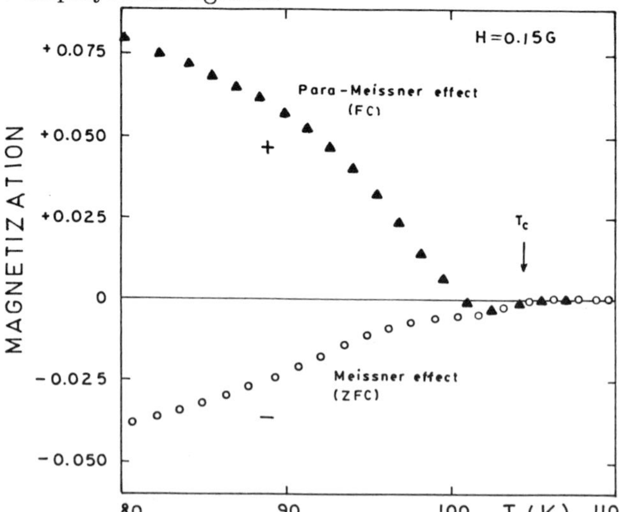

Fig. 4.2c. The magnetization as a function of temperature at a fixed field of $H = 0.15G$ for both the FC and ZFC samples of $Tl_2CaBa_2Cu_2O_8$ showing positive magnetization and hence the para-Meissner effect in the FC samples and negative magnetization corresponding to the Meissner effect in ZFC samples.

All of the curves have been obtained while cooling the samples. It is found that at low fields, the susceptibility becomes positive. As the field increases, the susceptibility becomes negative. This result is qualitatively in accord with calculated results. The magnetization is seen to become zero while passing from positive to negative values at a temperature of T_o. This characteristic temperature reduces as the field is increased. This experimental result is also in accord with the calculated result (4.44). We have plotted the inverse susceptibility as a function of temperature as shown in the inset of Fig. 4.2a, for $H = 0.1$G, which is represented by

$$\chi = \frac{17.2}{T} \tag{4.84}$$

This means that experimental measurement is consistent with a Curie type susceptibility term as in (4.43) which is positive and due to negative diamagnetic contribution gives $\chi = 0$ points as given by (4.44). The variation of the characteristic temperature T_o as shown in Fig. 4.3 is also consistent with the theory.

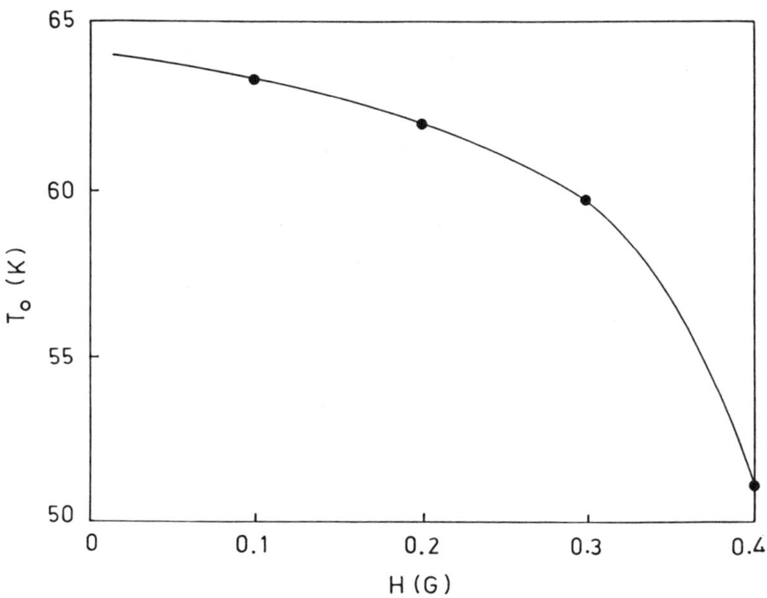

Fig. 4.3. The temperature T_o at which the susceptibility becomes zero while changing from positive to negative values as a function of magnetic field.

We assume that the sample consists of a single superconducting current loop with one Josephson junction. This system is solvable with a few approximations but not for the arbitrary shape of the loop. The solution for a linear branch of the loop in which the approximate diameter of the loop is treated as the size, can be obtained. The flux penetrating the loop is ϕ so that the energy of the loop is given by (4.53). The magnetic field arises in the above due to the flux quantization, $\phi = H.A$ with A as the area of cross section. The magnetization of this energy is given by $\partial\epsilon/\partial H$ in eq.(4.62), which is positive for $H_{o1} > H > H_{o2}$ and negative for $H_{o1} < H < H_{o2}$ where

$$\begin{aligned} H_{o1} &= \frac{1}{A}\left[\frac{\phi_o}{2}(2n+1) - LJ\right] \qquad \text{and} \\ H_{o2} &= \frac{1}{A}\left[n\phi_o - LJ\right] \end{aligned} \tag{4.85}$$

with n as an integer. The positive values give rise to the para-Meissner effect (PME), whereas the negative values give the Meissner effect. In general, the loop will be a random shape. If it is treated approximately as a circle, the diameter of the loop can be thought of as the size of the loop. In the case of a square loop, the length of the square may be treated as the size. In such a case, the problem is integrable along a length, l. We consider that the sample contains only one junction. Therefore, we assume that the loop size is l so that in order to determine the energy of the sample of length l, we multiply (4.53) by dz and integrate from 0 to l. The distance dependence of the magnetic field is taken to be $\tilde{H} = He^{-z/\lambda}$ where λ is the London penetration depth. The magnetization of the loop is assumed to decay as a function of distance as $e^{-z/\lambda}$ travelling along a sample of length l which may be linear along the length of a square or approximately equal to the diameter of a loop. The magnetization can then be written as,

$$M = AJ\int_o^l e^{-z/\lambda}\sin\left[\frac{2\pi}{\phi_o}(AHe^{-z/\lambda} + LJ)\right]dz \tag{4.86}$$

which is integrated to give,

$$M = \frac{J\phi_o\lambda}{2\pi H}\left[\cos\frac{2\pi}{\phi_o}(AHe^{-l/\lambda} + LJ) - \cos\frac{2\pi}{\phi_o}(AH + LJ)\right] . \tag{4.87}$$

We define two lengths,

$$l_{o1} = -\lambda \ln \left[\left\{ \frac{\phi_o}{2}(2n + \frac{1}{2}) - LJ \right\} \frac{1}{AH} \right] \tag{4.88}$$

$$l_{o2} = -\lambda \ln \left[\left\{ \frac{\phi_o}{2}(2n + \frac{3}{2}) - LJ \right\} \frac{1}{AH} \right] \tag{4.89}$$

for the size of the loop. The magnetization given by (4.87) shows negative sign only within the range of lengths,

$$l_{o2} > l > l_{o1} \ . \tag{4.90}$$

When the first term of (4.87) is negative and the second term is small but positive, the magnetization becomes negative. When the angle of the second cosine term in (4.87) is small,

$$\frac{2\pi}{\phi_o}(AH + LJ) \leq (2n + \frac{3}{2})\pi + \frac{\pi}{4} \ . \tag{4.91}$$

This is equivalent to a field restriction, $H \leq H_{o3}$, where

$$H_{o3} = \frac{1}{A}\left[\frac{\phi_o}{2}\left\{ 2n + \frac{3}{2} + \frac{1}{4} \right\} - LJ \right] .2 \tag{4.92}$$

For loops of lengths which do not satisfy (4.89), the magnetization becomes positive. For $l << \lambda$, $e^{-l/\lambda} \simeq 1 - l/\lambda$ so that the argument of the cosine can be expanded. For small angle the cosine term may also be expanded so that,

$$M = \frac{\pi l J H A^2}{\phi_o}\left[2 + \frac{LJ}{AH} - \frac{l}{\lambda} \right] \tag{4.93}$$

which gives positive (para-Meissner) magnetization. Therefore, depending on the loop size, either the Meissner effect or the PME would occur. The Josephson current varies with the gap of the superconductor, $J = c_1\Delta$. According to the B.C.S. theory the gap varies as $\Delta = \Delta_o(1 - T/T_c)^{1/2}$ so that,

$$J = c_1\Delta_o(1 - T/T_c)^{1/2} \ . \tag{4.94}$$

The current from the first of the equations of (4.85) is given by

$$J = \frac{\phi_o}{2L}(2n + 1) - \frac{AH_{o1}}{L} \tag{4.95}$$

so that the limiting field varies with temperature as,

$$H_{o1} = \frac{\phi_o}{2A}(2n+1) - \frac{Lc_1\Delta_o}{A}(1-T/T_c)^{1/2} \ . \tag{4.96}$$

This field is positive only if $T > T_{pM}$ where,

$$T_{pM} = T_c\left[1 - \{\phi_o(2n+1)/2Lc_1\Delta_o\}^2\right] \ . \tag{4.97}$$

Thus the Meissner effect occurs only above H_{o1} and above the temperature of T_{pm}. A transition temperature is thus generated below which the phase of the current is shifted compared with that in the usual Meissner state. In this state, the magnetization is positive and the resistivity is zero. Thermodynamically, the state has its own transition temperature, T_{pM} and its own law for the variation of susceptibility as a function of temperature, critical field etc. If the current and the magnetic field are treated as parallel to each other in the Meissner phase, then for the fixed direction of the magnetic field, the current is anti-parallel in the para-Meissner phase.

Buyers and Yang found that the flux is quantized for any value of charge, $\sum e$. There is no particular value of the charge for which flux is quantized. On the basis of experimental observation of Shapiro steps, it can be said that charge of the quasiparticles is $2e$ in the superconducting state so that the unit of flux becomes, $\phi_o = hc/2e$. Geshkenbein et al suggest that flux is quantized in units of $\phi_o(n+1/2)$ where n is an integer so that the difference between any two values is a multiple of ϕ_o only and the minimum value is $\phi_o/2$. For a flux-quantized Lorentz force work, equally spaced absorption lines we found so that the energy is determined by $J\phi(n+\frac{1}{2})$ where J is the current. This observation implies that the minimum flux may be $\phi_o/2$. Within the frame work of Ginzburg-Landau (GL) model, there is a phase which violates time reversal invariance, $\mathcal{T}$. In this case it is found that all values differenct from $\phi_o(n+\frac{1}{2})$ can occur only when the order parameter of the GL model violates $\mathcal{T}$ symmetry. Thus values less than $\phi_o/2$ are possible only in a $\mathcal{T}$ violating phase within the GL model. Such a $\mathcal{T}$ breaking phase arises in the superconductor, UPt_3. Buyers and Yang have found that the body current, I, in the superconductor is related to the partition function, Z, as,

$$I = ck_BT\partial \ln Z/\partial\phi \tag{4.98}$$

which can be written as

$$E = \frac{I}{c} d\phi = k_B T \, d\ln Z \tag{4.99}$$

replacing $d\phi$ by $n\phi_o$ we obtain,

$$E_n = \frac{I}{c} n\phi_o \tag{4.100}$$

2so that we obtain equally spaced levels. We replace n by $n + \frac{1}{2}$. Accordingly, $E_n = \frac{1}{c}\phi_o(n + \frac{1}{2})$, so that energy and current become linearly dependent. Buyers and Yang have also derived an expression for the quadratic dependence of ϕ on the energy. The magnetization is determined by (4.62) and the susceptibility of the system for $L = 0$ is given by $\partial M/\partial H$ so that

$$\chi = \frac{2\pi A^2 J}{c\phi_o} \cos\left(\frac{2\pi\phi}{\phi_o}\right) . \tag{4.101}$$

The cosine function changes sign at an angle of $\pi/2$ so that if there is a point at which the susceptibility changes sign from positive to negative values, then such a point occurs at $2\pi\phi/\phi_o = \pi/2$ or at $\phi = \phi_o/4$. Thus smallest flux becomes equal to $\phi_o/4$ whereas only $\phi_o/2$ is predicted by linear theory. The experimental measurements as well as the theoretical calculations are thus consistent with the interpretation that the minimum observable flux becomes $\phi_o/4$. The lines of force of the magnetic field are expelled out of the superconducting samples so that the induction field inside the sample is zero, $B_{in} = 0$. According to Maxwell equations, the $B_{in} = 0$ is completely consistent with the zero resistivity, $\rho = 0$. The phase ϕ/ϕ_o of the Josephson Hamiltonian is found to change to $(\phi/\phi_o)[1 + (2n + \frac{1}{2})\pi]$ till $(\phi/\phi_o)[1 + (2n + \frac{3}{2})\pi]$ so that the magnetization can be both positive (para-Meissner) or negative (Meissner) with zero resistivity in all the phases. Hence the para-Meissner phase is superconducting the same way as the Meissner phase is.

In the case of π-junctions, the phase change of π shows that the symmetry is that of the d wave. Thus the positive sign in the magnetization arises from the phase factor. The atomic paramagnetism arises from the 2 unfilled d-shell of the transition elements such as copper in the present case. The pair current is diamagnetic for all the phases and there are no unpaired single electrons which are paramagnetic. The phases of the two electrons of the Cooper pair are so correlated that there is a change in the phase in going from the s-wave to the d-wave leading to positive (para-Meissner) magnetization. When the field is changed the phase difference between the two electrons changes so that the para-Meissner effect changes to the Meissner effect. If $H_{c1} > H_{o2}$ as well as $H_{c1} > H_{o2}$ the PME region occurs for $0 < H < H_{o1}$ and ME is found for $H_{o1} < H < H_{o2}$ but a second PME region occurs for $H_{o2} < H < H_{c1}$. If $H_{c1} < H_{o2}$, the PME occurs for $0 < H < H_{o1}$, ME occurs for $H_{o1} < H < H_{c1}$ and then a para-Abrikosov region occurs for $H_{c1} < H < H_{o2}$. Just as the London penetration depth and the coherence lengths are given in terms of two critical fields, $\lambda^2 = n\phi_o/H_{c1}$ and $\xi^2 = n\phi_o/H_{c2}$, we find that H_{o1} and H_{o2} should also be associated with lengths i.e., $\xi_1^2 = n\phi_o/H_{o1}$ and $\xi_2^2 = n\phi_o/H_{o2}$ at which the current changes sign.

There are only certain size current loops which give the para-Meissner effect. Similarly, only in a certain temperature range, the para-Meissner effect is observable. Outside the range of the junction size and temperature interval needed for the para-Meissner effect, the magnetization becomes negative. The phase difference between the two electrons of a Cooper pair causes a change in the sign of the magnetization leading to para-Meissner effect. Usually in the B.C.S. theory, the difference between the free energy of a normal metal and that of a superconductor with a gap is compared with the electromagnetic energy to determine the critical field, the application of which destroys superconductivity. Consideration of other forces, such as the Lorentz force leads to two solutions for the critical field. These fields are defined as the H_{c1} which determines the London penetration depth and the H_{c2} which defines the coherence length. The consideration of Lorentz force and heat loss also determines the upper and the lower critical fields which define the boundaries between the normal and the superconducting phases but in addition to such critical fields we also find fields at which the free energy and hence the magnetization changes sign. Below and above these fields both the phases are superconducting. In a real material there may be a transition from type-I to type-II superconductivity upon doping or upon variation of a parameter such as temperature or field. There is no reason to believe that our sign reversal fields occur below H_{c1} and hence similar to a para-Meissner region there may also be a para-Abrikosov region. Similarly, in the case of Josephson vortices, we predict sign reversal fields so that para-Josephson regions are expected. Due to the flux-lattice melting we also expect a flux liquid region. We have thus discussed the conditions under which there is a change in sign of the magnetization of a superconductor so that para-Meissner effect is predicted. The experimental measurements performed also show that in field cooled samples at low fields, the magnetization becomes positive (para-Meissner). At large magnetic fields usual negative value (Meissner) is found. In view of the mesoscopic nature of the real samples, which have Josephson weak links we predict that para-Meissner regions occur in the phase diagram of a type-II superconductor as sketched qualitatively in Fig. 4.4.

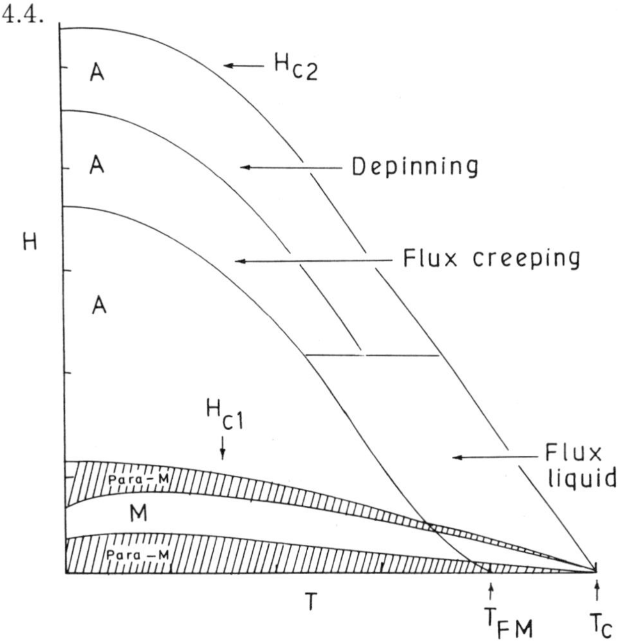

Fig. 4.4. Qualitative phase diagram of a type-II superconductor showing Abrikosov (A), Meissner (M) and para-Meissner (para-M) phases along with H_{c1}, H_{c2}, depinning, flux creeping and flux-liquid phases. The flux-lattice melting temperature, T_{FM} and T_c are also shown.

It has been found that the point at which the magnetization turns from negative towards positive values is located at $\phi_o/4$. According to the theory the flux is quantized in units of $\phi = \phi_o(n + \frac{1}{2})$ so that $\phi_o/2$ is the smallest value of the flux. Therefore $\phi_o(n+\frac{1}{2})$ may be called as the integer quantized flux. The values of the flux which do not fit into the expression $\phi_o(n + \frac{1}{2})$ may therefore be termed as the fractional quantized flux. The first fractional value of $\phi_o/4$ has thus been found. We now examine the experimental data of magnetization of superconducting aluminium discs to look for flux quantization. In one of the discs the magnetization is found to show jumps at the integer multiples of ϕ_o. In another disc it is found that for small fields mangetization is a continuous function of magnetic field. When the field is increased, a jump occurs at a certain value and for higher fields jumps occur at integer multiples of the flux but when the field is decreased from high fields to low fields the flux, which is quantized in units of ϕ_o is seen to quantize in units of $\phi = (10/3)\phi_o$ before reaching the normal continuous value.

The first term in the free energy is $H^2/8\pi$ which can be integrated over a given volume. The magnetization is defined by the field derivative of the free energy, $M = \partial F/\partial H$. Therefore, the first term in the magnetization depends linearly on the magnetic field. The flux quantization condition is linear in magnetic field, $SH = \phi_o$, so that the effective surface area S can be determined from the knowledge of the value of ϕ_o and the field, H. Once the area is ascertained, if there are more than one events, the value of $n\phi_o$ can be found. Therefore, we examine the magnetization of a sueprconducting disc of aluminium in Fig. 4.5, the radius of the disc is $R = 1.2\mu$m and the thickness is $d = 0.15\mu$m. In Fig. 4.5, a linear behaviour is seen for $H < 25$ Gauss as predicted by the theory. At 25.8 Gauss, the first $n = 6$, dip occurs so that we determine the area from the known value of $\phi_o = 2.07 \times 10^{-7}$ G cm^2. The field at $12\phi_o$ is 51.6 Gauss. We count the number of dips between $6\phi_o$ and $12\phi_o$ which gives 6 intervals so that the elementary flux at which dips occur is $1\phi_o$. The vlaues of the flux at the dips are 6, 7, 8, 9, 10, 11, 12, etc. For $n > 12$, the magnetization dips are seen at integer multiples of ϕ_o.

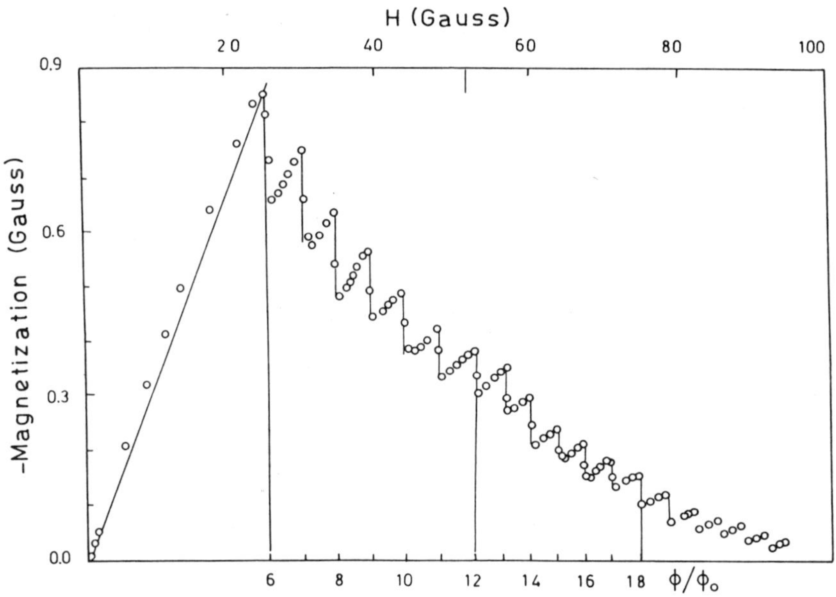

Fig. 4.5. The magnetization of superconducting discs of aluminium showing linear behaviour at small magnetic fields. For fields larger than that required for $6\phi_o$, dips are predicted at quantized values of the flux.

The quantization depends on the boundary conditions which are determined fromt he size of the disc. Therefore, we look at the experimental measurements of magnetization in one more disc of radius 2.5 μm. In the case of field cooled samples, the magnetization as a function of magnetic field is given in Fig. 4.6. In this case the initial rise corresponds to 2 flux quanta but there is no jump at $n = 3$ (15 Gauss) or at $n = 4$ (20 Gauss). For ϕ/ϕ_o larger than 6, single flux quantum is seen to move out of the disc. The quantization at integer values for $n > 6$ is quaite clear but there is an interesting phenomenon of fractional quantization of flux at $n = 10/3$. The flux quantization condition $HS = n\phi_o$ for the superconductors with $\phi_o = hc/2e$ at $n = 10/3$ shows that charge of the quasiparticles is 3/10 e. Such a possibility of charge is not contradicted by the theory of flux quantization. The paramagnetic Meissner effect (PME) depends on the particle size and hence on disc radius. In Fig. 4.5 PME occurs for $\phi < 6\phi_o$ whereas the point at which the magnetization turns from negative towards positive values, was found at $(1/4)\phi_o$ and in Fig. 4.6 this point is near $(10/3)\phi_o$.

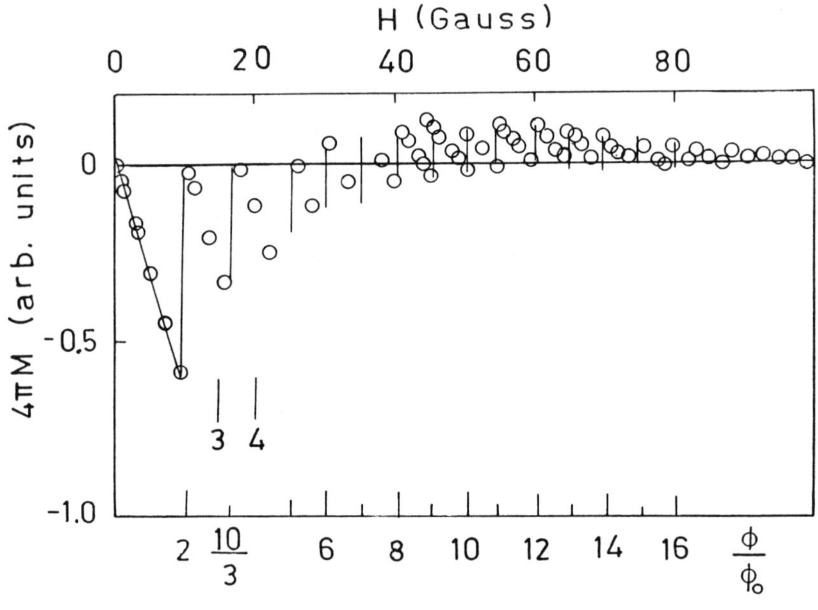

Fig. 4.6. The magnetization of a disc of aluminium showing continuous flux upto $2\phi_o$. For $n = 5$ and larger values, the flux is quantized in units of one ϕ_o. There is no quantization at $n = 3$ or 4 but there is one at (10/3) *of ϕ_o.*

Suggested References

4.1. V.G. Kogan, M.M. Fang and S. Mitra, Phys. Rev. B**38**, 11958 (1988).

4.2. L.J. Campbell, M.M. Doria and V.G. Kogan, Phys. Rev. B**38**, 2439 (1988).

4.3. K.N. Shrivastava, ref.0.17 and 0.18.

4.4. N. M. Krishna, L.S. Lingam, P.K. Ghosh and K.N. Shrivastava, Physica C**294**, 243 (1998).

4.5. D.J. Thompson, M.S.M. Minhaj, L.E. Wenger and J.T. Chen, Phys. Rev. Lett. **75**, 529 (1995).

4.6. M. Sigrist and T.M. Rice, Rev. Mod. Phys. **67**, 503 (1995).

4.7. V.V. Moshchalkov, X.G. Qiu and V. Bruyndoncx, Phys. Rev. B**55**, 11793 (1997).

4.8. V.B. Geshkenbein, A.I. Larkin and A. Barone, Phys. Rev. B**36**, 235 (1987).

4.9. K.N. Shrivastava, Phys. Today **51** [9] 15 (1998).

4.10. P.S. Deo, V.A. Schweigert, F.M. Peeters and A.K. Geim, Phys. Rev. Lett. **79**, 4653 (1997).

4.11. A.K. Geim, S.V. Dubonos, J.G.S. Lok, M. Henini and J.C. Mann, Nature **396**, 144 (1998).

4.12. A.K. Geim, et. al., Nature **390**, 259 (1997).

4.13. A.K. Geim, et. al., Appl. Phys. Lett. **71**, 2379 (1997).

4.14. K.N. Shrivastava, Physica B**281-282**, 1012 (2000).

Chapter 5

Microwave absorption

5.1 Electron spin resonance

The electron spin resonance is the absorption of microwaves by the energy levels due to splitting of states in a magnetic field. The absorption is affected by a small oscillating field applied normal to the large d.c. field. The Hamiltonian of the system is given by

$$\mathcal{H} = g\mu_B H.S \tag{5.1}$$

where g is the Lande's splitting factor,

$$g = 1 + \frac{J(J+1) + S(S+1) - L(L+1)}{2J(J+1)} \ . \tag{5.2}$$

Here L is the orbital angular momentum quantum number and S is the spin angular momentum quantum number with $J = L + S$. For $S = 1/2$ the value of g is equal to 2. However, the electron has a self field. Due to this electromagnetic correction, the value of g is found to be about 2.0023. The spin of the atom is S and μ_B is the Bohr magneton. Since the field is large compared with other interactions in the sample, it can be assumed that the field is along the z direction so that for $S = 1/2$ with projections along the field as $+1/2$ and $-1/2$. In this case the eigen values are given by

$$E_{\pm} = \pm\frac{1}{2}g\mu_B H \ . \tag{5.3}$$

A smal field, h_x, is applied normal to the field H so that one more term is added to the Hamiltonian, $\mathcal{H}' = g\mu_B h_x S_x$. Since $S_x \pm iS_y = S_{\pm}$, this term causes transitions. The absorption probability is given by the Fermi's golden rule,

$$P_{+\leftarrow -} = \frac{2\pi}{\hbar} | < +|\mathcal{H}'|- > |^2 \rho_f \tag{5.4}$$

where $P_{+\leftarrow -}$ is the probability of a transition from $| - 1/2 >$ to $| + I/2 >$ with ρ_f as the density of final states which is limited to the conservation of energy, i.e., the energy absorbed is equal to the energy level separation

$$\hbar\omega = g\mu_B H \tag{5.5}$$

as shown in Fig. 5.1a. Once the absorption occurs, the population of the upper level increases much more than is given by the Maxwell-Boltzmann distribution. Therefore, the absorption stops. For continuous observation of absorption, it is necessary that the atoms relax back to the lower energy level so that the population in the lower level is recuperated. This is achieved by the thermal contact of the spin with phonons. For this purpose, the simplest form of the interaction is written as,

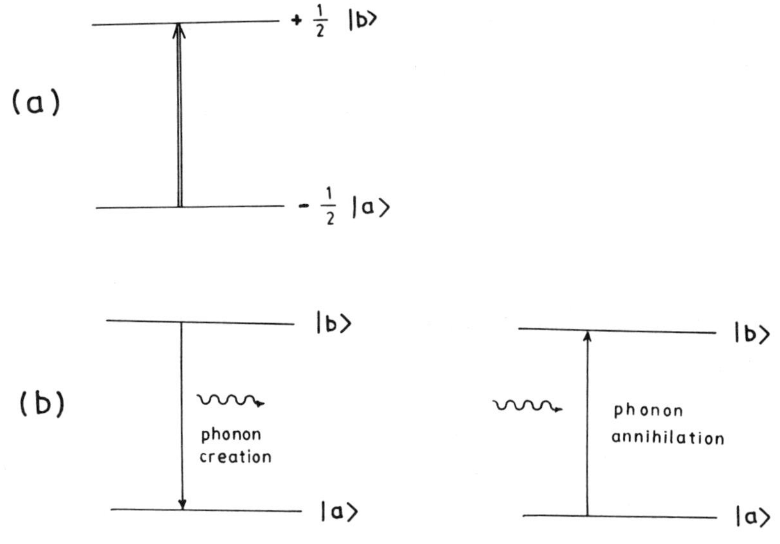

Fig.5.1 (a) The microwave is absorbed while a transition takes place from the state $|a >$ to $|b >$. The absorption is caused by small oscillating field along x-direction due to the separator S_x. The large magnetic field is along z direction.
(b) The emission of a phonon causes relaxation. The phonons are also absorbed due to hermiticity of the interaction.

$$\mathcal{H}'' = \sum_k \left(\frac{\partial V}{\partial R}\right)(S_+ + S_-)\left(\frac{\hbar}{2M\omega}\right)^{1/2}(a_k + a^\dagger_{-k}) \tag{5.6}$$

for which the unpertubed Hamiltonian is of the form,

$$\mathcal{H}_o = \Omega S_z + \sum_k \hbar\omega_k a^\dagger{}_k a_k \ . \tag{5.7}$$

The spin relaxes the lower level by emission of a phonon. Similarly, the spin may absorb a phonon as shown in Fig. 5.1b the sum of the probabilities of absorption, P_{ba} and emission P_{ab} defines the relaxation rate,

$$\frac{1}{\tau} = P_{ab} + P_{ba} \ . \tag{5.8}$$

In this way, we obtain the absorption of energy in the spin levels with the line width given by $\hbar/(2\tau)$.

5.2 E.S.R. in a superconductor

The ESR response due to pairs of electrons should be zero because ESR requires single particles with finite spins. In a real material single electrons with spin 1/2 occur but not in the B.C.S. theory of superconductivity, in which electrons are paired with zero spin and zero momentum. Therefore, the E.S.R. vanishes only at zero temperature. This means that ther eis a deviation from the B.C.S. theory which permits the existence of unpaired electrons so that there is a single in the electron spin resonance which vanishes only at very low temperature. The E.S.R response function is proportional to,

$$\chi'' = \frac{\delta\Omega}{(\omega-\omega_r)^2 - (\delta\Omega)^2} \tag{5.9}$$

where $\hbar\omega_r = g\mu_B H$ is the resonance frequency. At $\hbar\omega = \hbar\omega_r$, the resonance occurs and χ'' is peaked at this value of the frequency. In going from the normal to the superconducting state, the response function is to be multiplied by $l/\Lambda\xi_o$ where l is the mean free path of the electron and ξ_o is the superconducting coherence length with $1/\Lambda = ne^2x/mc^2$. At resonance,

$$\chi''_N = -\frac{1}{\delta\Omega} \tag{5.10}$$

in the normal state and

$$\chi_s'' = -\frac{1}{\Lambda\xi_o\delta\Omega/l} \tag{5.11}$$

in the superconducting state. This means that the line width broadens in going from normal to the superconducting state according to

$$\delta\Omega_s = \Delta\Omega_N(\xi_o/l) \ . \tag{5.12}$$

The E. S. R. of Cu^{2+} in $YBa_2Cu_3O_7$ at room temperature shows anisotropic exchange narrowed line with,

$$\begin{aligned} g_{\|} &= 2.2167 \pm 0.0010 \\ g_{\perp} &= 2.0475 \pm 0.0010 \ . \end{aligned} \tag{5.13}$$

This line shows that the sample is paramagnetic at room temperature. The temperature dependence of the g-shift above the transition temperature is shown in Fig.5.2. It is similar to what is expected from the Curie's law of susceptibility of a paramagnet. It is quite clear that the g shift is proportional to C/T. Therefore, $YBa_2Cu_3O_7$ for $T > T_c$ is said to be a paramagnetic insulator.

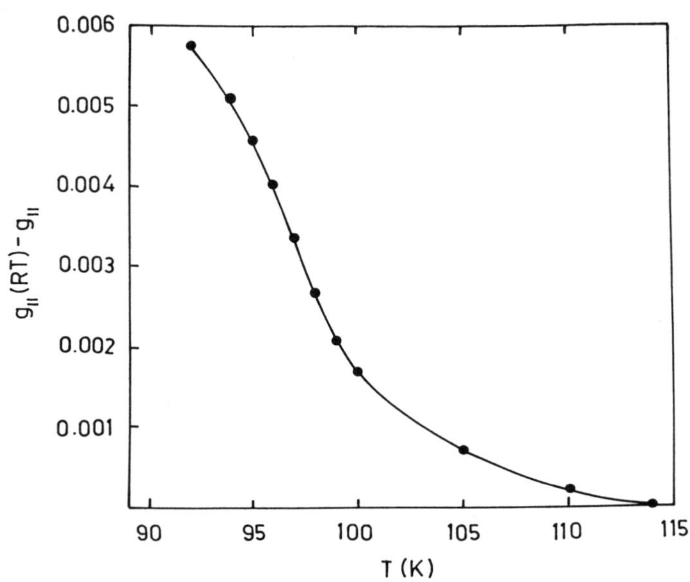

Fig.5.2 Temperature dependence of $g_{\|}$ of Cu^{2+} in $YBa_2Cu_3O_{7-\delta}$ for $T > T_c$ showing paramagnetic behaviour.

Below the T_c the g shift is shown in Fig.5.3. This shift is negative as expected from the susceptibility of a diamagnet. Therefore, the bulk of the material below T_c is diamagnetic.

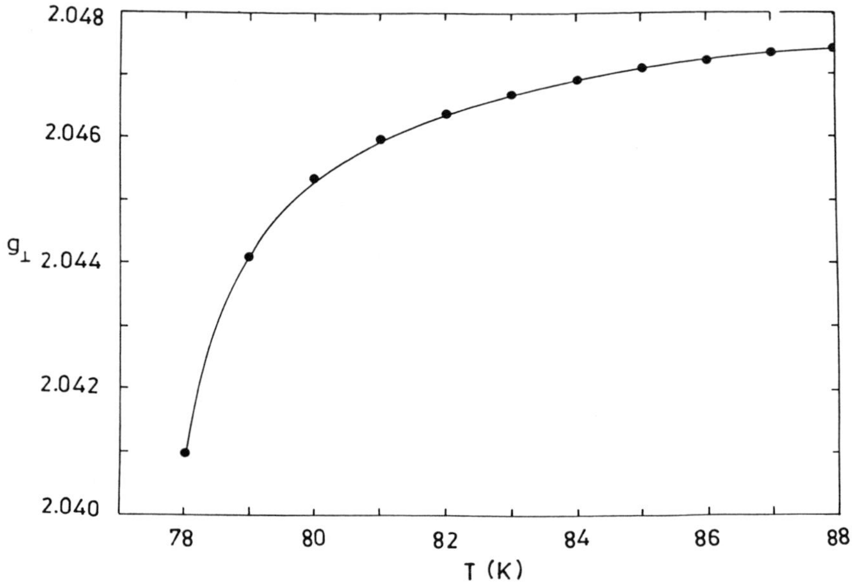

Fig.5.3 Temperature dependence of $g_\perp$ *of* Cu^{2+} *in* $YBa_2Cu_3O_{7-\delta}$ *for* $T < T_c$ *showing diamagnetic behaviour.*

The diamagnetic behaviour is seen more in $g_\perp$ than in $g_\parallel$. Thus the superconducting particles are in the ab plane and not along c axis. The intensity of the E.S.R. line as a function of temperature displaed in Fig. 5.4 approaches zero at $T = 0$. Thus we conclude that the number of single-particle unpaired electrons reduces upon cooling indicating that pairs of spin zero are formed. The electron spin resonance thus shows that paramagnetic electrons are surrounded by diamagnetic matter below T_c and upon cooling all paramagnetic electrons disappear or get paired.

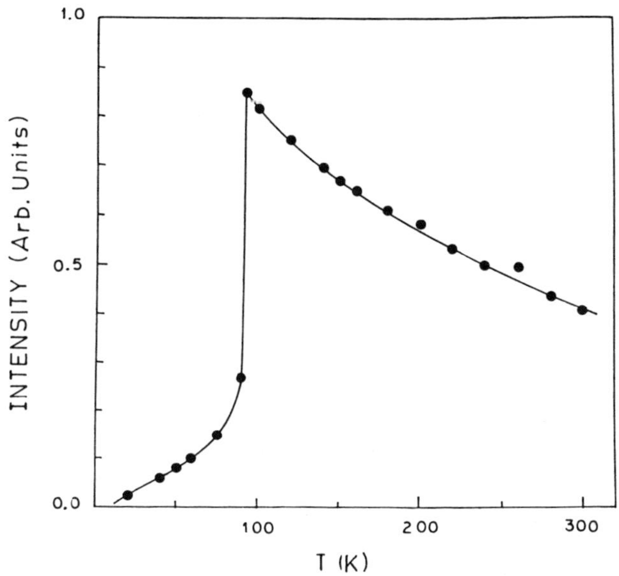

Fig.5.4 Temperature dependence of the intensity of the ESR line of Cu^{2+} showing the vanishing of the signal upon approaching towards zero temperature.

5.3 Microwave absorption

In the BCS theory there is a gap in the dispersion relation so that the microwaves can not be absorbed. In the present section we show that the B.C.S. theory does not account for the microwave absorption. The microwave signal is a manifestation of flux quantized energies. The maximum in the $d\chi''/dH$ corresponds to the maximum in the Josephson current occuring at the Josephson frequency, $\omega_J \simeq 2eV/\hbar$. Here $2e$ is the charge of the quasiparticles, and V is the applied voltage. The total current is given by the sum of the normal and the supercurrent, $J = J_n + J_s$. The Meissner effect is given by

$$c \text{ curl } \Lambda J_s + B = 0 \tag{5.14}$$

and the supercurrent at $E = 0$ is given by

$$\frac{\partial}{\partial t}\Lambda J_s - E = 0 \tag{5.15}$$

where

$$\frac{1}{\Lambda} = \frac{ne^2x}{m} \tag{5.16}$$

is which n is the charge density, e the electronic charge and m is the mass of the electron. The Josephson current is given by,

$$J = J_o \sin\left(\gamma_2 - \gamma_1 - \frac{2e}{h}\int \vec{A}.d\vec{l}\right) \tag{5.17}$$

where γ_1 and γ_2 are the phase factors, $\vec{A}$ is the vector potential of the electromagnetic wave and the integral is extended over the width of the Josephson junction. since $\vec{A} = \vec{B} \times \vec{r}$, we can write,

$$J = J_o \sin\left\{\delta(o) + \frac{2e}{h}Bx(t + 2\lambda)\right\} . \tag{5.18}$$

The maximum current is given by

$$I_{\max} = 2I_o \frac{\sin(\pi\phi_j/\phi_o)}{(\pi\phi_j/\phi_o)} \cos(\pi\phi/\phi_o) \tag{5.19}$$

where $\phi_j = Bx(2\lambda + t)$ with λ as the London penetration depth, t the thickness of the normal barrier, x the width, $\phi = n\phi_o$ and $\phi_o = hc/2e$. The amplitude in (5.17) is related to the gap, Δ of the superconductor and to the normal state resistivity R_N of the insulating layer, the surface area of which is S, is given by the relation

$$J_o = \frac{\pi\Delta}{2SR_N} \tag{5.20}$$

when there are a large number of Josephson junctions such as in an oxide material, we expect to find Gaussian behaviour rather than that of the spherical Bessel function as in (5.19). When a system is excited by a microwave of frequency, say 9.2 GHz, the resonance is given by

$$\chi'' = \frac{\delta\Omega}{(\omega - \omega_J)^2 - (\delta\Omega)^2} \tag{5.21}$$

where $\omega_J = 2eV/\hbar$ is the Josephson frequency. In the superconducting state we replace (5.15) by $\Lambda\xi_o/l$ where ξ_o is the coherence length and l is the mean free path of the electron in the normal state so that the resonance becomes

$$\chi'' = \frac{l\delta\Omega}{\Lambda\xi_o[(\omega-\omega_J)^2-(\delta\Omega)^2]} \ . \tag{5.22}$$

Using the relation $\xi_o = \hbar^2 k_F/2m\Delta$ for the coherence length, where the Fermi wave vector, $k_F = (3\pi^2 n)^{1/3}$ is determined from the electron density n, we find,

$$\chi'' = \frac{2lne^2x\Delta\ \delta\Omega}{\hbar^2 k_F[(\omega-\omega_J)^2-(\delta\Omega)^2]} \ . \tag{5.23}$$

In the approximation of $l = \tau(\hbar/m)(3\pi^2 n)^{1/3}$, we introduce the life time of the electron τ, so that

$$\chi'' = \frac{2\tau ne^2x\Delta\ \delta\Omega}{m\hbar[(\omega-\omega_J)^2-(\delta\Omega)^2]} \ . \tag{5.24}$$

The zeros of the current are located at

$$\delta(o) + \frac{2e}{\hbar}Bx(t+2\lambda) = p\pi \tag{5.25}$$

where p is an integer. At these fields the Josephson current is zero for finite integers and the superconductivity vanishes but on both sides, less or more, of such fields, the system remains superconducting. The maximum of the spherical Bessel function $(\sin x)/x$ occurs at $x = 0$, so that the maximum current occurs for a general solution of,

$$\delta(o) + \frac{2e}{\hbar}Bx(t+2\lambda) = 0 \ . \tag{5.26}$$

The phase difference $\delta(o)$ around a closed circuit which encompasses a total magnetic flux ϕ is given by

$$\delta(o) = \frac{2e}{hc}\phi \tag{5.27}$$

so that the maximum current occurs for a field of

$$Bx(t+2\lambda) = \frac{h}{2e} \ . \tag{5.28}$$

The Josephson frequency and voltage relationship is given by

$$\frac{\hbar\omega_J}{2\pi} = 2eV \ . \tag{5.29}$$

From (5.28) and (5.29), we find that

$$\frac{\omega_J}{2\pi} = \frac{V}{Bx(t+2\lambda)} \ . \tag{5.30}$$

The flux quantized energies occur in the form of spectroscopy as $d\chi''/dH$ is plotted as a function of magnetic field. The height of this signal at the maximum varies with temperature like the gap of the superconductor. The field at which the maximum occurs corresponds to the size of the insulating barriers in between the superconducting grains and resonances correspond to "flux quantized energies". The flux quantization leads to quantized energy levels with the use of Maxwell's equation relating curl of the magnetic field with the Josephson current and Lorentz force. The Maxwell equation needed is,

$$curl\vec{B} = -\frac{4\pi}{c}\vec{J_c} \ . \tag{5.31}$$

Since the external magnetic field is applied along a particular direction, we write the above equation in terms of components of the field as,

$$\begin{aligned} \frac{\partial}{\partial y}B_z - \frac{\partial}{\partial z}B_y &= -\frac{4\pi}{c}J_{cx} \\ -\frac{\partial}{\partial x}B_z + \frac{\partial}{\partial z}B_x &= -\frac{4\pi}{c}J_{cy} \\ \frac{\partial}{\partial x}B_y - \frac{\partial}{\partial y}B_x &= -\frac{4\pi}{c}J_{cz} \ . \end{aligned} \tag{5.32}$$

In case $B_x = B_y = 0$, $B = B_z$, the above equations assume a particularly simple form,

$$\frac{\partial}{\partial y}B_z = -\frac{4\pi}{c}J_{cx} \tag{5.33a}$$

$$\frac{\partial}{\partial x}B_z = \frac{4\pi}{c}J_{cy} \tag{5.33b}$$

$$J_{cz} = 0 \tag{5.33c}$$

which means that the component of the current along the direction of the magnetic field is zero. Multiplying (5.33a) by y and (5.33b) by x and integrating for flat distribution, we obtain

$$B_z = -\frac{4\pi}{c} y J_{cx} = +\frac{4\pi}{c} x J_{cy} \tag{5.34}$$

so that there are two solutions for B_z. The microwaves produce a supercurrent perpendicular to the direction of the magnetic field so that there is a Lorentz force,

$$\vec{F} = -\frac{V}{c}\vec{B} \times \vec{J_c} \tag{5.35}$$

since $B = B_z$ and $J_{cz} = 0$, the components become,

$$F_x = \frac{V}{c} B_z J_{cy} \tag{5.36a}$$

$$F_y = -\frac{V}{c} B_z J_{cx} \ . \tag{5.36b}$$

Taking the curl of (5.31),

$$curl\ curl \vec{B} = \frac{4\pi}{c}\ curl\ \vec{J_c} \tag{5.37}$$

which becomes,

$$-\nabla^2 B = \frac{4\pi}{c}\ curl\ J_c \ . \tag{5.38}$$

The London equation is,

$$curl\ J_c = -\frac{cB}{4\pi\lambda^2 L} \tag{5.39}$$

so that from (5.38) and (5.39) we find,

$$\nabla^2 B_z = B_z \lambda_L^{-2} \ . \tag{5.40}$$

The energy $E = \vec{F}.\vec{r}$ from (5.36) can be written as $xF_x + yF_y$ or

$$E = \frac{V}{c} B_z x J_{cy} - \frac{V}{c} B_z y J_{cx} \tag{5.41}$$

because $J_{cz} = 0$. As the applied field is H_z, the energy may be written as,

$$E = \frac{V}{c}[xJ_{cy} - yJ_{cx}][H_z \pm B_{oz} \exp(-z/\lambda_L)] \ . \tag{5.42}$$

Assuming $J_{cx} = J_{cy} = J_{cxy}$, (5.42) reduces to

$$E = \frac{V}{c}(x - y)J_{cxy}[H_z \pm B_{oz}\exp(-z/\lambda_L)] \tag{5.43}$$

which leads to quantized energy for flux quantized fields according to which $SB_{oz} = n\phi_o$ where S is the area normal to the field in the xy plane. The second term in (5.43) gives a small decaying field corresponding to Meissner effect. The first term has the area due to anisotropy, $S = V(x - y)/S'$ so that the quantized energy levels are predicted,

$$E_n = nJS\phi \ . \tag{5.44}$$

The Josephson relation is

$$\nu = 2eV/h \tag{5.45}$$

so that the voltage is related to the current

$$V = Jhc/(4e^2) \tag{5.46}$$

where J is the Josephson current, h is the Planck's constant, c is the velocity of light and e is the charge of the electron. The equation (5.46) is consistant with the Ohm's law except that the fundamental resistance unit is $hc/(4e^2)$. Substituting (5.46) in (5.45) gives,

$$\nu = Jc/(2e) \ . \tag{5.47}$$

Therefore,

$$h\nu = Jhc/2e \tag{5.48}$$

in which we introduce the flux quantization by using the value, $\phi_o = hc/2e$, so that

$$h\nu = nJS\phi_o \tag{5.49}$$

where S is the area of cross section. When the microwave frequency matches with the above, we expect resonance lines or oscillations. The magnetic field is contained in the current so that varying the magnetic field, H, the relation (5.49) can be tuned untill the field matches with the microwave frequency via the variation in the current so that there are oscillations in $d\chi''/dH$. Using the expression (5.49) with, $\nu = 9.4 \times 10^9$ Hz, $h\nu = 6.22863 \times 10^{-17}$ erg, $S = 0.01$ cm^2, $n = 1$, $\phi_o = 2.067784 \times 10^{-7}$ Gcm^2 and $e = 4.8035 \times 10^{-10}$ cm$^{3/2}$g$^{1/2}$s^{-1}, the current J is calculated to be 0.9×10^3A/cm^2. The current and the microwave frequency resonate causing the spectral lines so that $\phi_o = h\nu/JS$ can be determined from a measurement of frequency and current per unit area. When the microwave frequency agrees with $E_n - E_{n'}(n - n' = 1)$ we obtain a series of lines corresponding to $\omega = \omega_n$. Since the power is proportional to $H^2/8\pi$, we expect the field intervals to vary as the square root of power.

5.4 Power absorption in one d.c. and one a.c. coil

We assume that a magnetic field is applied normal to a Josephson junction and an oscillating current I_μ is also passing normal to the d.c. field. The current I_μ is induced as a surface current. The current due to the magnetic field is $I_c(B)$. In the case of strongly coupled long Josephson junctions the absorption occurs within the effective Josephson penetration depth λ_J. The supercurrent is given by,

$$I_s = I_c(B) \sin \phi \ . \tag{5.50}$$

The dynamical response of the Josephson junction is given by,

$$I_\mu \sin \omega t = I_c(B) \sin \phi(t) + \frac{\hbar}{2e2R_J(\phi)} \frac{d\phi}{dt} \tag{5.51}$$

where ω is the frequency of the microwave, $I_c(B)$ is the critical current of the junction at d.c. magnetic field, B. The subgap resistance is R_J and ϕ is the phase factor. The microwaves are absorbed at the relation $2eV_\omega = \hbar\omega$. The voltage produces a leakage current, $I_\omega = V_\omega/R_J$. We introduce the dimensionless currents, $i_c = I_c/I_\omega = I_cR_J/V_\omega$ and $i_\mu = I_\mu/I_\omega = I_\mu R_J/V_\omega$ so that (4.50) becomes,

$$\frac{d\phi}{dt} = i_\mu \sin \tau - i_c \sin \phi \tag{5.52}$$

where $\tau = \omega t$. The equation (5.52) can be plotted using four parameters i_μ, ω, i_c and R_J. In the linear approximation $\sin\phi \simeq \phi$, the solution is given by

$$i_s = \frac{i_\mu}{1 + (i/i_c)} \quad . \tag{5.53}$$

In the limit of i_c going to zero, $i_\mu < 1$, the Josephson current has a maximum value of i_c. The microwave power loss in the junction is described by

$$P(B) = \frac{1}{2} I_\mu^2 R_j [i_c^2(B) + 1]^{-1} \tag{5.54}$$

where the normal current is $iI_\omega(i_s/i_c)$ and $P = \frac{1}{2} I_m^2 R_J$. By taking numerical values for i_c, i_μ and τ it is seen that there are two values of ϕ for one value of τ. The system jumps from one value to another resulting into a dynamic flux quantization. The power loss of the junction can be expressed as,

$$P(B) = \frac{1}{2}(I_\omega)^2 R_J \left\langle \left(\frac{\Delta\phi}{\Delta\tau} \right) \right\rangle \quad . \tag{5.55}$$

As i_c goes to zero, $(\Delta\phi/\Delta\tau)^2$ approaches i_μ^2 so that the above equation can be written as,

$$P = \frac{1}{2} I_{rf}^2 R_J P_n(i_\mu, i_c) \tag{5.56}$$

where the normalized power loss is defined as $P_n \simeq < \dot{\phi}^2 > / i_\mu^2$. For a fixed value of i_μ steps corresponding to the dynamic flux quantization occur in $P_n(i_c)$ at the values of i_μ of the same order of magnitude as i_c.

5.5 Observation of Microwave Absorption

The equipment consists of a large electromagnet which provides the magnetic field along z direction. The microwaves are produced by a Gunn diode. The usual wave guide and the microwave cavity operating at 9 GHz is used to excite the sample. The modulation coil is provided in the microwave cavity such that the magnetic field of this coil is normal to the direction of the large d.c. field. The samples of $YBa_2Cu_3O_7$ when cooled below the transition temperature showed a line at low fields which is not found in the high temperature normal phase. Such an absorption is shown in Fig. 5.5. At large fields the usual ESR absorption of Cu^{2+} is visible. When the sample is cooled this absorption, first reduces in intensity and later vanishes as the temperature approaches zero due to pairing. On the other hand, the low field absorption occuring below T_c increases in intensity upon cooling. At $T = 0$ this absorption becomes very strong.

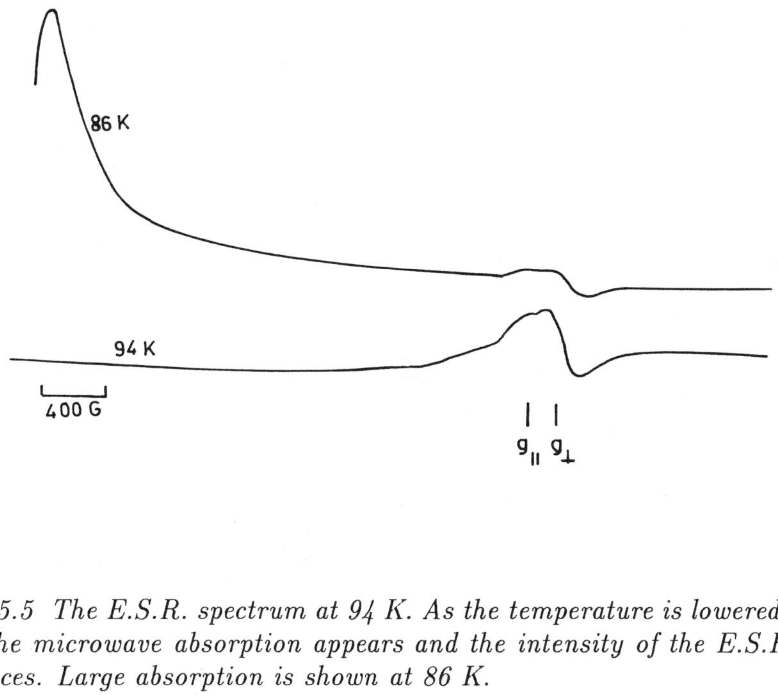

Fig.5.5 The E.S.R. spectrum at 94 K. As the temperature is lowered below T_c the microwave absorption appears and the intensity of the E.S.R. line reduces. Large absorption is shown at 86 K.

In Fig. 5.6 we show the low field absorption which shows small step like structures. The intensity of this absorption plotted as a function of temperature is shown in Fig. 5.7 The detailed microwave absorption as a function of temperature is also shown in Fig. 5.8. It is seen that the absorption is rich in structure and oscillations are developed in some of the spectra.

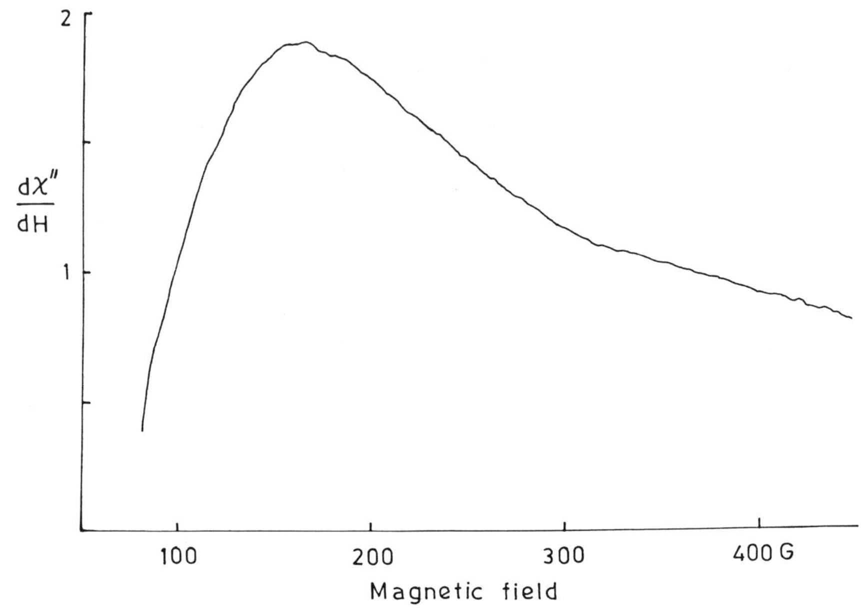

Fig.5.6 Low field micrwoave absroption of Fig. 4.5 as a function of magnetic field in $YBa_2Cu_3O_{7-\delta}(\delta \simeq 0.02)$ showing structure.

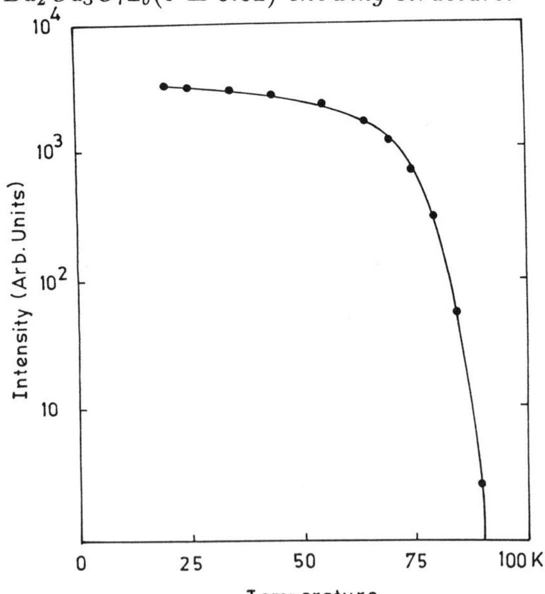

Fig.5.7 Temperature dependence of the intensity of the low field absorption showing that it vanishes at T_c.

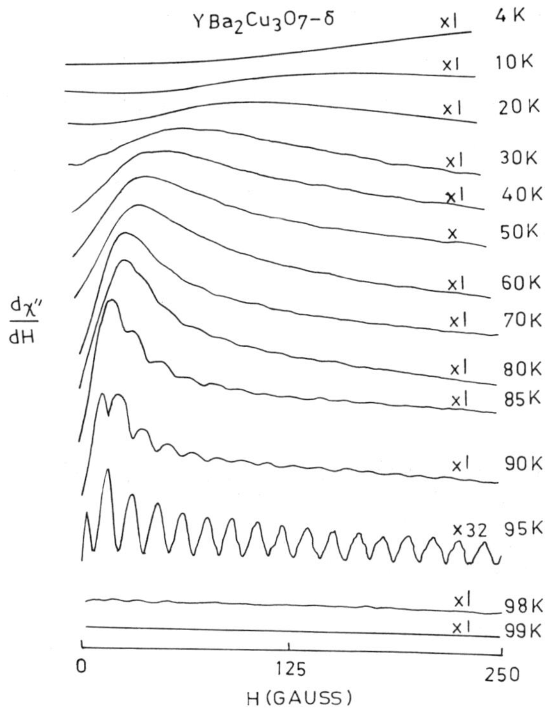

Fig.5.8 The microwave absorption in $YBa_2Cu_3O_{7-\delta}$ at several temperatures.

In a small chip, the microwave absorption is seen to display equally spaced lines. One of these spectra is shown in Fig. 5.9. When field changes from H to $-H$, the d/dH also changes sign. The scan for negative field is also seen. It is clear that due to sign change at both H and d/dH, the spectra should shift from the first quadrant to third quadrant. This is what is exactly observed. This also means that the system is invariant with respect to change of sign of the magnetic field which proves that there is time-reversal invariance. The lines also exhibit flux quantization according to (5.44). In one of the samples, when it is cooled below 60 K, there is a line splitting of about 2.3 G as shown in Fig. 5.10. This splitting is characteristic of emission from the centre of an absorption line. It is indicative of the saturation at which emission starts instead of absorption. Every line looks like a doublet. The property of equal spacing remains unchanged as the temperature is varied from 90 K to 60 K. In

(5.34) two values of the field B_z are seen, which means that the current in the x and y directions are so directed that if there is a current in the positive y direction then there is another component in the negative x direction and vice versa. We assume that the externally applied magnetic field is along the z direction, H_{oz}, then due to induction, the effective field has two solutions,

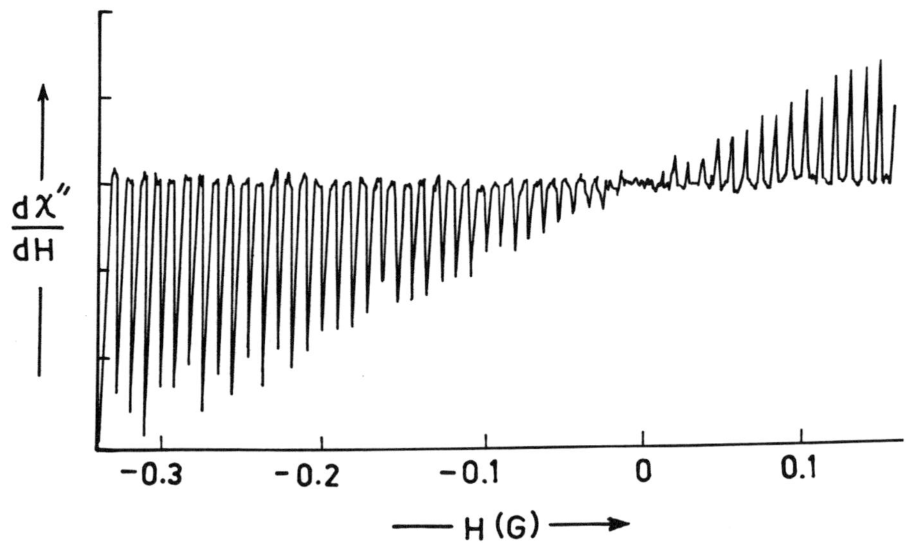

Fig.5.9 Equally spaced microwave absorption lines symmetrically situated near zero field showing time reversal invariance.

$$\tilde{H}_{oz} = H_{oz} - \frac{4\pi}{c} J_{cx} y$$
$$\tilde{H}_{oz} = H_{oz} + \frac{4\pi}{c} J_{cy} x \ . \tag{5.57}$$

The magnetic flux is transported against a macroscopic force in a volume V,

$$\vec{F} = -\frac{V}{c} \vec{B} \times \vec{J} \ . \tag{5.58}$$

Since $B = B_z$, the above equation becomes,

$$F_x = J_{cy} B_z V / c \tag{5.59a}$$
$$F_y = -J_{cx} B_z V / c \ . \tag{5.59b}$$

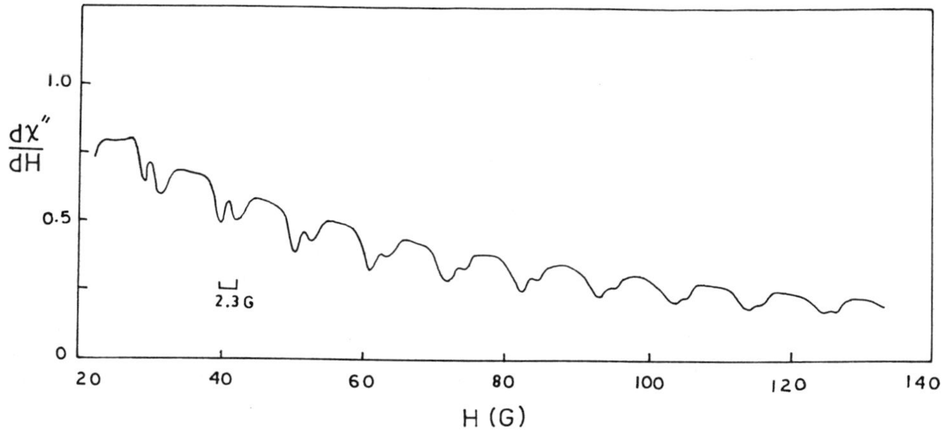

Fig.5.10 Low field micrwoave absorption having equally spaced lines which display splitting of about 2.3 G in a sample of $YBa_2Cu_3O_{7-\delta}$.

The energy associated with this force is given by

$$E = (J_{cy}x - J_{cx}y)B_zV/c \ . \tag{5.60}$$

Substituting (5.57) in (5.59) we find that

$$\begin{aligned} E_1 &= (J_{cy}x - J_{cx}y)(H_{oz} - \frac{4\pi}{c}J_{cx}y)V/c \\ E_2 &= (J_{cy}x - J_{cx}y)(H_{oz} + \frac{4\pi}{c}J_{cy}x)V/c \ . \end{aligned} \tag{5.61}$$

Assuming that $J_{cx} = J_{cy} = J_{cxy}$ eq.(5.61) reduces to

$$E_1 = \frac{1}{c}J_{cxy}V(x-y)H_{oz} - \frac{4\pi}{c^2}J_{cxy}^2V(x-y)y \tag{5.62a}$$

$$E_2 = \frac{1}{c}J_{cxy}V(x-y)H_{oz} + \frac{4\pi}{c^2}J_{cxy}^2V(x-y)x \tag{5.62b}$$

which shows line splitting in addition to what is given by (5.44) when we quantize the field given by the first term in (5.62a) and in (5.62b). This kind of line splitting is observed in $d\chi''/dH$ as shown in Fig. 5.10. Here S is the area of cross section perpendicular to the direction of the magnetic field. In the case of an isotropic system, the first term of (5.62) vanishes so that the equally spaced lines do not occur. The second term in the equation (5.62a) is negative while the corresponding term in (5.62b) is positive so that the lines shift in opposite directions resulting into doublets. This means that the equally spaced lines can be interpreted to arise from the quantized Lorentz force work. In the microwave absorption spectrum of a small sample of $ErBa_2Cu_3O_{7-\delta}$($\delta \sim= 0.2$) we find a large number of equally spaced lines at 9.2 GHz at 4.2 K at very low magnetic fields as shown in Fig. 5.11. These lines have a spacing of about 0.16 G at the microwave power of 0.03 mW. As the power is increased, the lines saturate as normally expected. Upon further increase of power to about 1 mW,

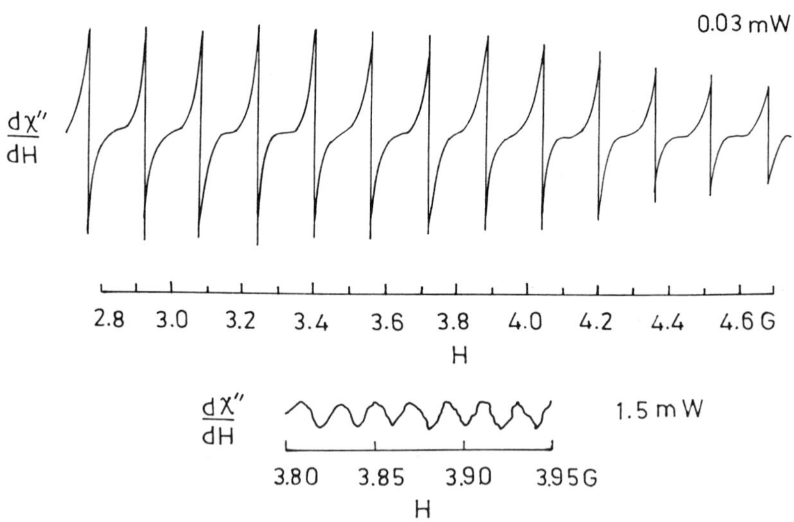

Fig.5.11 Low field microwave absorption in a small, ~ 0.01 cm size, single crystal of $ErBa_2Cu_3O_{7-\delta}$ at 9.2×10^9 Hz frequency for the magnetic field along the [110] direction. The upper spectrum is taken for 0.03 mW. The shape of one of the lines is recorded with increasing microwave power till it saturates. Upon further increase of the microwave power, the lines split into several lines.

every line splits and new lines arise within each line. This time the line spacings are about 0.02 G only. The large splitting of 0.16 G corresponds to a small Josephson vortex of size $\lambda_J = (\phi_o/\delta H)^{1/2} = 11.37 \times 10^{-4}$ cm. As the crystal is rotated away from the [110] direction, the lines spread apart. The position of two of the lines from the centre of the spectrum as a function of the angle of rotation is given by Fig. 5.12. The angular variation can be described by an anisotropic polynomial.

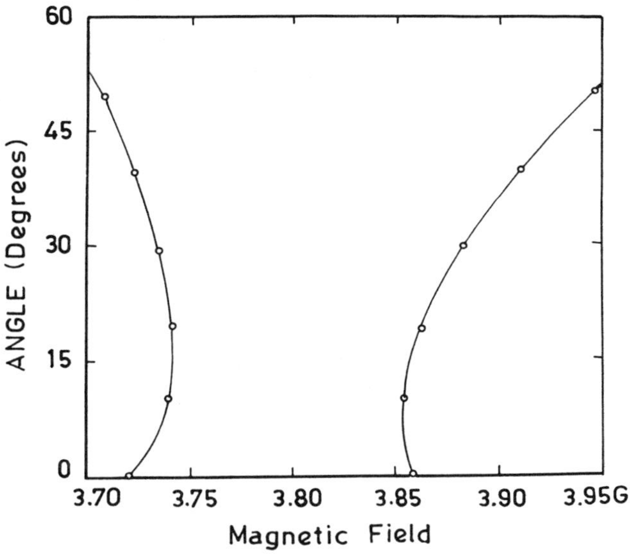

Fig.5.12 The dependence of the position of two of the lines from the low field microwave absorption from the previous figure on the angle between the mangetic field and the [110] direction in a plane showing anisotropic polynomial type behaviour.

Using the superconductor $Tl_2CaBa_2Cu_2O_8$, the low field microwave absorption at 9.2 GHz at a temperature of about 77 K is shown in Fig. 5.13. For a slow scan a large number of eqully spaced resonance lines are seen. However, as faster scan rates are used, the number of lines decreases, untill only one line is left. It is found that the appearance of one line requires a constant time of 2.8 s and the width of one line requires about 0.8 s. As large sweep times are used larger number of lines emerge. We interpret that the lines arise on account of the flux-quantized eigenvalues in the system. The microwave induced supercurrents on the sample tunnel through the Josephson like structures which are flux quantized. In this particular case, the sweep rate becomes comparable with the relaxation rate. Usually the relaxation time is small and the time taken to sweep the field equal to one line width is very large so that the spectra are completely independent of the sweep rate. When the relaxation times are long such as in glasses, then the sweap times become comparable to the relaxation time. Under the circumstances, we observe the phenomenon displayed in Fig. 5.13. Since the flux-quantized energies must have a relaxation mechanism it appears that the absorption of microwaves is associated with flux creep.

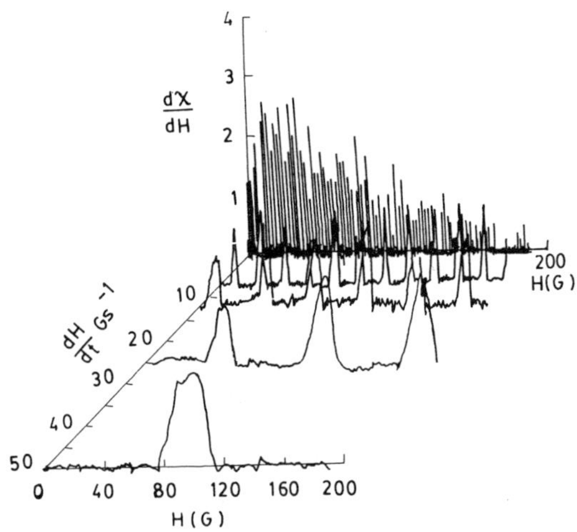

Fig.5.13 The field derivative of the microwave absorption, $d\chi''/dH$, in arbitrary units from the superconducting $Tl_2CaBa_2Cu_2O_8$ powder at 9.2 GHz at 77 K at different sweep rates. The time required to sweep the distance between two lines is a constant of about 2.8 s in all cases independent of the sweep rate. The spectra $d\chi''/dH$ are shown for several rates dH/dt in units of Gs^{-1}. The magnetic field is scanned from zero to 200 G in all the cases.

The energy absorbed from microwaves is associated with flux creep. The energy absorbed from microwaves is released by the flux creeping from inside to outside the sample. Thus in thermal equilibrium the relaxation rate is equal to the rate of flux creep. The flux quantized states release energy to relax to lower energy states by transferring energy to the flux lines which move out of the sample to creep some flux. In this way creation of microwave absorption lines enhances the Meissner effect.

In one of the studies of quantized microwave absorption lines in $YBa_2Cu_3O_{7-\delta}(\delta \sim 0.2)$ at about 4.2 K the relaxation time is $\leq$ 1 m-s and the sweep time is several minutes so that the spectra are completely independent of the sweep rate. In the present case the relaxation rate is of the order of the sweep rate so that we are able to observe the sweep rate dependent spectra. The low field microwave absorption of $YBa_2Cu_3O_{7-\delta}(\delta \sim 0.2)$ shows a large number of equally spaced lines as shown in Fig. 5.14. The critical temperature of these samples is about 98 K. The intensity of these lines as a function of temperature has been followed carefully. The lines are weak at low temperatures but their intensity increases to a maximum below the superconducting transition temperature, T_c. As the temperature is increased the intensity falls to zero at T_c indicating that the lines are characteristic of the superconducting phase. The rate at which a magnetic flux vortex hops from one site to another in a granular oxide superconductor is given by a thermally activated process. This rate is compared with the sweep rate of the spectrum. In Fig. 5.15, the two spectral scans show that the number of lines per unit time is a constant. The scan width of one line is about 1 s and the interline separation is about 3.4 s which is interpreted as the magnetic flux hopping or creep time.

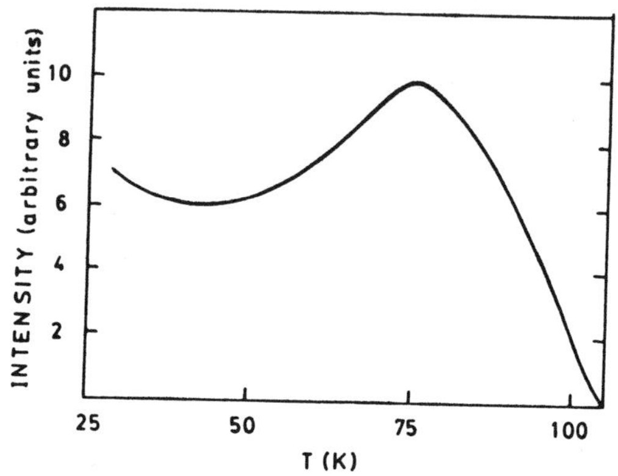

Fig.5.14 Low field microwave absorption from $YBa_2Cu_3O_{7-\delta}$($\delta \simeq 0.2$) at 95 K in an electron spin resonance spectrometer operating at 9.4 GHz. The gain in the spectrometer is (a) 1, (b) 16 and (c) 16, for three different samples. The overall shape of the lines is consistent with the Josephson diffraction formula. The equally spaced resonances are caused by magnetic flux quantization.

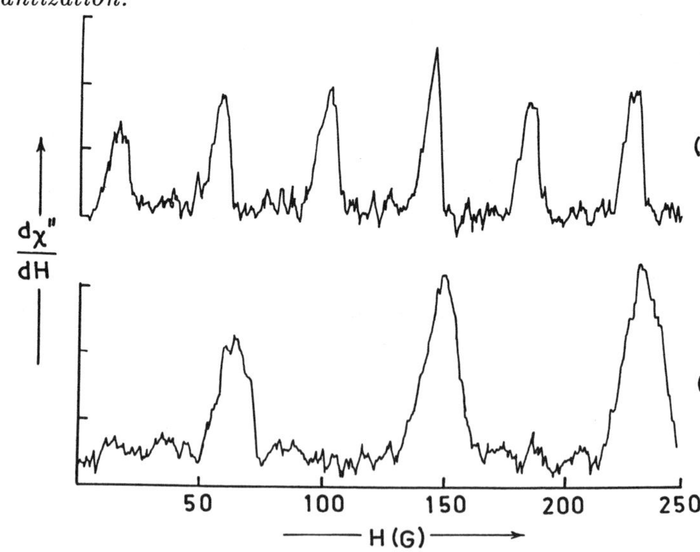

Fig.5.15 The magnetic field sweep rate dependence of the low-field microwave absorption at ~90 K from $YBa_2Cu_3O_{7-\delta}$($\delta \simeq 0.2$). The upper spectrum (a) is scanned in a time of 20 s while the lower spectrum (b) is scanned in 10 s showing that the number of lines per unit time is constant.

Since the intensities are related to a peaked function such as the diffraction type spherical Bessel function, $(\sin x)/x$, which is proportional to the Josephson current, the multiplet of lines is expected to have a phase inversion which was found in detailed spectra as the maximum of χ'' is equivalent to zero in $d\chi''/dH$. The lines are descrbed phenomenologically by,

$$H_n = H_o \pm \delta H(n + \frac{1}{2}) \ , \tag{5.63}$$

where n is an integer ($n = 0, 1, 2, \cdots$), δH is the line separation and the central field is $H_o \simeq 0$ G. The work described here is a manifestation of the quantization of the magnetic flux in granular oxide superconductors. We show the values of $d\chi''/dH$ as a function of magnetic field for a small sample of $YBa_2Cu_3O_{7-\delta}$ with $T_c \simeq 93$ K at the microwave frequency of 9.2 GHz and modulation frequency of 100 kHz in Fig. 5.16.

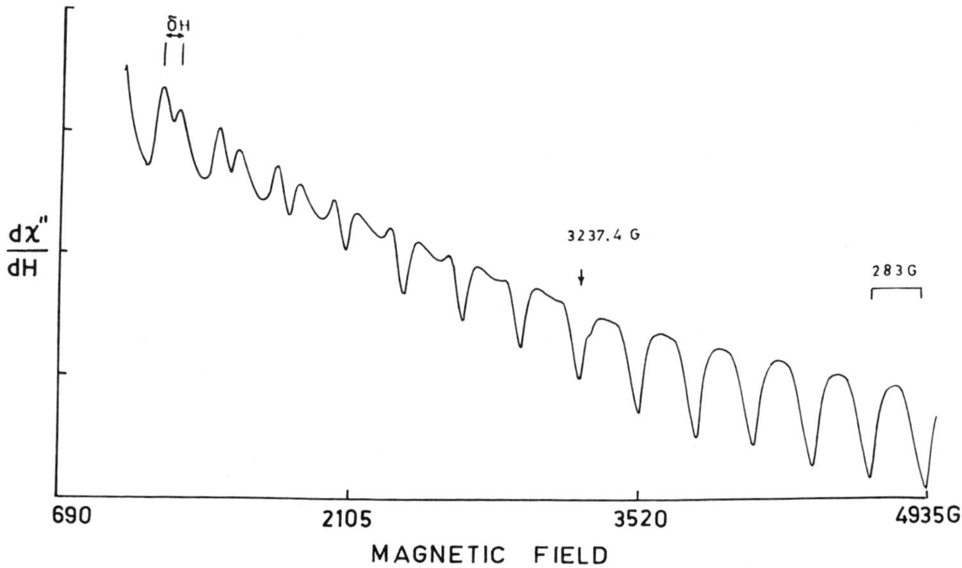

Fig.5.16 The microwave absorption $d\chi''/dH$, in a sample of $YBa_2Cu_3O_{7-\delta}$ ($\delta \simeq 0.2$) as a function of magnetic field showing equally spaced lines with spacing 283 G. An additional splitting of δH is also seen.

For temperatures larger than T_c usual electron spin resonance corresponding to the anisotropic Lande's splitting factor of Cu^{2+} is seen. Below T_c equally spaced oscillations are found. On the high field side, the spacing between lines is about 283 G which is resonably independent of the magnetic field. Since the spacing is constant it corresponds to the area of the grain size which is the normal region with a Josephson tunneling current carrying pairs of electrons. This means that peaks on the graph represent a change in the state of an individual micron size grain. This interpretation is consistent with the use of SQUIDS as gamma ray detectors in which flips of superconducting grains of tin are used to detect particles. There is an additional splitting of $\delta H \simeq 80$ G at 1237 G but this splitting slowly increases with increasing field to a value of 161 G at 2989 G as shown in Fig. 5.17. The large splitting corresponds to smaller areas of the Abrikosov triangular vortices as expected from the flux quantization. It appears that there are two features merging at about 1951 G. Usually the upper critical field varies with temperature from a large value at $T = 0$ to zero at T_c as $H_{c2} \propto [1 - (T/T_c)]^{1/2}$. The region of large upper critical field cannot be reached by ordinary coil type electromagnets but the region near $T = T_c$ and $H_{c2} \to 0$ can be approached. For $YBa_2Cu_3O_{7-\delta}(\delta \simeq 0.2)$, the upper critical field $H_{c2}(T = 92K) = 3125$ Oe and $H_{c2}(T = 90K) \simeq 10 \times 10^4$ Oe. Very near the T_c it is possible to measure the upper critical field by using the microwave signal. As this signal occurs only in the superconducting state, increasing the field beyond H_{c2} leads to vanishing of the signal. Thus upon scanning the magnetic field we can measure the H_{c2} from the point where the intensity of the mcirowave signal vanishes. In Fig. 5.18 we show the microwave signal as a function of magnetic field at $T = 92$ K for $T_c \simeq 93$ K. It is found that the signal vanishes at about 3125 Oe which we identify with H_{c2}. This leads to a new method for determining H_{c2} In Fig. 5.18 we show the microwave absorption in a sample of $YBa_2Cu_3O_{7-\delta}$ as a function of temperature at high fields. From this measurement $H_{c2}(T)$ can well be determined.

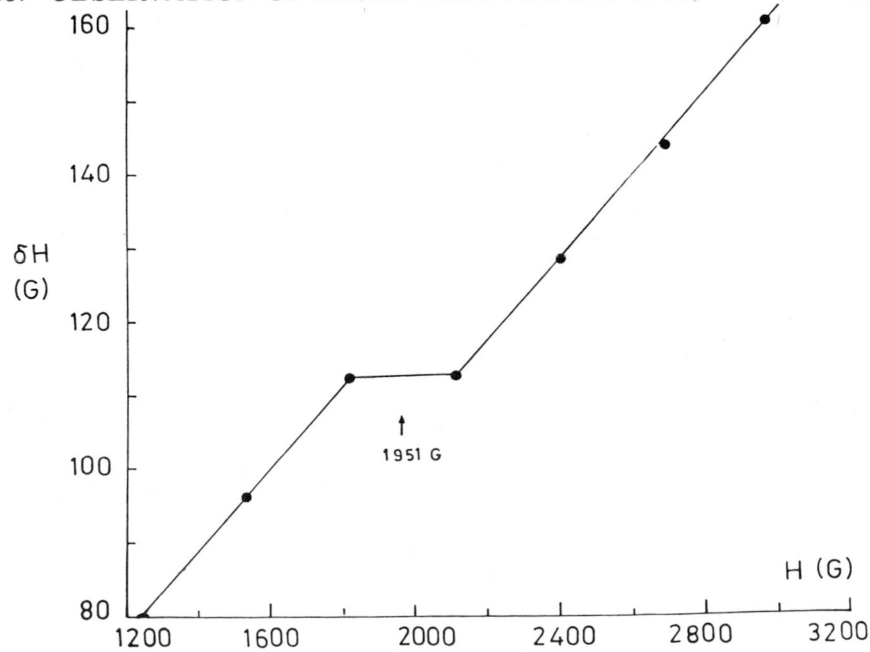

Fig.5.17 *The splitting δH deduced from the data of Fig. 5.16 is shown as a function of magnetic field. There is some indication of vortex-vortex interaction near the field of 1951 G.*

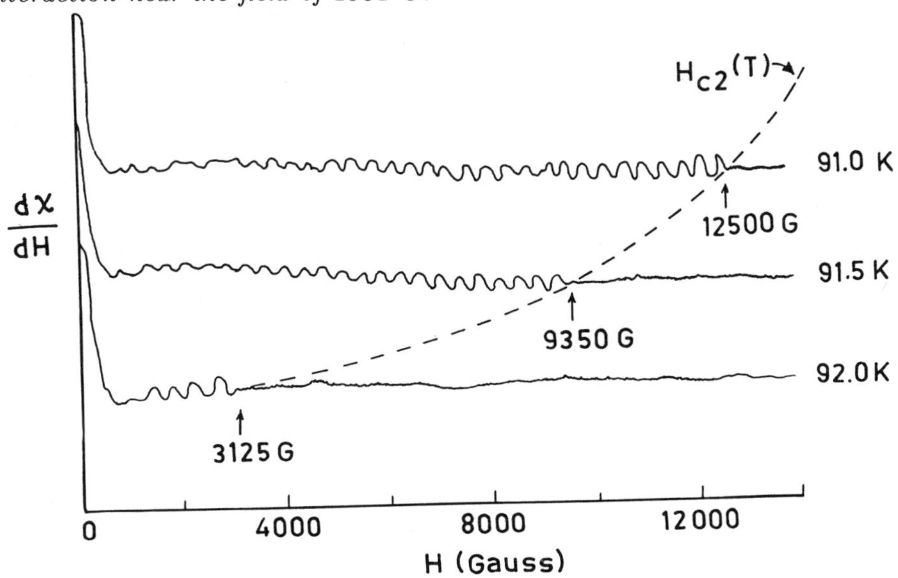

Fig.5.18 *The microwave absorption in $YBa_2Cu_3O_{7-\delta}$($\delta \simeq 0.2$) at 9.4 GHz with a modulation of 100 kHz showing a low field microwave absorption peak and oscillations extending upto high fields terminating at H_{c2}. Then $H_{c2}(T)$ dashed line can be delineated.*

We describe a transient state in the electron spin resonance experiment of $Tl_2CaBa_2Cu_2O_8$ which has a superconducting transition temperature of about 105 K. At room temperature, there is an exchange narrowed line with $g_{\|} \simeq 2.217 \pm 0.002$ and $g_{\perp} = 2.048 \pm 0.002$. As the sample is cooled, the intensity of the Cu^{2+} line reduces and the flux-quantized oscillations appear. The flux quantized oscillations and the Cu^{2+} ESR as followed through a temperature cycle seem to be complementary to each other. Below a temperature of $T_c \simeq 66.5 \pm 1.0$K only flux quantized oscillations appear while above T_c only Cu^{2+} appears. In the intervening region, $T_c > T > T_o$, both the signals appear; flux quantized signal on the heating branch and Cu^{2+} on the cooling branch of the temperature cycle as shown in Fig. 5.19. The flux quantized oscillations are equally spaced as expected. At a temperature of 90 K, we find that the separation of two nearest flux lines is 257.5 ± 0.5 G. The line separation is precisely constant as the field varies from about 1000 G to 5000 G. The reproducibility of the line separation is as accurate as the measurement of the field. The lines are truely equally spaced. The Cu^{2+} $g_{\perp}$ dominating the line sits exactly on top of a flux line so that the nearest flux lines are equally separated from the Cu^{2+} line as shown in Fig. 5.20. Thus the flux wings occur symmetrically spaced with respect to the Cu^{2+} line. As the system is cooled, the Cu^{2+} slowly disappears and then only the flux lines remain. In view of the two dimensional character of the problem, the electrons leading to $g_{\perp}$ in the ab plane contribute to the flux quantization in the superconducting state which mostly occupies the ab plane. This means that the pairs of electrons in the superconducting phase are generated directly from the insulator without a single particle metallic state.

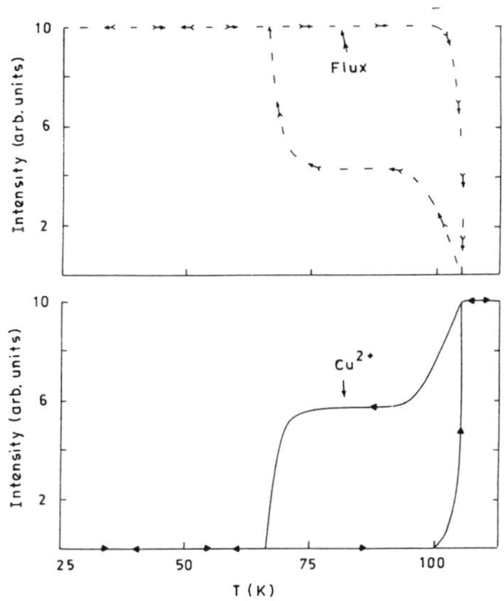

Fig.5.19 The relative intensity of one of the flux lines as compared with that of Cu^{2+} E.S.R. line as a fucntion of temperature from powdered samples of the superconductor $Tl_2CaBa_2Cu_2O_8$ (T_c =105 K) showing oscillatory state. Below $T_o = 66.5 \pm 1.0K$, the state is superconducting while above 105 K, it is paramagnetic and insulating.

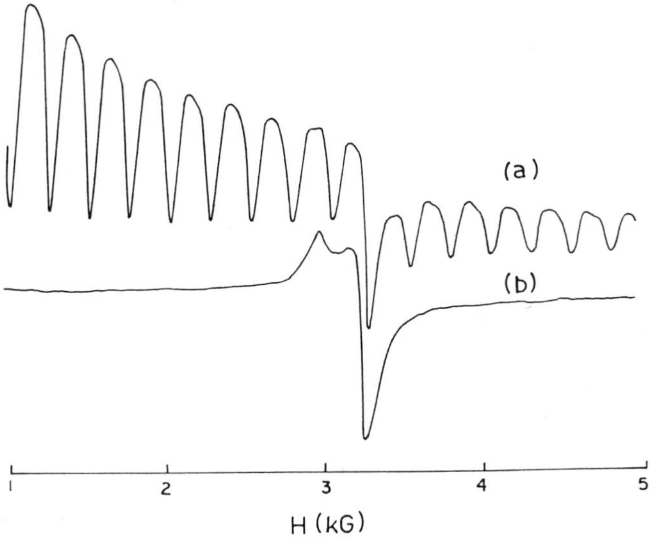

Fig.5.20 The E.S.R. response of the powdered superconductor $Tl_2CaBa_2Cu_2O_8$ *in cooling cycle showing flux quantized lines occuring in the superconducting state. In figure (a) the temeprature is 90 K while in (b) it is 110 K showing normal E.S.R. due to Cu^{2+} in an insulator.*

Suggested References

5.1 K.N. Shrivastava, J. Phys. C **20**,L 789 (1987).

5.2 A. Abragam and B. Bleaney, Electron Paramagnetic Resonance of Transition Ions, Oxford University Press, 1970; Dover edition, New York, 1986.

5.3 K.N. Shrivastava, Phys. Stat. Solidi B**117**, 437 (1983).

5.4 K.N. Shrivastava, J. Phys. (Paris) Colloq. **49**, C8-2239 (1988).

5.5 K.N. Shrivastava, Solid Stat. Commun. **68**, 259 (1988).

5.6 K.N. Shrivastava, Solid Stat. Commun. **68**, 1019 (1988).

5.7 K.N. Shrivastava, Solid Stat. Commun. **77**, 433 (1991).

5.8 K.N. Shrivastava, Supercond. Sci. Technol. **4**,S 430 (1991).

5.9 K.N. Shrivastava, Solid Stat. Commun. **78**, 403 (1991).

5.10 K.N. Shrivastava, Phys. Stat. Solidi B**164**, K 51 (1991).

5.11 K.N. Shrivastava, Solid State Commun. **80**, 933 (1991).

5.12 K.N. Shrivastava, M. Puri, J. Bear and L. Kevan, J. Chem. Soc. Faraday Trans. **87**, 3893 (1991).

5.13 K.N. Shrivastava, M. Puri, J. Bear and L. Kevan, Solid State Commu. **81**, 613 (1992).

5.14 K.N. Shrivastava, Solid Stat. Commun. **85**, 227 (1993).

5.15 K.N. Shrivastava, Bull. Mater Sci. (India) **14**, 625 (1991).

5.16 I. Ciccarel, Europhys. Lett. **7**, 185 (1988); Physica C**161**, 39 (1789).

5.17 D.K. De, J. Phys. C**21**, 4481 (1988).

5.18 L. Kevan, ACS Symp. **377**, 223 (1988).

5.19 F. Mehran, S.E. Barnes, E.A. Giess and T.R. McGuire, Solid State commun. **67**, 55 (1988).

5.20 S. Ducharme, R. Durny, J. Hautala, D.J. Zheng, P.C. Taylor, O.G. Symko and S. Kulkarni, J. Appl. Phys. **66**, 1252 (1989).

5.21 R. Durny, S. Ducharme, J. Hautala, D.J. Zheng, O.G. Symko, P.C. Taylor and S. Kulkarni, J. Opt. Soc. Am. **6**, 465 (1989).

5.22 S.K. Misra and L.E. Misiak, Solid State Commun. **72**, 117 (1989); **72**, 351 (1989); Phys. Rev. B**42**, 4776 (1990).

5.23 G.A. Petrakovskii, G.S. Patrin, Yu. N. Ustyuzhanin, K.A. Sablina and G.N. Stepanov, Solid State Commun. **72**, 759 (1989).

5.24 S. Curvier, M. Puri, J. Bear and L. Kevan, J. Phys. Chem. **94**, 3864 (1990).

5.25 R. Jones, R. Janes, R. Armstrong, K.K. Singh, P.P. Edwards, D.J. Keeble and M.R. Harrison, J. Chem. Soc. Faraday Trans. **86**, 683 (1990).

5.26 F.J. Owens, J. Phys. Condens. Matter **2**, 8345 (1990).

5.27 C. Yu, C. Lai, S.A. Marshall, D.R. Yoder-Short and Y.N. Zhang, Phys. Stat. Solidi B**157**, 379 (1990).

5.28 R. Janes, R.S. Liu, P.P. Edwards and J.L. Tallon, Physica C**167**, 520 (1990).

5.29 R. Janes, R.S. Liu, P.P. Edwards, A.D. Stevens and M.C.R. Symons, J. Chem. Soc. Faraday Trans. **87**, 1209 (1991).

5.30 A.A. Romanyukha, Yu. N. Shavachko and V.V. Ustinov, Usp. Fiz. Nauk **161**, 37 (1991); Sov. Phys. Uspekhi **34**, 862 (1991).

5.31 F. Mehran, Solid State Commun. **67**, 55 (1988).

5.32 N.E. Alekseevski, J. Low Temp. Phys. **77**, 87 (1989).

5.33 S. Hwang, Bull. Korean Chem. **10**, 23 (1989).

5.34 T. Kato, Jpn. J. Appl. Phys. (Pt.2) **27**, 2336 (1988).

5.35 K. Tagaya, Jpn. J. Appl. Phys. (Pt.2) **29**, L757 (1990).

5.36 R.J. Barham and D.C. Doetschman, J. Mat. Res. **7**, 565 (1992).

5.37 F. Mehran, J. Mag. Mag. Mat. **104**, 637 (1992).

5.38 H. Schinz and F. Schwarbl, J. Low Temp. Phys. **88**, 347 (1992).

5.39 H.G. Chi and A.D.S. Nagi, J. Lot Temp. Phys. **86**, 139 (1992).

5.40 L.E. Misiak and M. Subotowicz, Solid State commun. **80**, 761 (1991).

5.41 D.S. Babu, G.N. Rao, L. Brohan, Int. J. Mod. Phys. B**7**, 157 (1993).

5.42 S.L. Cooper, D. Reznik, A. Kotz, Phys. Rev. B**47**, 8233 (1993).

5.43 H. Muto, K. Matsurra, K. Yasuda, Physica C**211**, 147 (1993).

5.44 T. Endo and H.D. Yan, Jpn. J. Appl. Phys. (Pt.1) **33**, 103 (1994).

5.45 N. Guskos, V. Likodimos, J. Kuriata, Phys. Stat. Solidi B**180**, 491 (1993).

5.46 M.B. Suvasini, W.M. Temmerman and B.L. Gyorffy, Phys. Rev. B**48**, 1202 (1993).

5.47 P. Subhash, G.N. Rao and D.S. Babu, Mod. Phys. Lett. B**10**, 133 (1996).

5.48 T. Endo, H. Yan and S. Nagase, J. Supercond **8**, 259 (1995).

5.49 B.I. Kochelaev, R.G. Demminov and L.R. Tagirov, Fiz. Tverd. Tela **38**, 2293 (1996).

5.50 S.V. Sharma, T.K. Nath and A.K. Majumdar, Physica C**290**, 229 (1997).

5.51 T. Endo, S. Nagase and S. Sugiura, Physica C**282**, 1591 (1997).

5.52 T. Endo and H. Yan, "Studies in High Temperature Superconductors" edited by A.V. Narlikar, Nova Science Publisher, Inc. **14**, 1994, pp. 65-106 and 81, 82.

5.53 K. N. Shrivastava, Phys. Rep.**200**, 51 (1991).

5.54 K. N. Shrivastava and S. Koka, Solid State Commun.**78**,735 (1991).

Chapter 6

Surface resistance

6.1 Surface resistance

The electrons are scattered by phonons or by magnetic atoms so that there is a finite conductivity. When an electric field E of frequency ω is applied on the electrons, they experience a force which determines their velocity, $mv/\tau = -eE$. The current is determined by $j = nev$ where n is the concentration of particles, e the charge and v is the velocity. The electrical conductivity, σ, is determined by $j = \sigma E$ and the resistivity is the inverse conductivity, $\rho = 1/\sigma$. Here m is the mass of the electron. Substituting for the velocity in terms of electric field in the expression for the current and using the Ohm's law with $\omega = 1/\tau$, we find the normal state conductivity as,

$$\sigma_n = \frac{ne^2}{m\omega} \tag{6.1}$$

Here n is the concentration of electrons in the normal state. The normal state resistivity varies linearly with temperature above T_c and vanishes at T_c. We will see that a.c. resistivity does not vanish below T_c. Similarly, we define the superfluid number density, n_s and hence the conductivity of the superconducting state becomes

$$\sigma_s = \frac{n_s e^2}{m\omega} = \frac{1}{\mu_o \omega \lambda^2} \tag{6.2}$$

where λ is the London penetration depth,

$$\lambda^2 = \frac{m}{\mu_o n_s e^2} \ . \tag{6.3}$$

However, it should be noted that compared with (6.1) we use $2m$, $2e$ and $n_s/2$ in going from the normal to the superconducting state for m, e and n, respectively. The number of pairs is clearly half the number of electrons and hence $n_s/2$ is the density of pairs assuming that the volume is fixed and the charge is constant. The complex conductivity is given by

$$\sigma_s = \frac{i\mu_o\omega}{(R_s + iX_s)^2} \tag{6.4}$$

where the surface impedance is $Z_s = R_s + iX_s$ so that the penetration depth is $\lambda = X_s/(\mu_o\omega)$ and μ_o is the permeability. The complex surface impedance is related to the complex conductivity and penetration depth as,

$$Z_s = R_s + iX_s = [i\mu_o\omega/(\sigma_1 - i\sigma_2)]^{1/2} \ . \tag{6.5}$$

We can express σ_1 and σ_2 in terms of R_s and X_s as,

$$\sigma_1 = \frac{2\mu_o\omega R_s X_s}{(2R_sX_s)^2 + (R_s^2 - X_s^2)^2} \tag{6.6a}$$

$$\sigma_2 = \frac{\mu_o\omega(X_s^2 - R_s^2)}{(2R_sX_s)^2 + (R_s^2 - X_s^2)^2} \tag{6.6b}$$

so that a measurement of permeability and surface resistivity at a known frequency leads to the measurement of the London penetration depth.

We assume that fluxoids are subject to a harmonic force with the force constant, k, the viscous force ηv proportional to the velocity, v and the ordinary fluxoid mass, m, multiplied by the acceleration is determined by the quantized flux,

$$M\frac{dv}{dt} + \eta v + kx = \frac{1}{c}J\phi_o \ . \tag{6.7}$$

The fluxoid velocity has the time dependence of $v = v_o e^{-i\omega t}$ which substituted in (6.7) above gives,

$$v = \frac{J\phi_o}{c(\eta - i\omega M + ik/\omega)} \ . \tag{6.8}$$

The moving fluxoid produces an electric field, $E_\varphi = -\frac{1}{c}vB$, that opposes the current so that

$$E_\varphi = -\frac{J\phi_o B}{c^2(\eta - i\omega M + ik/\omega)} \ . \tag{6.9}$$

Actually $v \times B$ can be written as $vB \sin\theta$ where θ is the angle between v and B. We can put the angle dependant factor when necessary. The diamagnetic current is expressed in terms of a vector potential,

$$J = -\frac{c}{4\pi\lambda_L^2} A \ . \tag{6.10}$$

Differentiating (6.10) with respect to time and replacing the time derivative of the vector potential, $-\partial A/dt = E + \nabla\varphi$, we can write,

$$\frac{dJ}{dt} = \frac{c^2}{4\lambda_L^2}(E + E_\varphi) \ . \tag{6.11}$$

Substituting (6.9) into (6.11) and taking the time dependence of the current as $J = J_o e^{-i\omega t}$, we find,

$$J = E\left[\frac{\phi_o B}{c^2(\eta - i\omega M + ik/\omega)} - \frac{4\pi i\omega\lambda_L^2}{c^2}\right]^{-1} \tag{6.12}$$

so that the complex resistivity may be defined by,

$$\rho = \frac{E}{J} = \frac{\phi_o B}{c^2(\eta - i\omega M + ik/\omega)} - \frac{4\pi i\omega\lambda_L^2}{c^2} \ . \tag{6.13}$$

Separating the real and imaginary parts of (6.13) we find that the complex resistivity is given by,

$$\rho = \frac{\phi_o B\eta}{c^2(\eta^2 + m^2)} - i\left[\frac{m(\phi_o B/c^2)}{\eta^2 + m^2} + \frac{4\pi\omega\lambda_L^2}{c^2}\right] \tag{6.14}$$

where $m = (k/\omega) - \omega M$. Replacing ϕ_o by $n\phi_o$ where n is an integer, it is found that the resistivity oscillates and the values of n give quantized resistivity. At $B = 0$, the resistivity determines the penetration depth. We write the Maxwell equations as,

$$\begin{aligned} \vec{\nabla} \times \vec{E} &= -(1/c)\partial\vec{B}/\partial t \\ \vec{\nabla} \times \vec{H} &= (4\pi/c)\vec{J} + (1/c)\partial\vec{D}/\partial t \end{aligned} \tag{6.15}$$

where $B = \mu H$ with μ the magnetic permeability at the frequency ω. For the plain wave $e^{i(qz-\omega t)}$ neglecting the displacement current, we obtain,

$$q = (4\pi i\omega\mu/\rho c^2)^{1/2} \ . \tag{6.16}$$

The surface impedance is found to be,

$$Z = \frac{4\pi}{c}\frac{E}{H} = \frac{4\pi\omega\mu}{c^2 q} = \left(-\frac{4\pi i\omega\mu\rho}{c}\right)^{1/2} . \tag{6.17}$$

Substituting (6.13) for ρ by neglecting the term containing the mass of the vortex and vortex oscillations,

$$Z = R - iX = Z_o(1 + 2iB/B_o)^{1/2} \tag{6.18}$$

with

$$Z_o = -iX_o = -(4\pi i\omega\lambda_L/c^2)\mu^{1/2} \tag{6.19}$$

and

$$B_o = 8\pi\omega\eta\lambda_L^2/\phi_o . \tag{6.20}$$

The real and imaginary parts of the surface impedance are given by,

$$\begin{aligned} R &= X_o\left[\frac{1}{2}\{-1 + (1 + 4B^2/B_o^2)^{1/2}\}\right]^{1/2} \\ X &= X_o\left[\frac{1}{2}\{1 + (1 + 4B^2/B_o^2)^{1/2}\}\right]^{1/2} . \end{aligned} \tag{6.21}$$

The flux viscocity is estimated to be about $\eta = 7.5 \times 10^{-10}$ cgs units. This value used to estimate the conductivity of the fluxoid core,

$$\sigma \simeq \frac{\eta c^2}{\phi_o H)c2} \tag{6.22}$$

is about 5 (ohm-cm)$^{-1}$.

6.2 Penetration depth

The equation (6.14) shows that the London penetration depth increases upon the application of an external magnetic field. Therefore, the effective penetration depth can be defined as

$$\lambda_{eff}^2 = \lambda^2(o)\left[1 + \frac{i\phi_o B}{4\pi\omega\{\eta - i\omega M + (ik/\omega)\}\lambda_L^2(o)}\right] \tag{6.23}$$

in which $\lambda(o)$ gives the penetration depth at zero field. The above expression can be written as

$$\lambda_{eff} = \lambda_L(o)\left[1 + \frac{\phi_o B}{4\pi\omega\lambda_L^2(o)(m - i\eta)}\right]^{1/2} \tag{6.24}$$

from which it is clearly seen that the complexity is caused by the magnetic field. We separate the real and imaginary parts to define the absolute value of the London penetration depth as,

$$\lambda_{eff} = \lambda_L(o)\left[\left\{1 + \frac{\phi_o Bm}{4\pi\omega\lambda_L^2(o)(m^2 + \eta^2)}\right\}^2 + \left\{\frac{\phi_o Bn}{4\pi\omega\lambda_L^2(o)(m^2 + \eta^2)}\right\}^2\right]^{\frac{1}{4}} . \tag{6.25}$$

This expression is of general nature. It can give rise to the square root of the magnetic field dependence in the penetration depth. When expanded according to binomial theorem, it gives a linear function of magnetic field. The full phase factor introduced in ϕ_o due to Josephson tunneling can bend the $\lambda(H)$ curve from the linear behaviour and introduction of quantized flux, $n\phi_o$, instead of ϕ_o, leads to quantized resistivity. We consider several special cases.

Case I: At large field,

$$\lambda_{eff} = B^{1/2}(\phi_o/4\pi\omega)^{1/2}\left[\left(\frac{k}{\omega} - \omega M\right)^2 + \eta^2\right]^{-1/4} \tag{6.26}$$

and for large η,

$$\lambda_{eff} = B^{1/2}\left(\frac{\phi_o}{4\pi\omega}\right)^{1/2}\eta^{-1/2} \tag{6.27}$$

which shows that the effective penetration depth depends on the square root of the magnetic field.

Case II: For small fields, we obtain the binomial expansion,

$$\lambda_{eff} = \lambda_L(o)\left[1 + \frac{\phi_o B}{8\pi\omega\lambda_L^2(m - i\eta)}\right] \tag{6.28}$$

which for small viscosity, $\eta = 0$, becomes,

$$\lambda_{eff} = \lambda_L(o)\left[1 + \frac{\phi_o B}{8\pi\omega\lambda_L^2 m}\right] . \tag{6.29}$$

For zero vortex mass, $M = 0, m = k/\omega$, so that,

$$\lambda_{eff} = \lambda(o)\left[1 + \frac{\phi_o B_{c2}}{8\pi\lambda_L^2 k}\left(\frac{B}{B_{c2}}\right)\right] . \tag{6.30}$$

Introducing dimensionless variable $h = B/B_{c2}$ and

$$\beta = \frac{\phi_o B_{c2}}{8\pi k \lambda_L^2(o)} \tag{6.31}$$

we find that the penetration depth depends linearly on the magnetic field,

$$\lambda_{eff} = \lambda(o)\left[1 + \beta h\right] . \tag{6.32}$$

Case III: In the superconductors with weak links, the critical current is replaced by the Josephson current with quantized phase so that on the right hand side of (6.7), we can add a phase factor. For small magnetic fields, we expand the cosine term, retaining only two terms,

$$\frac{J\phi_o}{c}\cos\left(\frac{2\pi}{\phi_o}\phi_x\right) \approx \frac{J\phi_o}{c}\left[1 - \frac{2\pi^2\phi_x^2}{\phi_o^2}\right] \tag{6.33}$$

so that the vortex velocity becomes,

$$v = \frac{J\phi_x - 2J\pi^2 A^2 B^2 \phi_o^{-1}}{c[\eta - i\omega M + (ik/\omega)]} \tag{6.34}$$

where flux is quantized within the area A such that $BA = \phi_x$. This velocity produces the electric field of

$$E_\varphi = \frac{-J\phi_o B + J(2\pi^2 A^2 \phi_o^{-1})B^3}{c^2[\eta - i\omega M + (ik/\omega)]} \tag{6.35}$$

for which the effective penetration depth is found to become,

$$-\frac{4\pi i\omega}{c^2}\lambda_{eff}^2 = -\frac{4\pi i\omega}{c^2}\lambda_L^2 + \frac{\phi_o B}{c^2[\eta - i\omega M + (\frac{ik}{\omega})]} - \frac{(2\pi^2 A^2 \phi_o^{-1})B^3}{c^2[\eta - i\omega M + (\frac{ik}{\omega})]} . \tag{6.36}$$

Upon comparing (6.23), we find that the field is,

$$B_{eff} = B\left[1 - \frac{2\pi^2 A^2 B^2}{\phi_o^2}\right] \tag{6.37}$$

so that in the approximation which leads to the square root of the field as in (6.27), the penetration dpeth varies as,

$$\lambda_{eff} \propto B^{1/2}\left[1 - aB^2\right]^{1/2} \tag{6.38}$$

where $a = 2\pi^2 A^2/\phi_o^2$ and in the approximation which gives penetration depth linear in field,

$$\lambda_{eff} \propto B\left[1 - aB^2\right] \tag{6.39}$$

in which the bending in $\lambda_{eff}(B)$ varies as B^3. Retaining the full cosine term in the resistivity, it is seen that

$$\rho = -\frac{4\pi i\omega\lambda_L^2}{c^2} + \frac{\phi_o B}{c^2\eta}\cos\left[\frac{2\pi\phi_x}{\phi_o}\right] \tag{6.40}$$

which oscillates as a function of magnetic field. We observe that ρ is proportional to the square of the penetration depth. Thus oscillations occurs in the square of the penetration depth as a function of magnetic field. In the eq.(6.40), ϕ_o may be replaced by $n\phi_o$ where n is an integer. Then it is seen that resistivity is quantized,

$$\rho \approx nB \tag{6.41}$$

in units of $\phi_o/c^2\eta$ leaving out a phase factor as equal to unity and the first term which does not depend on the field. Similarly, the square of the effective London penetration depth $\lambda_{eff}^2 \propto nB$, gets quantized.

Case IV: We determine the voltage as a function of field from the above theory. The electric field generated by the motion of fluxoids in the magnetic field,

$$E = -\frac{J\phi_o B(\eta - im)}{c^2(\eta^2 + m^2)} \ . \tag{6.42}$$

The voltage is applied across a distance r so that, $V = Er$, is the voltage. Differentiating this voltage with respect to the magnetic field B, we find the effective vortex velocity, $v = \partial V/\partial B$ as,

$$v = \frac{-J\phi_o r(\eta - im)}{c^2(\eta^2 + m^2)} = -v_1 + iv_2 \tag{6.43}$$

with

$$v_1 = \frac{J\phi_o \eta r}{c^2(\eta^2 + m^2)} \tag{6.44}$$

$$v_2 = \frac{J\phi_o m r}{c^2(\eta^2 + m^2)} \; . \tag{6.45}$$

The real part v_1 is nonzero even if $m = 0$. This is the velocity which can be greater than the velocity of light in the medium for

$$\frac{J\phi_o \eta r}{\eta^2 + m^2} > c^3 \; . \tag{6.46}$$

For small m, $(J\phi_o r/\eta)^{1/3} > c$ in a medium is allowed and v_2 may also be larger than c. The ratio of the two velocities $v_1/v_2 = \eta/m$, indicates that one of the two velocities may be greater than c depending on the ratio η/m. The system is thus in a dynamic equilibrium and can switch from one velocity to another. For example, the vortex has the velocity v_1 and a parameter such as the voltage is varied, then at a certain value of the voltage, the velocity can change to v_2. Since one depends on η which is defined in (6.7) and the other depends on $(k/\omega) - \omega M$ the two quantities are independent of each other and hence can be quite different.

6.3 Experimental results

(a) The change in the penetration depth in a single crystal of $ErNi_2B_2C$ as a function of magnetic field has been obtained by Eskildsen et al from the measurement of r.f. kinetic inductance. Except at small fields, the measured values are in accord with the predicted dependence of the penetration depth on the square root of magnetic field as shown by the expression (6.26), $\delta\lambda_L \propto B^{1/2}$, for large viscosity. In Fig. 6.1 we show the measured values of the penetration depth as a function of the square root of the magnetic field which agree with the theory.

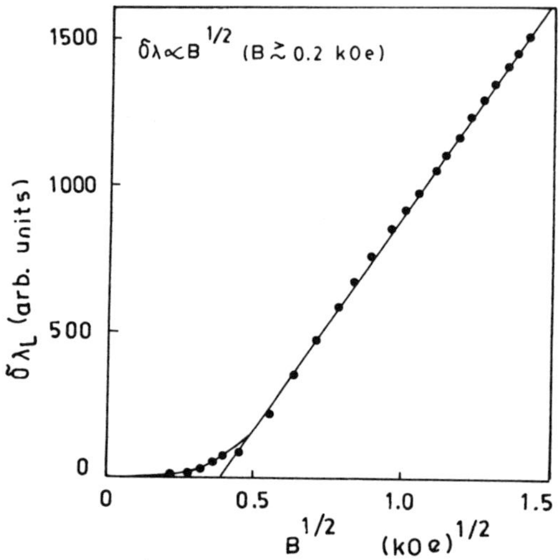

Fig.6.1: The measured changes in the penetration depth in single crystals of $ErNi_2B_2C$ shown by dots are plotted against the square root of the magnetic field. It is found that except at small fields, the change in penetration depth is proportional to the square root of the magnetic field. The measured values are taken from Eskildsen et al.

(b) In the case of small viscosity, the equations (6.30) and (6.32) predict that the penetration depth is linearly proportional to the magnetic field. Sonier et al have measured the magnetic field dependence of the penetration depth $\lambda_{ab}(H)$ in the vortex state of $NbSe_2$ at 0.6 T_c and also at 0.33 T_c. In both the cases, the penetration depth is seen to linearly increase with increasing magnetic field. In particular the data agrees with eq.(6.32) for

$$\begin{aligned} \lambda_{ab}(o) &= 1323\mathring{A} \ , \ \beta = 1.61 \ \text{ at } \ T = 0.33T_c \ \text{ and} \\ \lambda_{ab}(o) &= 1436\mathring{A} \ , \ \beta = 1.56 \ \text{ at } \ T = 0.60T_c \ . \end{aligned}$$

The data for $T = 0.33T_c$ are displayed in Fig.6.2 showing the amount of agreement between the theory and the experimental measurements.

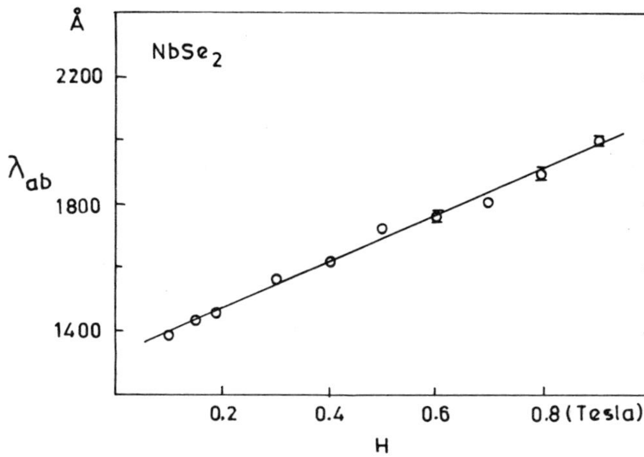

Fig.6.2: The measured values of the penetration depth in $NbSe_2$ are seen to be linearly proportional to the magnetic field for $T = 0.33T_c$. The data are taken from Sonier et al and the calculation is given in the text.

Sonier et al have also measured the $\lambda_{ab}(H)$ in $YBa_2Cu_3O_{6.6}$ in both the twinned as well as the detwinned samples. For twinned samples, $H_{c2} = 70T$, $\lambda_{ab}(o) = 1586\mathring{A}$ and $\beta = 6.6$ and for the determined samples $\lambda_{ab}(o) = 1699\mathring{A}$ and $\beta = 5.0$. In both the cases eq.(6.32) is well obeyed.

(c) The magnetic field penetration depth λ_L in $CeRu_2$ has been obtained using transverse field μSR by Yamashita et al. At 2K for a field of 0.9846 T the penetration depth is found to be 1983±15 Å. As the field is increased to 1.9621 T, the penetration depth increases to 2101 ± 11 Å but upon further increase of field to 2.9312 T the penetration depth reduces but increases to 2937±24 Å upon further increase of field to 3.9003 T. This type of oscillatory behaviour is qualitatively in accord with that predicted by (6.29) and is caused by the phase factor dependent term in the tunneling current. Considering the phase factor (6.29) becomes,

$$\lambda_{eff} = \lambda(o)\left[1 + \frac{\phi_o B}{8\pi\omega\lambda_L^2 m}\cos\frac{2\pi\phi_x}{\phi_o}\right] . \tag{6.47}$$

Between two fields B_1 and B_2, the cosine term shows one half oscillation when,

$$\left(\frac{2\pi\phi_{x1}}{\phi_o} - \frac{\pi}{2}\right) - \left(\frac{2\pi\phi_{x2}}{\phi_o} - \frac{\pi}{2}\right) = \pi . \tag{6.48}$$

Since the field is quantized in an area A, we write $\phi_{x1} = AB_1$ and $\phi_{x2} = AB_2$ so that one half oscillation is obtained when

$$B_1 - B_2 = \phi_o/\pi A . \tag{6.49}$$

In Fig. 6.3 we have shown the experimental data points of penetration depth as a function of field in $CeRu_2$ along with a qualitative curve drawn as a guide to the eyes. It is found that the data are in qualitative agreement with the interpretation given above. The period of oscillations in Fig. 6.3 is about 2×10^4 Gauss. Substituting this value for $B_1 - B_2$ and the value of ϕ_o we find that $A^{1/2} \simeq (10/\pi)^{1/2}10^{-6}$cm which is quite reasonable for the size of the current loops.

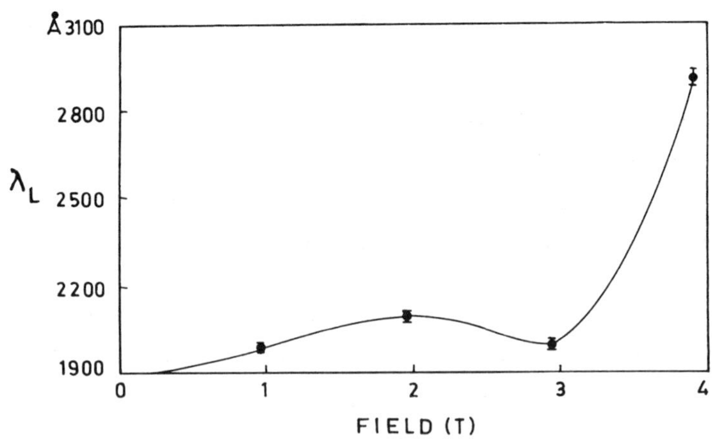

Fig.6.3: Oscillation of period of about 2T is found in $CeRu_2$. The experimental data is taken from Yamashita et al and the theory is given by (6.47).

(d) For Josephson junctions the voltage frequency relation is $2eV = h\nu$ from which the voltage is found to be,

$$V = \frac{hc}{2e}\frac{\nu}{c} = \phi_o\frac{\nu}{c} \tag{6.50}$$

where V is the applied voltage and ν is the microwave frequency. The voltage can be written in terms of the unit flux, $\phi_o = hc/2e$. For a stack of N intrinsic Josephson junctions,

$$V = N\phi_o\frac{\nu}{c} \ . \tag{6.51}$$

The voltage is being applied across the thickness t of the junction so that,

$$2eV = 2eB\bar{v}t \tag{6.52}$$

where $\bar{v}$ is the standard Swihart velocity,

$$V = B\bar{v}t \tag{6.53}$$

so that for N junctions

$$V = N\bar{v}Bt \ . \tag{6.54}$$

The plot of voltage as a function of field in a stack of N Josephson junctions is thus predicted to be linear with only one slope. It has been shown in (6.43) that the velocity becomes complex so that the system can jump from one state with velocity v_1 to another state of velocity v_2. The c-axis in $Bi_2Sr_2CaCu_2O_{8+y}$ is analogous to a stack of N josephson junctions. The tunneling voltage along this direction as a function of field measured by Hechtfischer et al shows two slopes as expected for a system which switches from v_1 to v_2 at a field of 17 kOe. Therefore, the measurements are in accord with the theory. Thus the penetration depth in type-II superconductors is determined by viscous oscillatory forces. For large viscosity the change in penetration depth as a function of magnetic field depends on the square root of the magnetic field. For small viscosity, the penetration depth depends linearly on the magnetic field. For intermediate cases due to quantized phase factor we expect oscillations in the penetration depth as a function of magnetic field. The vortex velocity becomes complex when viscous forces become comparable with those of oscillatory motion of the vortex. In such a case a transition from one vortex velocity to another vortex velocity is predicted.

6.4 Electric field effect

We describe the mechanism of change of resistivity upon the application of an electric field in a polarizable superconducting device near the onset temperature. As the system is cooled, the resistivity approaches zero. We find that in the case of a finite electric dipole moment, the resistivity is quantized. We assume that there are vacanicies of oxygen ions, so when the electric field E is applied, the system develops an electric dipole moment,

$$p = -e \sum_i r_i \tag{6.55}$$

where r_i is the radius vector of the i-th ion. The potential energy of the ion due to the dipole moment is given by

$$V^{(1)} = -\vec{E}. < \vec{p} > \quad . \tag{6.56}$$

The quadratic polarization effects give rise to a change in 2energy expressed by the usual second-order perturbation theory,

$$V^{(2)} = -\sum_n \frac{< 0|E.p|n >< n|E.p|0 >}{E_n - E_o} \quad . \tag{6.57}$$

The change in energy due to the electric field is then given by the sum of the previous two terms,

$$V = -|E| \left[< p > + \sum_n \frac{< 0|E.p|n >< n|p|0 >}{E_n - E_o} \right] \quad . \tag{6.58}$$

The current, j, is related to the electric field E through the conductivity, σ per unit length, as

$$j = \sigma E. \tag{6.59}$$

The flux is quantized in units of $\phi_o = hc/2e$ so that $nj\phi_o/c$, where n is an integer, represents the current in terms of energy. We have taken current in units of e/s, ϕ_o in terms of G cm^2 and c is the velocity of light in cm s^{-1}. Here e is the electronic charge, so $j\phi_o/c$ is measured in erg. Accordingly $n\sigma|E|\phi_o/c$ represents the energy of a flux quantized system with electric field E, so that,

$$-\frac{n\sigma|E|\phi_o}{c} = -|E|\left[<p> + \sum_n \frac{<0|E.p|n><n|p|0>}{E_n - E_o}\right]. \tag{6.60}$$

Therefore, we can write the conductivity as,

$$\sigma = \frac{c<p>}{n\phi_o}\left[1 + \sum_n \frac{<0|E.p|n><n|p|0> / <p>}{E_n - E_o}\right]. \tag{6.61}$$

When the sign of E is changed the second term in the above changes sign, so for one direction of E we have,

$$\sigma_+ = \sigma_o(1+\gamma) \tag{6.62}$$

whereas for the direction of E reversed we get

$$\sigma_- = \sigma_o(1-\gamma). \tag{6.63}$$

Thus the ratio of the conductivity for the two directions of E is

$$\frac{\sigma_+}{\sigma_-} = \frac{1+\gamma}{1-\gamma}. \tag{6.64}$$

Similarly, the ratio of resistivity for the two directions of the electric field is, $(1-\gamma)/(1+\gamma)$ where

$$\sigma_o = \frac{c<p>}{n\phi_o} \tag{6.65}$$

$$\gamma = \sum_n \frac{<0|E.p|n><n|p|0> / <p>}{E_n - E_o}. \tag{6.66}$$

The conductivity is the inverse resistivity, so

$$\rho_+ = \frac{1}{\sigma_+} \quad ; \quad \rho_- = \frac{1}{\sigma_-}. \tag{6.67}$$

The theory may be compared with the experiments. At 90 K, the increase in the drain source resistance upon the application of a gate voltage +10 V is 0.11 kΩ while the decrease in the same resistance upon reversing the gate voltage to -10 V is 0.18 kΩ so the ratio $\rho_+/\rho_- = 0.61$, when compared with the calculated conductivity expression (6.64) gives $(1-\gamma)/(1+\gamma)$=0.61, and hence $\gamma = 0.24$ indicating that there is a 24% correction to the energy of the system due to second-order dipole moment. When the sign of E changes, the sign of γ also changes, so that resistivity is either increased or reduced depending upon the direction of the electric field. This feature of the experimental measurements carried out by Mannhart et al is also fully in accord with the expressions (6.62) and (6.63). The quantized resistivity

$$\rho_+ = \rho_o(1+\gamma)^{-1} \tag{6.68}$$

$$\rho_- = \rho_o(1-\gamma)^{-1} \tag{6.69}$$

has a prefactor of $\rho_o = n\phi_o/c < p >$, so when $< p >= 0$, the system has infinite resistivity, i.e., it becomes an insulator. Thus for the theory a non-zero $< p >$ is needed. This requirement is met in a polarizable material where $\langle \sum_i r_i \rangle \neq 0$. Once the system is polarized, the polarization decays unless there are pinning centres. In this way pinning becomes necessary to hold the dipole moments.

The displacement vector in terms of the dielectric constant and the electric field is given by

$$\vec{D} = \epsilon \vec{E}, \tag{6.70}$$

and hence the polarization is identified by

$$\vec{P} = \left(\frac{\epsilon - 1}{4\pi}\right) \vec{E}. \tag{6.71}$$

The solid consists of positive and negative charge concentrations. When the electric field is applied perpendicular to the plane of the film which is the xy plane, the positively charged background is deformed so that as an example, the charge density along z direction, which is the direction of the applied gate field is given by,

$$\rho^+(z) = n_o e + \rho_{ext} \sin(k_z z) \ . \tag{6.72}$$

Similarly, the negative charge density is given by

$$\rho^-(z) = -n_o e + \rho_{ind} \sin(k_z z). \tag{6.73}$$

Since E is applied perpendicular to the film, k_z is the wave vector with its minimum value determined by the inverse depth of the film. Here $\rho_{ext}(k_z)\sin(k_z z)$ gives an electrostatic field, that is, the external field applied to the electron gas and $\rho_{ind}(k_z)$ is the amplitude of the charge density variation induced in the electron gas. The Thomas-Fermi screening length is given by $1/k_s$. This length is related to the static dielectric function as,

$$\epsilon(0,k) = 1 + \frac{k_x^2}{k^2}. \tag{6.74}$$

Eliminating the dielectric constant from (6.71) and (6.74), we find that the polarization can be expressed as

$$4\pi P = E\frac{k_s^2}{k^2}. \tag{6.75}$$

We define

$$k_s = (3\pi^2 n_o)^{1/3} = \left(\frac{4\pi P k^2}{|E|}\right)^{1/2}. \tag{6.76}$$

The current along the z direction due to polarization caused by the gate voltage is given by,

$$nev = \sigma E \tag{6.77}$$

where v is the velocity of the electrons and σ the change in conductivity due to polarization, $\sigma = 1/R$, which is the inverse of the change in resistivity along the z direction. From (6.76) we can write the electron concentration as,

$$n_o = \frac{1}{3\pi^2}\left(\frac{4\pi P k^2}{E}\right)^{3/2} \tag{6.78}$$

so that from (6.77), the resistance is found to become,

$$R = \frac{3\pi^2 E^{5/2}}{ev(4\pi P k^2)^{3/2}}. \tag{6.79}$$

The energy imparted by the electric field to the electrons is $e\epsilon_o E d$ where d is the thickness of the film along the z direction. Therefore, the velocity of the electrons along the z direction may be determined from the relation $mv^2/2 = e\epsilon_o E d$ and substituted in (6.79) so that the resistance becomes,

$$R = \frac{3\pi^2 m^{1/2} E^2}{\sqrt{2}\,(4\pi e P)^{3/2} k^3 \epsilon_o^{1/2} d^{1/2}}. \tag{6.80}$$

which varies as E^2. The maximum value of k is π/d so that the change in resistance becomes,

$$R = \frac{3\pi^2 m^{1/2} E^2 d^{5/2}}{\sqrt{2}\,(4\pi e P)^{3/2} \epsilon_o^{1/2} \pi^3}. \tag{6.81}$$

From which the change in resistance due to the applied field is found to be proportional to the (5/2)th power of the depth of the film. A static approximation of (6.81) also exists. Since $\epsilon_o \simeq 25$ may be treated as a constant, we can ignore the dynamics of the system and write, $4\pi P = \epsilon_o E$ which substituted in (6.81) gives,

$$R = \frac{3\pi^2 m^{1/2} E^{1/2} d^{5/2}}{\sqrt{2} e^{3/2} \epsilon_o^2 \pi^3} \tag{6.82}$$

which shows that the resistance varies as the square root of the applied electric field. In (6.60), the first term is linear in $|E|$ and the second term is proportional to E^2. In the normal state this energy gives rise to a current determined by eRJ so that from (6.60) we write,

$$eRJ = |E| \left[< p > + \sum_n \frac{< 0|E.p|n >< n|p|0 >}{E_n - E_o} \right] \tag{6.83}$$

where $< p >$ is the average dipole moment. Since the voltage, $V = |E|d$, the first term gives the Ohm's law type correction to the resistance and the second term gives the resistance proportional to $|E|$.

The resistivity of $YBa_2Cu_3O_{7-x}$ as a function of electric field has been measured by Frey et al. For small values of the electric field, E, the curve is linear as expected from (6.83). For the fields larger than about half of the breakdown, (BD), voltage, the variation of resistivity as a function of the electric field is shown in Fig.6.4. We find that $R \propto E^{1/2}$ which is in agreement with that predicted approximately from the field driven movement of the mobile charge carriers in the normal state of the superconductor as long as the dielectric constant is independent of the electric field as in (6.82). The experimentally measured variation of resistivity as a function of increasing electric field shows two separate mechanisms. (a) For small electric fields, the resistivity is a linear function of electric field, and (b) for $E > E_{BD}/2$ the resistivity varies as $E^{1/2}$. These experimental results are in agreement with the theory. Indeed the theoretical interpretation is so simple that the experiments appear to belong to ordinary metals.

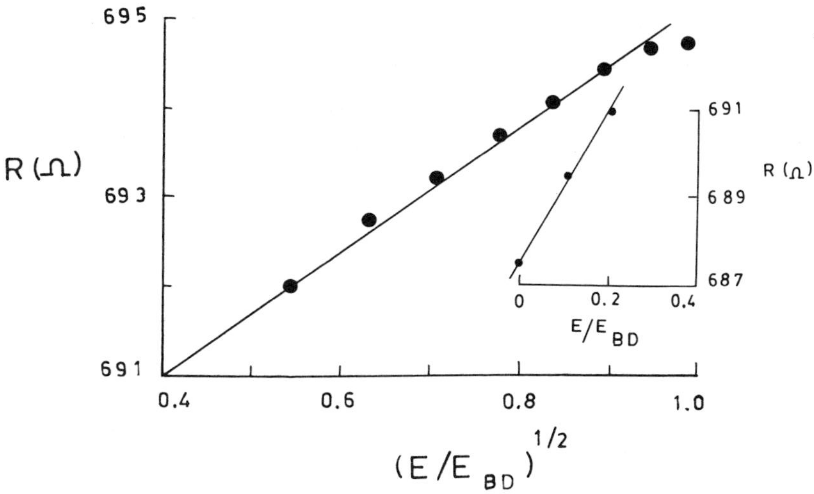

Fig.6.4: The resistivity as a function of square root of the electric field at a temperature of 100 K. The experimental values are taken from Frey et al. The resistivity is seen to vary as the square root of the electric field. For small fields, $E/E_{BD} \leq 0.2$, the resistivity shown in the inset is linear in E.

6.5 Electric fields in Ginzburg-Landau model

We assume that the carriers have no kinetic energy and the entire kinetic energy is caused by the application of the electric field only. There is a change δn in the carrier density by the application of a field, the zero-field value of which is n_o. In the Thomas-Fermi approximation, the relative change in the carrier concentration upon the application of the electric field is given by

$$\frac{\delta n}{n_o} = \left(\frac{E}{E^*}\right) \exp\left(-\frac{x}{l_D}\right) \tag{6.84}$$

where $E^* = 4\pi e l_D n_o/\epsilon$ with l_D the Debye screening length,

$$l_D = \left(\frac{4\pi N(o)e^2}{\epsilon}\right)^{-1/2} . \tag{6.85}$$

Here $N(o)$ is the density of states and ϵ is the dielectric constant. There is a shift in the critical temperature which is small because $l_D/\xi << 1$. We consider the order parameter of the Ginzburg-Landau free energy which is Δ_o in the absence of the electric field becoming $\Delta_o + \delta\Delta$ upon the application of the electric field. The term which has the square of the order parameter is $\alpha|\Delta|^2$. For

$$\tau = 1 - \frac{T}{T_c} \tag{6.86}$$

the coefficient of the square term in the presence of an electric field becomes,

$$\alpha = \tau + \frac{1}{g}\frac{\delta n}{n_o} \tag{6.87}$$

where g is the coupling constant of the B.C.S. theory. The free energy with the application of electric field then becomes,

$$\begin{aligned} G \;=\; & \int \left[\frac{(B-H)^2}{8\pi} + N(o) \left\{ \left| \xi_j \left(\nabla - \frac{2ie}{c}A\right)_j \Delta \right|^2 \right.\right. \\ & \left.\left. -\alpha|\Delta|^2 + \frac{\beta}{2}|\Delta|^4 \right\} \right] d^3r \end{aligned} \tag{6.88}$$

where Δ is the superconducting energy gap which is being used in place of the order parameter, B is the magnetic induction inside the sample and H is the external field. The electric field is along x-direction, perpendicular to the yz plane. A vortex is located at (x_o, y_o). The superconducting gap is space dependent being zero at the site of the vortex and increasing by going away from the vortex centre. In absence of external electric field,

$$|\Delta_o(x,y)|^2 = \frac{[(x-x_o)/\xi_\perp]^2 + [(y-y_o)/\xi_{||}]^2}{1 + [(x-x_o)/\xi_\perp]^2 + [(y-y_o)/\xi_{||}]^2} \left(\frac{\tau}{2\beta}\right) \tag{6.89}$$

obtained by minimizing the last two terms of (6.88). The substitution of (6.86) into (6.87) and the resulting term into the last two terms of (6.88), we find the free energy due to the electric field as,

$$\delta G = -\frac{N(o)E}{gE^*}\int_o^\infty dx \int_o^\infty dy \exp\left(-\frac{x}{l_D}\right)\left(|\Delta_o(x,y)|^2 - |\Delta_o(\infty)|^2\right) \tag{6.90}$$

where $\Delta_o^2(\infty) = \tau/2\beta$ is the order parameter far from the vortex. Upon integration the above becomes,

$$\delta G = \frac{N(o)\tau\pi E}{2\beta g E^*}\xi_{\|} I\left(\frac{\xi_\perp}{l_D}\right) \tag{6.91}$$

where the integral is given by

$$I(a) = \int_o^\infty \frac{\exp(-xa/\xi_\perp)dx}{[1 + \{(x - x_o)/\xi_\perp\}^2]^{1/2}}. \tag{6.92}$$

The change in free energy due to the application of an electric field as a function of $x_o/\xi_\perp$ has a maximum value, $\delta G(x_o^m)$ and it decreases upon increasing $x_o/\xi_\perp$. Therefore $\delta G(x_o^m)$ works like a potential barrier. For small values of $x_o/\xi_\perp$ where $\xi_\perp/l_D \simeq 1$, the effect of electric field on the free energy is the largest. The lower critical field can be obtained from

$$\frac{\phi_o H_{c1}}{4\pi} = \epsilon_o + \delta G \tag{6.93}$$

where

$$\epsilon_o = \left(\frac{\phi_o}{4\pi}\right)^2 \left[\ln(\kappa)/\lambda_\perp\lambda_{\|}\right]. \tag{6.94}$$

Here $\lambda_\perp$ is the penetration depth in the direction in which field is applied and $\lambda_{\|}$ is in the plane. Therefore,

$$H_{c1} = H_{c1}^o + \frac{4\pi}{\phi_o}\frac{N(o)\tau}{2\beta}\frac{\xi_{\|}}{g}\frac{E}{E^*} I(\xi_\perp/l_D) \tag{6.95}$$

where

$$H_{c1}^o = \frac{4\pi\epsilon_o}{\phi_o} \tag{6.96}$$

in the lower critical field when applied electric field is zero and I is determined by (6.92). Using the above, we find that H_{c1} changes upon the application of the electric field,

$$\delta h = \frac{H_{c1} - H_{c1}^o}{H_{c1}^o} = \frac{N(o)\tau E}{2\beta E^*}\frac{\xi_{\|}}{\epsilon_o g} I(\xi_\perp/l_D)\,. \tag{6.97}$$

This field is zero at $\tau = 0$. Thus H_{c1} is enhanced upon the application of the electric field. The current flowing in the $y - z$ plane of a sample of thickness d along the direction of the electric field produces a field which at $H = H_{c1}$ and $J = J_o$ is lower with,

$$J_o = \frac{H_{c1}c}{2\pi d} \approx 1.6\frac{H_{c1}}{d}. \tag{6.98}$$

For $d >> \lambda$, the current flows only along the surfaces of the slab and decreases as $\exp(-x/\lambda_L)$. For $d << \lambda$, the current is almost constant over the slab. The enhancement of H_{c1} by δh of (6.97) enhances the current.

Suggested References

6.1. A.M. Portis, K.W. Blazey, K.A. Müller and J.G. Bednorz, Europhys. Lett. **5**, 467 (1988).

6.2. P.K. Ghosh, N.M. Krishna and K.N. Shrivastava, J. Phys. Condens. Matter **9**, L663 (1997).

6.3. M.R. Eskildsen, P.L. Gammel, B.P. Barber, U. Yaron, A.P. Ramirez, D.A. Huse, D.J. Bishop, C. Bolle, C.M. Lieber, S. Oxx, S. Sridhar, N.H. Andersen, K. Mortensen and P.C. Canfield, Phys. Rev. Lett. **78**, 1968 (1997).

6.4. J.E. Sonier, R.F. Kiefl, J.H. Brewer, J. Chakhalian, S.R. Dunsiger, W.H. Hardy, W.A. MacFarlane, R.I. Miller, A. Wong, G.M. Luke and J.W. Brill, Phys. Rev. Lett. **79**, 1742 (1997).

6.5. J.E. Sonier, J.H. Brewer, R.F. Kiefl, D.A. Bonn, S.R. Dunsiger, W.H. Hardy, R. Liang, W.A. MacFarlane, R.I. Miller, T.M. Riseman, D.R. Noakes, C.E. Stronach and M.F. White, Jr., Phys. Rev. Lett. **79**, 2875 (1997).

6.6. A. Yamashita, K. Ishii, T. Yokoo, J. Akimitsu, M. Hedo, Y. Inada, Y. Onuki, E. Yamamoto, Y. Haga and R. Kadono, Phys. Rev. Lett. **79**, 3771 (1997).

6.7. G. Hechtfischer, R. Kleiner, K. Schlenga, W. Walkenhorst, P. Muller and H.H. Johnson, Phys. Rev. B**55**, 14638 (1997).

6.8. A. Beck, O.M. Frochlich, D. Koelle, R. Gross, H. Sato and M. Naito, Appl. Phys. Lett. **68**, 3341 (1996).

6.9. J.C. Swihart, J. Appl. Phys. **32**, 461 (1961).

6.10. G. Hechtfischer, R. Kleiner, A.V. Ustinov and P. Muller, Phys. Rev. Lett. **79**, 1365 (1997).

6.11. S.K. Yip and J.A. Sauls, Phys. Rev. Lett. **69**, 2264 (1992).

6.12. M. Franz, I. Affleck and M.H.S. Amin, Phys. Rev. Lett. **79**, 1555 (1997).

6.13. I. Affleck, M. Franz and M.H.S. Amin, Phys. Rev. B**55**, 704 (1997).

6.14. I. Kosztin and A.J. Leggett, Phys. Rev. Lett. **79**, 135 (1997).

6.15. M.W. Coffey and J.R. Clem, Phys. Rev. Lett. **67**, 386 (1991).

6.16. K.N. Shrivastava, J. Phys.: Condens. Matter **5**, L 597 (1993).

6.17. L.S. Lingam and K.N. Shrivastava, Mod. Phys. Lett. B**23**, 1123 (1996); Physica B**223**,577 (1996).

6.18. J. Mannhart, Supercond. Sci. Technol **9**, 49 (1996).

6.19. T. Frey, J. Mannhart, J.G. Bednorz and E.J. Williams, Phys. Rev. B**51**, 3257 (1995).

6.20. L. Burlachkov, I.B. Khalfin and B.Ya. Shapiro, Phys. Rev. B**48**, 1156 (1993).

Chapter 7

Mössbauer effect

7.1 The γ-decay schemes

A nucleus is identified by the number of neutrons, N and the number of protons Z. The mass number is $A = Z + N$. The charge on the nucleus determines completely the electronic structure and hence the chemical properties of the element. Thus not only the neutron number N but also the atomic number Z becomes unnecessary,

$$^{A}_{Z}X_N =^A X \ .$$

The energy levels of ^{57}Fe are shown in Fig. 7.1. The electron capture by ^{57}Co whose life time is 270 days produces 136.32 keV state with 99.84% efficiency which decays to two levels. It produces a γ-ray of 121.9 keV energy with 85% efficiency and another γ-ray of energy 136.32 keV with 11% efficiency. The energy of the first excited state is thus 14.41 keV. The resonance from 136.32 keV is much weaker than that from 14.41 keV. The 14.41 keV level is associated with nuclear spin 3/2 whereas the lowest level belongs to nuclear spin 1/2. The parity is conserved in the transition from 3/2 to 1/2. The amount of $E2$ admixture being only about 0.0006% is very small. The life time of the 14.41 keV level is $\sim$ 99 ns which corresponds to the width of 0.192 ms^{-1}.

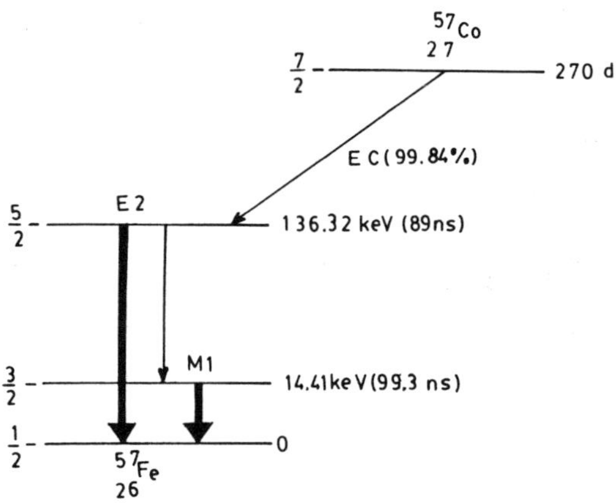

Fig. 7.1: The γ-decay scheme of ^{57}Co by electron capture showing Mošsbauer transitions of ^{57}Fe at 14.4 keV and 136.32 keV.

There are two isotopes of iodine which give low energy γ-rays. The ^{129}I has a γ-ray transition at 27.72 keV and ^{127}I has 57.60 keV. The ground state of ^{129}I is radio active with a half life of 1.7×10^7 y whereas natural iodine ^{127}I is stable. The first excited state of ^{129}I can be populated by decay of 33d ^{129m}Te or 70 minute ^{129}Te. Both parents are produced by the ^{128}Te(n,γ) reaction. $^{129m}_{52}$Te produces $^{129}_{52}$Te by isomeric or electromagnetic transition as shown in Fig. 7.2.

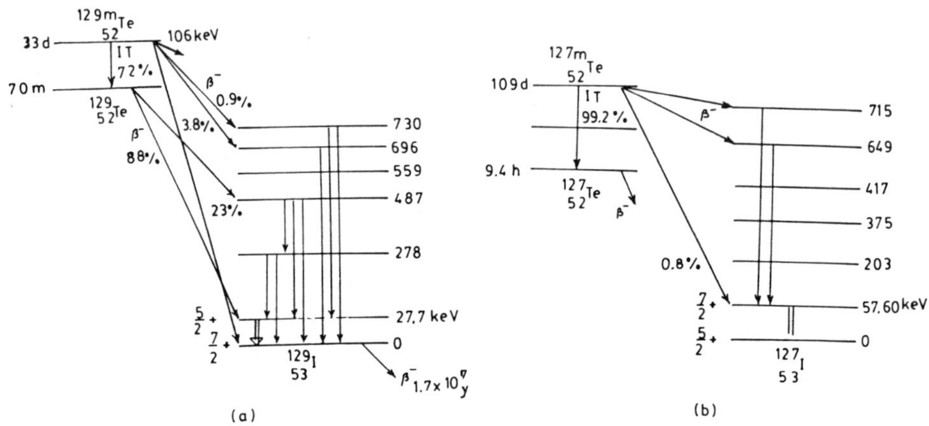

Fig.7.2: (a) The decay of ^{129}I and (b) the decay of ^{127}I by isomeric transitions.

The 23.875 keV γ-ray is produced by $^{119}_{50}$Sn. The 65.66 keV γ-ray is more strongly internally converted than the 23.875 keV γ-ray and hence is weak in intensity. The decay scheme of ^{119}Sn is shown in Fig. 7.3.

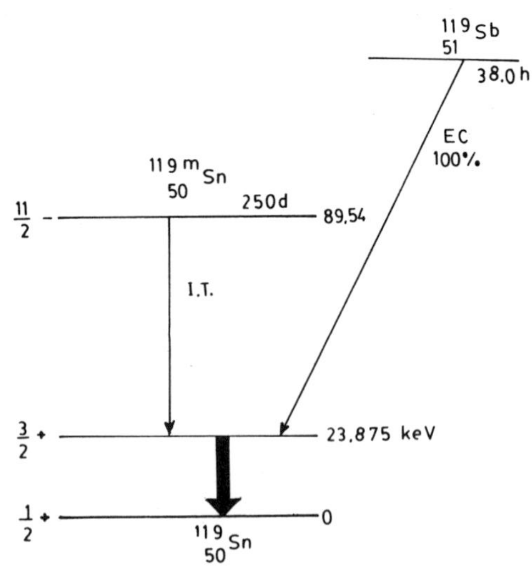

Fig.7.3: The decay of ^{119}Sn by isomeric transitions and electron capture.

The 93.26 keV excited state of ^{67}Zn has a life time of 9.4×10^3 ns. This gives an unusually narrow γ-ray transition of line width 3.12×10^{-4}mms^{-1}. It offers high precision which can be utilized for the measurement of gravitational red shift. The energy levels of ^{67}Zn are shown in Fig. 7.4.

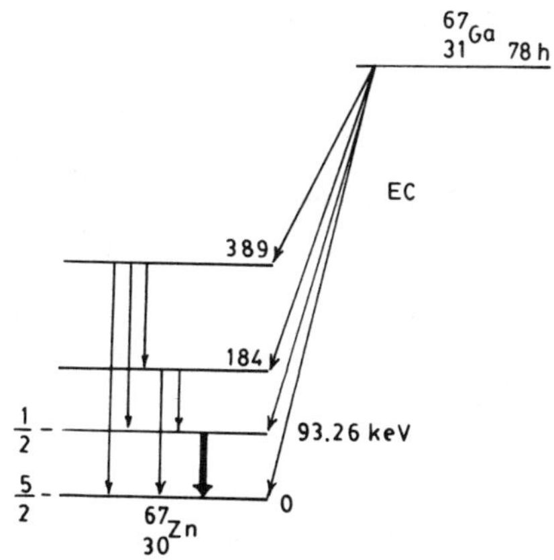

Fig.7.4: Decay shceme of ^{67}Zn.

The ^{191}Ir and ^{193}Ir decay schemes are given in Fig. 7.5. There are several γ-ray transitions. The half life of 129.4 keV γ transition is about $T_{1/2} = 0.1$ns. The ^{191}Ir has a γ-ray of energy 82.3 keV produced by electron capture of ^{191}Pt. The half life of 82.3 keV excited state is about $T_{1/2} \simeq 3.8$ ns. The half life of 138.9 keV transition of ^{193}Ir is 0.08 ns.

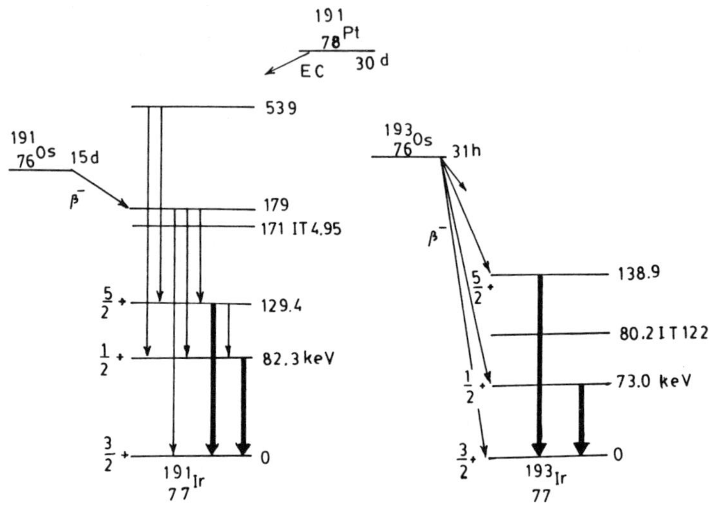

Fig. 7.5: Decay shceme of ^{191}Ir *and* ^{193}Ir.

7.2 Moššbauer effect

The phenomenon of emission or absorption of a γ-ray photon without loss of energy due to recoil of the nucleus is known as the Mössbauer effect and to understand the importance of recoilless emission of γ-rays, we consider the interplay of a variety of factors such as the energetics of free-atom recoil and thermal broadening, energy and momentum transfer to the lattice, recoil free fraction, Debye-Waller factor, etc. We consider an isolated atom in the gas phase and define the energy difference between the ground state of the nucleus E_g and its excited state E_e as $E = E_e - E_g$. If the photon is emitted from a nucleus of mass M moving with an initial velocity V_x along the positive x driection, at the moment of emission, the total energy of the nucleus at rest is $E + \frac{1}{2}MV_x^2$. After emission, the γ-ray has the energy E_γ and the nucleus, a shifted velocity, $V_x \pm v$ due to recoil. It can be $V_x \pm v$ depending on the direction of v relative to V_x. The total energy of the system is $E_\gamma + \frac{1}{2}M(V_x + v)^2$. By conservation of energy,

$$E + \frac{1}{2}MV_x^2 = E_\gamma + \frac{1}{2}M(V_x + v)^2 \ . \tag{7.1}$$

Hence

$$\delta E = E - E_\gamma = \frac{1}{2}Mv^2 + MvV_x \tag{7.2}$$

or

$$\delta E = E_R + E_D \ . \tag{7.3}$$

The γ-ray energy is thus seen to differ from the nuclear energy level separation by an amount which depends on the recoil kinetic energy, $E_R = \frac{1}{2}Mv^2$ and on $E_D = MvV_x$ which is proportional to the velocity of the atom. The mean kinetic energy per translational degree of freedom of a free atom in a gas with random thermal motion is,

$$E_k = \frac{1}{2}MV_x^2 \tag{7.4}$$

where V_x^2 is the mean square velocity of the atoms. Hence $(V_x^2)^{1/2} = (2E_k/M)^{1/2}$ and the mean broadening

$$E_D = Mv(V_x^2)^{1/2} = \sqrt{2E_kMv^2} = 2\sqrt{E_kE_R} \ . \tag{7.5}$$

The values of E_R and E_D can be more conveniently expressed in terms of the γ-ray energy, E_γ. Thus,

$$E_R = \frac{1}{2}Mv^2 = \frac{(Mv)^2}{2M} = \frac{p^2}{2M} \tag{7.6}$$

where p is the recoil momentum of the atom. Since momentum must be conserved, this momentum will be equal and opposite to the momentum of the γ-ray photon, p_γ,

$$p = -p_\gamma = \frac{-E_\gamma}{c} \ . \tag{7.7}$$

Hence,

$$E_R = \frac{E_\gamma^2}{2Mc^2} \quad \text{and} \tag{7.8}$$

$$E_D = 2\sqrt{E_kE_R} = E_\gamma\left(\frac{2E_k}{Mc^2}\right)^{1/2} \ . \tag{7.9}$$

The recoil energy is usually of the order of $10^{-2} - 1$ kJmol^{-1}. This energy is much less than the chemical binding energy which is in the range of $10^2 - 10^3$ kJmol^{-1}, but of the same order of magnitude as the lattice vibrational energy.

If the recoil energy is transferred directly to the vibrational energy, the γ-ray energy is lowered. However, the phonon energies are quantized and the recoil energy can only be transferred to the lattice if it corresponds closely to an allowed quantum jump. The transfer of energy to the lattice can take place only in integral multiples of $\hbar\omega$ $(0, \pm\hbar\omega,\ \pm 2\hbar\omega, \cdots)$. If a fraction, f, of emission events results in no transfer of energy to the lattice (zero-phonon transitions), and a fraction (1-f) transfer one phonon of energy $\hbar\omega$, then we have $E_R = (1-f)\hbar\omega$ or $f = 1 - \frac{E_R}{\hbar\omega}$. Since many emission processes are considered, the average energy transferred per event must equal E_R. In a zero-phonon emission, the whole crystal rather than a single nucleus recoils. The equations (7.8) and (7.9) for E_R and E_D contain the reciprocal mass, $1/M$. This would make E_R and E_D almost negligible. However, the nucleus is not bound rigigily in the crystal as assumed, but is free to vibrate. Hence the recoil momentum is transferred to the crystal as a whole. Since both the recoil energy and Doppler broadening become much less in a crystal than in a gas, the zero-phonon transitions are referred to as "recoilless". The parameter f, is known as the recoilless or recoil-free fraction. The decay schemes of the γ-transitions of several nuclei are given in the previous section. The ^{119m}Sn has a half-life of 250 days. The energy of the transition observed in the Mössbauer spectra is 23.875 keV. It arises from the decay of $^{119}Sb_{51}$ which is made by neutron capture in ^{118}Sn. The Fig.7.1 shows the γ-decay scheme of ^{57}Co and the energy levels of the excited state. The ^{57}Co has a half life of 270 days. In the Mössbauer experiment the emitting and absorbing nuclei are embedded in a solid lattice. The total energy of the γ-transition E, must be conserved and can be shared between,

(a) energy of the γ-ray photon, E_γ,

(b) the lattice vibrations,

(c) the translational kinetic energy of the individual atom and

(d) the translational kinetic energy of the solid as a whole.

Due to high chemical binding energy and large mass, the translational kinetic energies can be neglected. The γ-transition energy is thus shared between the γ photons and the lattice vibrations. The recoil-free fraction depends on

(a) the free-atom recoil energy, which is proportional to E_γ^2,

(b) the properties of the solid and

(c) the temperature.

Thus, the smaller the probability of exciting lattice vibrations, the larger will be the value of f, i.e., the smaller the γ-ray energy, the stronger is the binding of the atom in the lattice. A lowering of the temperature increases the recoil-free fraction and lowers the counting rate of the transmitted γ-photons. The probabiliy, W of the zero-phonon γ emission from a nucleus in a solid, is proportional to the square of the matrix element connecting the initial $|i>$ and final $<f|$ states,

$$W \propto | < f|\mathcal{H}|i > |^2 \tag{7.10}$$

where $\mathcal{H}$ is the interaction Hamiltonian operator and depends upon the positional coordinates of the atom and the momenta and spins of the particles within the nucleus. The forces acting within the nucleus are extremely short range whereas those holding the lattice together are of much longer range. Hence the nuclear decay and the vibrational state are independent of each other. We consider the interaction of a radioactive nucleus bound in a crystal with a radiation field. For a single nucleus,

$$\mathcal{H} = -\left(\frac{e}{m_p}\right)\sum_i A(x_i)p_i \tag{7.11}$$

where, x_i and p_i are the proton coordinates and momenta, m_p is the mass of the proton and summation includes all protons in the nucleus. Consider a plane electromagnetic wave of wave vector k_γ so that the vector potential, may be written as, $A(x_i) \propto e^{ik_\gamma.x_i}$. When the nucleus makes a transition from $|i>$ to $<f|$, the state of the lattice may change from n_i to n_f in one of the phonon modes. The matrix element for the transition is,

$$M =< n_f(q)|\exp(ik_\gamma.x_\mu)|n_i(q) >< f|a(k_\gamma)|i > \tag{7.12}$$

where

$$a(k_\gamma) \simeq \sum_j \exp(ik_\gamma.r_j)A_o.p_j \ . \tag{7.13}$$

Here A_o is the amplitude of the radiation field and transition rate is given by

$$p(n_f, n_i) = |< n_f| \exp(ik_\gamma.x_\mu)|n_i|^2 \ . \tag{7.14}$$

The probability of a nuclear transition is independent of the phonon states. Hence, only the lattice matrix element is needed. The recoilless fraction, f is the relative probability that a transition will occur without change in energy of the phonon distribution,

$$f = \sum_f |< n_i| \exp(ik_\gamma.x_\mu)|n_f > |^2\delta(E_f - E_i) \ . \tag{7.15}$$

Averaging over the probability of a particular distribution of occupation number, multiplying by $P_{ni}(T)$, the probability of obtaining a particular phonon distribution at a temperature T, and taking the sum over all states n_i determines the recoilless fraction. This fraction f can also be written as follows: For zero-phonon emission, the lattice modes are unchanged and hence the probability for recoilless emission is

$$f = \text{const}| < n_i|e^{ik.u}|n_i > |^2 \ . \tag{7.16}$$

Since n_i is normalized,

$$f = e^{-k^2.u^2} \tag{7.17}$$

and u is a displacement vector, u^2 can be replaced by $< u^2 >$, the component of the mean square vibrational amplitudes of the emitting atom in the direction of the γ-ray. Since $k^2 = \frac{4\pi^2}{\lambda^2} = E_\gamma^2/(\hbar c)^2$, λ is the wave length of the γ-ray, we obtain,

$$f = \exp\left(\frac{-4\pi^2 < u^2 >}{\lambda^2}\right) = \exp\left(-\frac{E_\gamma^2 < u^2 >}{(\hbar c)^2}\right) \ . \tag{7.18}$$

This indicates that the probability of zero-phonon emission decreases exponentially with the square of the γ-ray energy. Also, f increases exponentially with decrease in $< u^2 >$ which in turn depends on the firmness of binding and on the temperature. The fraction f gives the intensity of the Mošsbauer line. For optical phonons, the displacement of the Mošsbauer atom in a harmonic approximation is given by

$$\delta R = (\hbar/2M\omega_k)^{1/2}(a_k + a^{\dagger}_{-k}) \ . \tag{7.19}$$

Therefore, in the dimensionality, d, the average value of the square of the atomic displacement is found to be,

$$< u^2 >= \frac{3(2\pi)^{1-d}\hbar}{\rho v^d}\left(\frac{k_B T}{\hbar}\right)^{d-1}\int(2n+1)x^{d-2}dx \tag{7.20}$$

where $n = [e^{\hbar\omega/k_B T} - 1]^{-1} = [e^x - 1]^{-1}$. For $d = 3$,

$$< u^2 >= \frac{3\hbar}{2(2\pi)^2\rho v^3}\left(\frac{k_B\theta_D}{\hbar}\right) + \frac{3}{2\pi^2\rho v^3}\left(\frac{k_B T}{\hbar}\right)^2\int\frac{x dx}{e^x - 1} \tag{7.21}$$

Here M is the mass per atom, k_γ is the wave vector of the γ ray and θ_D is the Debye cut-off temperature. The quantity $\hbar^2 k_\gamma^2/2M$ gives the recoil energy and $k_B\theta_D$ the Debye energy. Thus,

$$f \simeq \exp\left[-\frac{3\hbar^2 k_\gamma^2}{4Mk_B\theta_D}\right] \ . \tag{7.22}$$

At absolute zero of temperature, in the Debye model, the idea of a single vibrational frequency is abandoned and a continuum of oscillator frequencies ranging from zero upto a maximum ω_D is assumed. A characteirstic temperature called the Debye temperature θ_D is defined as $\hbar\omega_D = k_B\theta_D$ and the average frequency is $\hbar\omega = \frac{3}{4}\hbar\omega_D$. The Debye model leads to the expression,

$$f = \exp\left[-\frac{6E_R}{k_B\theta_D}\left\{\frac{1}{4} + \left(\frac{T}{\theta_D}\right)^2\int_o^{\theta_D/T}\frac{x dx}{e^x - 1}\right\}\right] \ . \tag{7.23}$$

This is often written as $f = e^{-2W}$. The factor W is called the Debye-Waller factor. Thus, recoilless emission or absorption is optimized for a low energy γ-ray with the nucleus strongly bound in a crystal lattice at low temperatures.

7.3 Second-order Doppler shift

The emission of γ-radiation from the radioactive nuclei embedded in solids is considered to be recoilless because the recoil is absorbed by a large number of atoms in the solid. However, the atoms in the solid are at a finite temperature so that they execute small oscillations due to thermal energy. Alternatively, we may say that as the nucleus emits the γ-ray, it looses some mass which causes a small shift in the frequency of the emitted radiation. The existence of a relativistic temperature dependent contribution to the shift was pointed out by Josephson who proposed a frequency shift in the Mössbauer spectrum due to the relativistic decrease in the mass of the oscillator which emits radiation. The frequency shift was derived from the expression of the relativistic time dilation. It was found that this frequency shift is precisely that due to the Doppler effect. The mean velocity of the nuclei due to thermal vibrations is zero. However, in the second order the average of a square of the thermal velocity of atoms in solids is not zero, so that there is a net second-order Doppler shift. The relativistic equation for the Doppler effect for an emitted photon gives the observed frequency ν' for a velocity v as,

$$\nu' = \nu\left(1-\frac{v}{c}\right)\left(1-\frac{v^2}{c^2}\right)^{-1/2} . \tag{7.24}$$

Since $v^2/c^2 << 1$, the second bracketed quantity can be expanded upto the second term in the binomial series so that

$$\nu' \simeq \nu\left(1-\frac{v}{c}\right)\left(1+\frac{v^2}{2c^2}\right) \tag{7.25}$$

where ν is the frequency of the stationary system. The first-order term in velocity is a function of the velocity of the atom vibrating on its lattice site in the direction of the γ-ray and will average to zero as it is a term linear in v. It is usually referred to as a second-order Doppler shift, (SODS), and for a Mössbauer resonance,

$$\nu' \simeq \nu\left(1+\frac{< v^2 >}{2c^2}\right) . \tag{7.26}$$

where a nucleus of one of the atoms in a crystal lattice emits a γ-ray photon, its mass changes by an amount $h\nu/c^2$, where ν is the frequency of the γ-ray. The change in the mass of the atom causes a corresponding change in the energy of the interacting system to which emitting atom belongs.

Consider the interaction Hamiltonian

$$\mathcal{H} = \sum_i \frac{p_i^2}{2m_i} + V(r_1, r_2, \cdots) \tag{7.27}$$

as the nucleus emits a γ ray, its mass changes by the amount, $\delta m_i = \frac{h\nu}{c^2}$. As a consequence, the change in energy becomes,

$$\begin{aligned} \delta E &= \frac{\delta \sum_i p_i^2}{2m_i} \\ &= -\frac{< p_i^2 >}{2m_i}\frac{\delta m_i}{2m_i} \\ &= -\frac{h\nu}{m_i c^2}\left\langle \frac{p_i^2}{2m_i} \right\rangle \\ &= -\frac{h\nu}{m_i c^2 \left\langle \frac{1}{2m_i v_i^2} \right\rangle} \quad . \end{aligned} \tag{7.28}$$

This change is reflected as a corresponding change in the γ-ray photon. Thus the energy of the γ-ray photon is increased by $\Delta E = -\delta E$. Hence,

$$h\Delta\nu = \Delta E = -\delta E = \frac{h\nu}{m_i c^2}\left\langle \frac{1}{2} m_i v_i^2 \right\rangle = \frac{h\nu}{2c^2} < v_i^2 > \quad . \tag{7.29}$$

Dropping the suffix, $\frac{\Delta\nu}{\nu} = \frac{<v^2>}{2c^2}$. The mean square velocity of an atom changes with temperature. Therefore, the second-order Doppler shift, (SODS), depends on temperature. Accordingly, the Mössbauer resonance in an absorber moves towards negative velocity as the temperature is raised. The SODS contribution to the observed chemical shift is smallest at absolute zero of temperature but is finite because of a zero-point motion of the nucleus.

We can calculate the second-order Doppler shift in the Debye model, which assumes a contribution with a cutoff frequency, $\hbar\omega_D$. The SODS of the Mössbauer line, in velocity units, is given by,

$$\delta_{SOD} = \frac{9k_B\theta_D}{16mc} + \frac{9k_BT}{2mc}\left(\frac{T}{\theta}\right)^3 \int_o^{\theta/T} \frac{x^3dx}{e^x - 1} \quad . \tag{7.30}$$

Since the thermally averaged velocities of the atoms in the solid determine the SOD, this shift is related to the specific heat of the solid,

$$\frac{\partial \nu}{\partial T} = \frac{\nu}{2mc^2}\frac{\partial}{\partial T}\left(\frac{p^2}{2m}\right) = \frac{\nu}{2mc^2}C_p \quad .$$

Hence

$$\frac{1}{\nu}\left(\frac{\partial \nu}{\partial T}\right) = \frac{C_p}{2mc^2} \quad . \tag{7.31}$$

There has been considerable work on the Moššbauer effect of atoms in the high temperature superconductors but the effect of the superconducting lattice on the spectra has only recently been reported.

7.4 Isomer shift

We now discuss briefly the effect of the finite size of the nucleus on the shift of the Mössbauer line. Out side the nucleus the electric field due to the nucleus is that of a point charge Ze while inside it flattens off so that the potential reaches a finite value as $r \to 0$. An electron occupies an orbit with a radius much larger than that of the nucleus. Since the nuclear radius is of the order of 10^{-13}cm whereas the radius of the first Bohr orbit is 0.529 Å, the wave function of the electron is determined primarily by the electric field outside the nucleus. The wave function of the s electron will be approximately that computed under the assumption of a point nucleus and hence we calculate the change in energy of this state by computing,

$$\delta E = \int_o^\infty \psi_s^*(r)\delta V(r)\psi_s(r)4\pi r^2 dr \tag{7.32}$$

where $\delta V(r)$ is the difference between the potential due to a distribution of charge and the potential due to a point charge

$$\psi_s(r) = \frac{1}{\sqrt{\pi}}\left(\frac{Z}{\alpha_B}\right)^{3/2} e^{-Zr/\alpha_B} \tag{7.33}$$

where α_B is the Bohr radius. We ignore the effect of finite size of the nucleus on this wave function of the atomic electron to calculate the energy shift of the s state due to a finite nuclear radius. We assume that the nuclear charge is uniformly distributed. Then the potential inside the nucleus is,

$$V(r) = -\frac{Ze^2}{R}\left\{\frac{3}{2} - \frac{1}{2}\left(\frac{r}{R}\right)^2\right\} \qquad (r \leq R) \qquad (7.34)$$

and outside,

$$V(r) = -\frac{Ze^2}{r} \qquad (r > R) \qquad (7.35)$$

as shown in Fig.7.6,

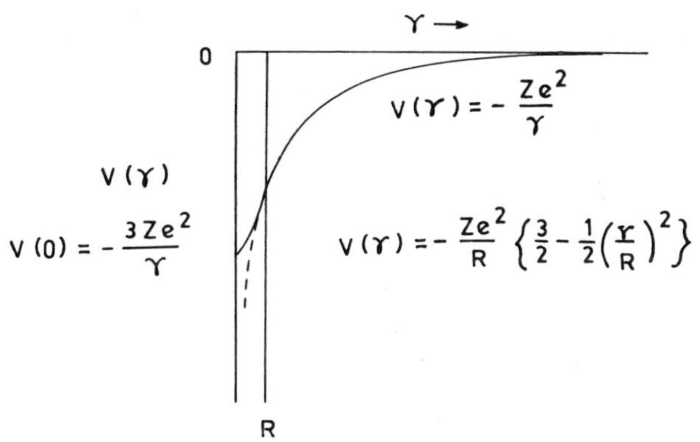

Fig.7.6: Nuclear potential for a finite size nucleus.

$$\delta E = \int_o^R \psi_s^*(r)\left\{\frac{Ze^2}{r} - \frac{Ze^2}{R}\left[\frac{3}{2} - \frac{1}{2}\left(\frac{r}{R}\right)^2\right]\right\}\psi_s(r)4\pi r^2 dr \ . \qquad (7.36a)$$

Using the hydrogenic wave function (7.33) in the above, we find,

$$\delta E = \left(\frac{Z}{\alpha_B}\right)^3 4Ze^2 \int_o^R e^{-2Zr/\alpha_B}\left\{\frac{1}{r} - \frac{3}{2R} + \frac{1}{2}\frac{r^2}{R^3}\right\} r^2 dr \ . \qquad (6.36b)$$

This change in energy is to be compared with the total potential energy E of

$$E \simeq -\left(\frac{Z}{\alpha_B}\right)^3 4Ze^2 \int_o^\infty e^{-2Zr/\alpha_B} r dr = -\frac{z^2 e^2}{\alpha_B} \ . \tag{7.37}$$

Since $R << \alpha_B$,

$$\begin{aligned} \delta E &\simeq \left(\frac{Z}{\alpha_B}\right)^3 4Ze^2 \int_o^R \left\{ r - \frac{3}{2}\frac{r^2}{R} + \frac{1}{2}\frac{r^4}{R^3} \right\} dr \\ &= \left(\frac{Z}{\alpha_B}\right)^3 \frac{2}{5} Ze^2 R^2 \end{aligned} \tag{7.38}$$

The fractional shift of the energy of the s electron is,

$$\frac{\delta E}{\frac{1}{2}E} = \frac{4}{5} Z^2 \left(\frac{R}{\alpha_B}\right)^2 . \tag{7.39}$$

As an example, the first s state in lead is reached from the next p state by emission of a K x rays. Since the p wave function is not concentrated in the nucleus and falls to zero as $r \to 0$, the shift in p state energy is small in comparison with the shift in s state energy and the shift in the energy of the K x ray $\delta E_K / E_K$ ($p \to s$ transition) is given by

$$\frac{\delta E_K}{E_K} = \frac{4}{3}\frac{\delta E}{E} \simeq \frac{16}{15} Z^2 \left(\frac{R}{\alpha_B}\right)^2 . \tag{7.40}$$

For Pb, $R \simeq 8 \times 10^{-13}$cm, $Z = 82$, so

$$\frac{\delta E_K}{E_K} \simeq 10^{-4} \ . \tag{7.41}$$

This is not a small effect but the quantity that has been studied is the splitting between $2p_{1/2}$ and $2p_{3/2}$ levels which is affected strongly. In the Mössbauer spectra the changes in the shift of this order are always detected. We need not use the hydrogen type wave function in (7.36). Instead of hydrogen type wave functions in (7.36) we use the effective charge density, $\rho = -e|\psi(o)|^2$ so that the energy shift becomes,

$$\delta E = \frac{2\pi}{5} Ze^2 R^2 |\psi(o)|^2 . \tag{7.42}$$

If the energy of the excited nuclear state is E_e and that of the ground state is E_g, the change in energy of the γ ray due to the nuclear transition from the excited state to the ground state owing to the electrostatic interaction is,

$$\delta E_e - \delta E_g = \frac{2\pi}{5} Ze^2 |\psi(o)|^2 (R_e^2 - R_g^2) \tag{7.43}$$

where R_e and R_g represent the nuclear radii in the excited state before the emission and in the ground state after the emission of the γ ray. We compare the energy of the source E_g with that of the absorber E_a, so that the net isomershift becomes,

$$\Delta S^{(i)} = \frac{2\pi}{5} Ze^2 \left[|\psi_s(o)|^2 - |\psi_a(o)|^2 \right] \left[R_e^2 - R_g^2 \right] \tag{7.44}$$

where $\psi_a(o)$ is the electronic wave function at the site of the nucleus in the absorber and $\psi_s(o)$ in the source. If we know the electronic wave functions, a measurement of the isomershift (7.44) can lead to an estimate of change in size of the nucleus in going from the excited to the ground state.

7.5 Recoilless fraction in a superconductor

The intensity of the Moššbauer line is determined by the recoilless fraction given by (7.22). It may be written as,

$$f = \exp\left(-\frac{1}{3} k_\gamma^2 < u^2 > \right) \tag{7.45}$$

where $< u^2 >$ is the mean-square distance between atoms. We now calculate the effect of optical phonons on the recoilless fraction of the Mössbauer spectrum in a superconductor. We use a B.C.S. type theory to find the change in the single phonon energy which we use to determine the number density of the phonons in the paired state. Since the single particle energy decreases, the number density increase so that the mean square displacement of an atom, $< u^2 >$, increases leading to a decrease in the f factor of the γ-ray in going from the normal to the paired state. We consider the interaction of conduction electrons with lattice vibrations so that the Hamiltonian becomes,

$$\mathcal{H} = \sum_{k,\sigma} E_{k,\sigma} c_{k\sigma}^\dagger c_{k\sigma} + \sum_q \hbar\omega_q a_q^\dagger a_q + \sum_{k,q,\sigma} D_q c_{k+q,\sigma}^\dagger c_{k,\sigma} a_q$$

$$+\sum_{k,q,\sigma} D_{-q} c^{\dagger}_{k-q,\sigma} c_{k,\sigma} a^{\dagger}_{q} + h.c. \tag{7.46}$$

where D_q is the matrix element of the electron-phonon interaction and h.c. stands for the hermitian conjugate of the previous terms. The operators $c^{\dagger}_{k\sigma}(c_{k\sigma})$ create (destroy) electrons in the conduction band and $a^{\dagger}_{q}(a_q)$ create (destroy) phonons. $E_{k\sigma}$ is the single electron energy and $\hbar\omega_q$ is the single particle phonon energy. Using the anticommutators for electrons and commutators for phonons, the electron-phonon interaction transformed upto second order may be written as,

$$\mathcal{H}' = \frac{1}{2} \sum_{k,k',\sigma,\sigma'} V_{k,k'} c^{\dagger}_{k',\sigma'} c^{\dagger}_{-k',\sigma} c_{-k,\sigma} c_{k,\sigma'} \tag{7.47}$$

where

$$V_{k,k'} = 2|D_{k-k'}|^2 \hbar\omega_{k-k'} [(E_k - E_{k'})^2 - (\hbar\omega_{k-k'})^2]^{-1} \ . \tag{7.48}$$

We introduce the gap as in the B.C.S. theory as,

$$\Delta_k = \sum_{k'} V_{k,k'} c^{\dagger}_{k'} c^{\dagger}_{-k'} \tag{7.49}$$

where k is associated with spin up while $-k$ is with spin down so that the gap corresponding to spin singlet state. Substituting (7.49) in (7.47) we find,

$$\mathcal{H}' = \frac{1}{2} \sum_{k,\sigma,\sigma'} \Delta_k c_{-k\sigma} c_{k,\sigma'} + \frac{1}{2} \sum_{k',\sigma,\sigma'} \Delta_{k'} c^{\dagger}_{k',\sigma'} c^{\dagger}_{-k',\sigma} + \frac{1}{2V_{k,k'}} \Delta_k \Delta_{k'} \ . \tag{7.50}$$

The potential (7.48) is attractive for $E_k - E_{k'} << \hbar\omega_{k-k'}$, otherwise it is repulsive. The attractive potential is,

$$v_{k-k'} = -2|D_{k-k'}|^2 / \hbar\omega_{k-k'} \ . \tag{7.51}$$

The average value of the interaction (7.50) is thus found to be,

$$< H' >= -2\Delta^2 / v_{k-k'} \tag{7.52}$$

where $\Delta^2 = \Delta_k \Delta_{-k}$ which means that the square of the gap energy is used. The result (1.65) was obtained for zero excitation of phonons which is correct at $T = 0$ only. A finite temperature calculation gives a value which depends on the phonon number density. Therefore, we calculate the second-order energy of the system from the Hamiltonian (7.46) from which we evaluate the derivative with respect to the phonon number density to determine the self-energy of the phonon which upon use of the pairing operator gives the modified single particle energy of the phonon as,

$$\hbar\omega' = \hbar\omega_q - 2\Delta^2/v_q \tag{7.53}$$

which shows that the frequency of the phonons is reduced upon pairing of electrons by the amount $-2\Delta^2/v_q = -(\Delta^2/D_q^2)\hbar\omega$. The phonon number density at the modified frequency then becomes,

$$n_q = \left[\exp\left\{(\hbar\omega_q - 2\Delta^2/v_q)/k_BT\right\} - 1\right]^{-1} \tag{7.54}$$

so that the average value of the square of the atomic displacement becomes,

$$< u^2 >= \frac{3(2\pi)^{1-d}\hbar}{\rho v^d}\left(\frac{k_BT}{\hbar}\right)^{d-1}\int(2n+1)x^{d-2}dx \tag{7.55}$$

where ρ is the mass density of the solid, v the sound velocity and d the dimensionality. For $d = 3$, it has two terms, a zero-point contribution, $u^2(o)$ and a temperature dependent term $< u^2(T) >$ as,

$$< u^2 >=< u^2(o) > + < u^2(T) > \tag{7.56a}$$

with

$$< u^2(o) >= \frac{3\hbar}{2(2\pi)^2\rho v^3}\left(\frac{k_B\theta_D}{\hbar}\right)^2 \tag{7.56b}$$

and

$$< u^2(T) >= \frac{3}{2\pi^2\rho v^3}\int\left[\exp\left\{\frac{(\hbar\omega_q - 2\Delta^2/v_q)}{k_BT}\right\} - 1\right]^{-1}(\hbar\omega_q - \frac{2\Delta^2}{v_q})d\omega_q\,. \tag{7.56c}$$

In order to understand the integral further, we need the wave vector dependence of v_q. For optical phonons, we assume,

$$D_q = D_o\left(\frac{\hbar}{2M\omega_q}\right)^{1/2} \tag{7.57}$$

where $D_o = (\partial V/\partial r)_o$ is the distance derivative of the crystal potential evaluated at the equilibrium. We define a parameter,

$$s_o = \Delta^2/(D_o^2\hbar^2/2M) \tag{7.58}$$

so that using (7.51) and (7.57), the temperature dependent part of the amplitude of the oscillation becomes,

$$\begin{aligned} < u^2(T) > &= \frac{3}{2\pi^2\rho v^3}\int\left[\exp\{(\hbar\omega_q - s_o\hbar^2\omega_q^2)/k_BT\} - 1\right]^{-1} \\ &\times (\hbar\omega_q - \frac{2\Delta^2}{v_q})d\omega_q\,. \end{aligned} \tag{7.59}$$

Substituting $\hbar\omega/k_BT = x$, we find that,

$$\begin{aligned} < u^2(T) >= \frac{3\hbar}{2\pi^2\rho v^3}\left(\frac{k_BT}{\hbar}\right)^2\int\left[\exp\{x - s_o x^2 k_BT\} - 1\right]^{-1} x dx \\ -\frac{3s_o\hbar^2}{2\pi^2\rho v^3}\left(\frac{k_BT}{\hbar}\right)^3\int\left[\exp\{x - s_o x^2 k_BT\} - 1\right]^{-1} x^2 dx \end{aligned} \tag{7.60}$$

which describes the reduction of the logarithm of the f factor upon pairing. The larger the gap energy, the larger is the reduction. Upon cooling, further reduction is predicted. In the case of ^{119}Sn in $EuBa_2Cu_{2.98}Sn_{0.02}O_{7-\delta}$ the value of f at $T = 0$ extrapolated from the normal state is about 0.868 whereas the actually measured value in the superconducting state is 0.800. There is a reduction in the f value in going from the normal value to the superconducting value. This reduction of 0.068 amounts to about 8.5 per cent in going from the normal to the superconducting state. The calculation shows that the recoil-free fraction is indeed reduced in going from the normal to the superconducting state by about this amount. There is a large reduction in the recoilless fraction of ^{57}Fe in $YBa_2(Cu_{0.98}Fe_{0.02})_4O_8$ superconductor but the data are masked by a structural distortion. However, we have found that the isomer shift in this system is correlated with the diamagnetic susceptibility. The integrals in (7.60) at temperatures lower than the Debye cutoff temperature are independent of temperature. Therefore, the logarithm of the recoilless fraction varies as,

$$-\frac{1}{3}k_\gamma^2 < u^2 >= -a_1T^2 + b_1T^3\left(1 - \frac{T}{T_c}\right) + c \tag{7.61}$$

where

$$a_1 = \frac{k_B^2(k_\gamma^2/3)}{4\rho v^3 \hbar} \qquad (7.62a)$$

$$b_o = \frac{6Mk_B^3\zeta(3)(k_\gamma^2/3)}{\pi^2\rho v^3\hbar^3 D_o^2} \qquad (7.62b)$$

and

$$b_1 = b_o\Delta_o^2 \ . \qquad (7.62c)$$

The quantity c is caused by the quantum mechanical corrections such as those arising from the first term of (7.56a). Here $\zeta(3)$ is the Riemann zeta function and we have used the temperature dependent gap,

$$\Delta^2 = \Delta_o^2(1 - T/T_c) \qquad (7.63)$$

for the B.C.S. singlet state. This equation (7.61) describes a peaked function with the value at T_c deeper than at $T = 0$. Such a dependence is not comparable with the experimental data of the recoilless fraction of ^{57}Fe in $Bi_4Sr_3Ca_3Cu_{3.92}Fe_{0.08}O_{16.36}$. For this purpose, we use a complex or a triplet type gap,

$$\Delta^2 = \Delta_1^2 + \Delta_o(1 - T/T_c) \ , \qquad (7.64)$$

where Δ_1 corresponds to a gap energy corresponding to structural distortion whereas Δ_o corresponds to the superconducting gap. The structural distortions have been detected in experimental measurements by Boolchand et al. The result (7.61) with structural distortions may thus be written in the form,

$$-\frac{1}{3}k_\gamma^2 < u^2 >= -a_1T^2 + b_1T^3\left(1 - \frac{T}{T_c}\right) + b_2T^3 + c \qquad (7.65)$$

where $b_2 = b_o\Delta_1^2$. In Fig. 7.7, we show the predicted value of the logarithm of the recoilless fraction of ^{57}Fe in $Bi_4Sr_3Ca_3Cu_{3.92}Fe_{0.08}O_{16.36}$

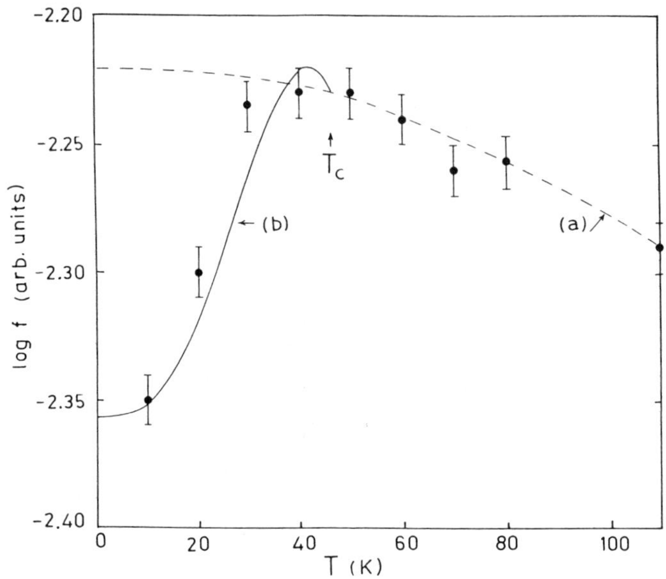

Fig. 7.7: The recoilless fraction of ^{57}Fe *in* $Bi_4Sr_3Ca_3Cu_{3.92}Fe_{0.08}O_{16.36}$ *(a) in the normal state and (b) in the superconducting state. The curves are calculated as given in the text and the dots are the values of the logarithm of the absorption area experimentally measured by Lin and Lin.*

in arbitrary units, for $a_1 = 5 \times 10^{-6}\mathrm{K}^{-2}$, $b_1 = 6.4 \times 10^{-6}\mathrm{K}^{-3}$, $b_2 = 1.4134 \times 10^{-6}\mathrm{K}^{-3}$, $c = -2.357$ and $T_c = 46\mathrm{K}$. The normal state value is also shown. We find that the predicted temperature dependence of the recoilless fraction in the superconducting state is in reasonable agreement with the experimental data only with the complex gap as in (7.64). The resistivity of the sample drops to 50% at 53K and becomes zero at $T_c' = 33\mathrm{K}$. Thus the value of $T \simeq 46\mathrm{K}$ is in accord with the value at which the resistivity is about one half of the normal state value. From the measured value of a_1 and the expression (7.62a) we measure the velocity of sound. Using the known values of the mass desnity, $\rho = 6\ \mathrm{gcm}^{-3}$ and the γ ray wave vector $k_\gamma = 7.29 \times 10^8\ \mathrm{cm}^{-1}$ corresponding to $E_\gamma \simeq 14.4$ keV, we calculate $v = 2.98 \times 10^5\ \mathrm{cms}^{-1}$ for the superconductor. This value of the velocity of sound in a pellet is quite reasonable. Thus the recoilless fraction is used as a method of measuring the sound velocity. The parameters b_1 and b_2 are dependent on a_1 and on the ratio of the gap energy to the matrix element of the electron-phonon interaction. To demonstrate this aspect of the theory we write,

$$b_1 = \frac{12\zeta(3) a_1 \delta_c^2}{\pi^2 T_c} \tag{7.66}$$

where

$$\delta_c = \frac{\Delta_o}{\hbar D_o/(2Mk_B T_c)^{1/2}} \ . \tag{7.67}$$

For the measured values of a_1, b_1, T_c and $\zeta(3) = 1.2$, we calculate $\delta_c = 6.35$ which shows that the strength of the electron-phonon interaction is about one sixth of the gap energy which is quite reasonable for the B.C.S. theory. Since the structureal-distortion temperature is very near the transition temperature, we expect b_1 and b_2 to be of the same order of magnitude as they are. The value of Δ_o derived from the measured value of b_1 is in accord with the B.C.S. relation $2\Delta_o = 3.5k_B T_c$ for the given value of δ_c. In Fig.7.8 we compare the experimental measurements of the recoilless fraction of ^{119}Sn in $EuBa_2Cu_{2.98}Sn_{0.02}O_{7-\delta}$. For $T > T_c$ in the normal

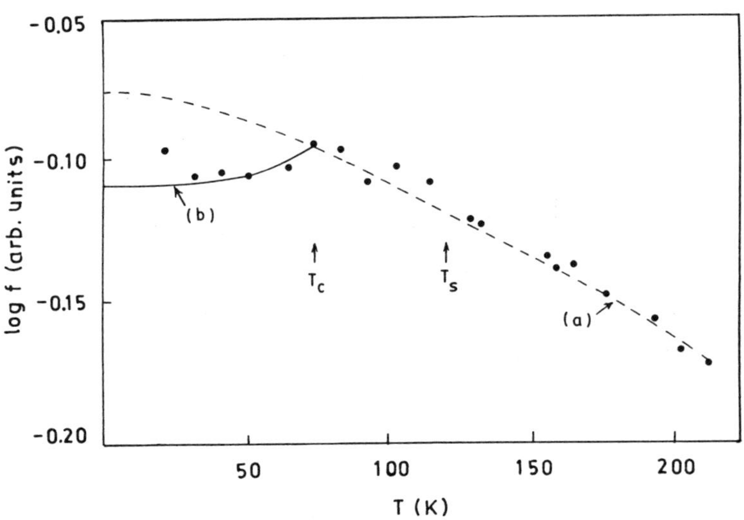

Fig.7.8: The recoilless fraction of ^{119}Sn in $EuBa_2Cu_{2.98}Sn_{0.02}O_{7-\delta}$. For $T < T_c$ the calculated value is drawn by dashed curve along with experimental points. For $T > T_c$ the normal behaviour is seen. There is a structural distortion at T_s. The superconducting region is shown by the calculated curve (b) along with experimental points.

state the theory fits the data well for $\theta_D = 320$K. For $T < T_c$ the agreement between the calculated curve and the measured points is reasonable for $a_1 = 5 \times 10^{-7}\text{K}^{-2}$, $b_1 = 0.1 \times 10^{-7}\text{K}^{-3}$, $b_2 = 0.413 \times 10^{-7}\text{K}^{-3}$, $T_c = 74$K and $c = -0.11$. The value of the transition temperature determined from the resistivity measurements for a slowly cooled smaple is about 85K. Therefore our value of 74K found from the recoilless fraction is in reasonable agreement with the value found from the resistivity in consideration of quenching effects. The wave vector of the γ-photon corresponding to the energy of 23.875 keV is $12.1 \times 10^8\text{cm}^{-1}$. Therefore we predict the velocity of sound to be,

$$v = \left(\frac{k_B^2 k_\gamma^2}{12\rho\hbar a_1}\right)^{1/3} \simeq 9 \times 10^5 \text{ cm s}^{-1} \tag{7.68}$$

for $\rho = 6\ \text{gcm}^{-3}$ this value of the velocity of sound is quite reasonable for a dense material. Thus the recoilless fraction of a Mössbauer line is reduced in going from the normal to the superconducting state. The reduction is related to the gap energy of the electron dispersion relation.

7.6 Second-order Doppler shift

We calculate the shift in the Mössbauer spectra in going from the normal to the superconducting state due to the change in the velocity of an atom owing to the gap in the dispersion relation of the superconductor. This shift is caused by the relativistic second-order Doppler effect because of the change in velocity of an atom in going from the normal to the superconducting state. The resonance energy of the γ-ray is reduced due to the second-order Doppler effect by the amount,

$$\delta_{SOD} = \frac{1}{2}\frac{< v^2 >}{c^2}E_\gamma \tag{7.69}$$

where v is the velocity of the atom which emits the γ-ray. E_γ is the γ-ray energy in the absence of the second-order Doppler effect and c is the velocity of light. Since γ-ray looses energy, the gain in energy of the lattice is given by,

$$\delta E = \delta(p^2/2m) = -\frac{p^2}{2}\frac{\delta m}{m^2} = -\frac{v^2}{2}\delta m = -\frac{v^2}{2c^2}E_\gamma \tag{7.70}$$

where $c^2\delta m = E_\gamma$ is the γ-ray energy. In order to calculate the change in the second-order Doppler shift when the Debye lattice becomes superconducting, we consider the electron-phonon interaction. The electrons in the conduction band interact with the phonons so that the Hamiltonian of the system is given by (7.46) where the coupling constant of the electron-phonon interaction may be written as in (7.57). The electron-phonon interaction upto second order is given by (7.47) with the coupling constant (7.48). In a zero-temperature theory the gap of the system is given by $\Delta_k = \sum'_{k'} V_{kk'} < 0|c^\dagger_{k'}c^\dagger_{-k'}|0 >$ which is a c number because of the vacuum expectation value. The reduced Hamiltonian is then given by (7.50). We take the attractive part of the potential for $E_k - E_{k'} << \hbar\omega_{k-k'}$ so that the average value of (7.50) can be written as

$$< \mathcal{H} >= 2\Delta^2/V_{k,k'} \tag{7.71}$$

due to different permutations of the last term of (7.50). Here $V_{k,k'} = -2D^2/\hbar\omega$ which upon using (7.57) becomes,

$$V_q = -D_o^2/(M\omega_q^2) \tag{7.72}$$

which is the attractive potential between electrons below the transition temperature. Substituting (7.72) in (7.71), we obtain

$$< \mathcal{H}' >= -\frac{2\Delta^2 M\omega_q^2}{D_o^2} \ . \tag{7.73}$$

This is the amount of attractive energy by which the energy of the system is reduced in going from the normal- to the superconducting state. We use the phonon dispersion relation as $\hbar\omega_q = \hbar qv$ where v is the velocity of sound so that the above energy becomes,

$$< \mathcal{H} >= -\frac{2\Delta^2 Mv^2q^2}{D_o^2} \ . \tag{7.74}$$

The contribution of this energy to the velocity of an atom is given by

$$\hbar^{-1}\partial < \mathcal{H}' > /\partial q = -\frac{4\Delta^2 Mv\omega}{\hbar D_o^2} \ . \tag{7.75}$$

The velocity of an atom in a superconductor, v_{SC}, thus becomes,

$$v_{SC} = v_N \left(1 - \frac{4\Delta^2 M\omega}{\hbar D_o^2}\right) \tag{7.76}$$

where v_N is the velocity of an atom in the normal state. The velocity of an atom in a superconductor thus varies as the square of the gap energy. It is of interest to find a numerical estimate of the change in the velocity in going from the normal to the superconducting state whether it is observable. For this purpose numerical values of Δ and $D_q = D_o(\hbar/2M\omega_q)^{1/2}$ are needed. We note that for $T_c \simeq 85\text{K}$ $\Delta = 1.75k_BT_c = 205.4 \times 10^{-16}\,\text{erg}$. The phonon frequency may be estimated from Raman and infrared measurements. Therefore, for the Bi containing superconductors we assume, $\hbar\omega \simeq 480\ \text{cm}^{-1}$, corresponding to Cu-O-Bi bonds and about 500 cm^{-1} for the $YBa_2Cu_3O_{7-\delta}$. The order of magnitude of D_q may be estimated from the attractive interaction, $D_q \simeq 320 \times 10^{-16}$ erg for $\hbar\omega \simeq 500$ cm^{-1} so that $2\Delta^2/(D_o^2\hbar/2M\omega) \simeq 0.82$. Thus the second term on the right hand side of (7.76) is about 82 per cent of the first term so that $v_{SC}/v_N = 0.18$. Therefore, the velocity of an atom in a superconductor is predicted to be much less than in the normal state. This reduction in the velocity being proportional to T_c^2 is much larger for the high-temperature superconductors than for the low-temperature superconductors. In general, the temperature dependence of the gap of the superconductor is a numerically involved problem. However, near the transition temperature, $\frac{1}{2}T_c < T < T_c$, it may be described by the mean field theory, according to which $\Delta = \Delta_o(1 - T/T_c)^{1/2}$ so that the velocity of an atom in a superconductor becomes,

$$v_{SC} = v_N \left[1 - \frac{4M\omega\Delta_o^2}{\hbar D_o^2}\left(1 - \frac{T}{T_c}\right)\right] . \tag{7.77}$$

At $T = 0$, the velocity of an atom is given by (7.76) whereas at $T = T_c$, v_{SC} becomes equal to v_N. The shift of the Mössbauer line in a superconductor is thus found to be

$$\delta E_{sc} = -\frac{E_\gamma v_N^2}{2c^2}\left[1 - \delta_o^2\left(1 - \frac{T}{T_c}\right)\right]^2 \tag{7.78}$$

where

$$\delta_o = \frac{2\Delta_o}{[D_o(\hbar/M\omega)^{1/2}]} \ . \tag{7.79}$$

The value of the parameter δ_o is of the order of the matrix element of the electron-phonon interaction divided by the phonon frequency and hence is always less than unity. The shift as a function of temperature describes slightly parabolic behaviour in the superconducting state owing to the fourth power of gap energy $\delta E_{SC} \propto \Delta^4$. For a reasonably strong electron-phonon interaction, we have plotted the shift, δE_{SC} as a function of temperature in Fig. 7.9 for three different values of $\delta_o = 0.3, 0.4$ and 0.5. For a Debye temperature of $\theta_D \simeq 500$K, the mass of one ^{57}Fe atom as equal to 57 times the mass of the proton, $m_p = 1.6725 \times 10^{-24}$ g, we find that $9k_B\theta_D/16mc \simeq 0.15$ mm/s. The second term of (7.23) is smaller than the first term at all temperatures. At low temperatures, $T < \theta_D/10$, the shift varies as T^4 similar to the lattice energy. At high temperatures $T > \theta_D/2$, the shift varies linearly with temperature. Thus for a given value of $\theta_D = 500$K with m, c and k_B as constants, the δ_{SOD} from (7.23) is calculated to be smaller than 0.3 mm/s at all temperatures. For $T > T_c$, in addition to the second-order Doppler shift, there is some isomer shift. Whereas for $T < T_c$, additional contribution to the shift arises from the gap of the superconductor. In the Mössbauer spectra of ^{57}Fe in $Bi_4Sr_3Cu_{3.92}Fe_{0.08}O_{16.36}$ the shift in the normal state at 156 K is about 0.31 mm/s so that it is mostly of the type of the second-order Doppler shift. As the system is cooled below $T_c \simeq 53K$, it shows superconducting behaviour. The shift has a value of about 0.353 mm/s at 41 K reducing to 0.346 mm/s at 20 K. Thus the change in the shift in going from 41 K to 20 K is only about 0.007 mm/s caused by the superconducting behaviour so that the superconducting contribution to the line shift in going from the transition temperature to zero is 0.017 mm/s.

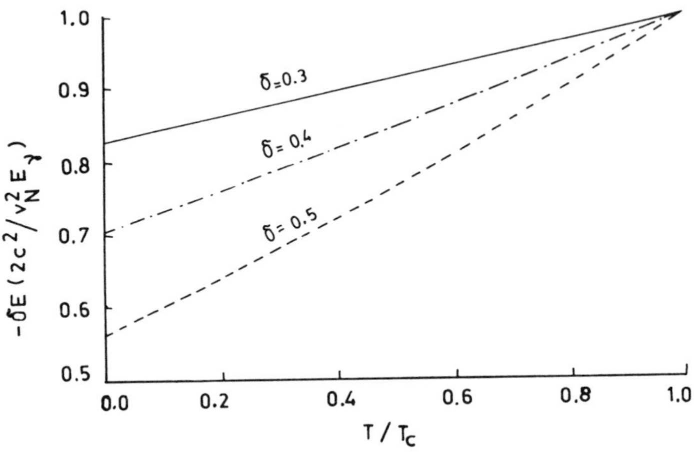

Fig.7.9: The second-order Doppler shift in the Mössbauer effect in a superconductor in units of $2c^2/(v_N^2 E_\gamma)$ as a function of temperature for three different values of the strength of the electron-phonon interaction and constant phonon frequency as for optical phonons from (7.78).

The comparison of the experimental data of ^{57}Fe in $YBa_2(Cu_{0.98}Fe_{0.02})_4O_8$ with a simple phenomenological model incorporating shifts and susceptibilities was found to be sufficient. Further details are needed for the separation of the isomershift and the second-order Doppler shift. In the normal state the first term of (7.23) is 0.03 mm/s for $\theta_D \simeq 100$K. Thus at 300 K, part of the shift $\simeq$0.06 mm/s is due to the SODS and the remaining $\sim$0.06 mm/s may be assigned to the isomer shift so that the total shift comes out to be about 0.12 mm/s. About 0.04 mm/s is due to the superconductivity in the system below T_c. The error bars in these measurements are of about $\pm$0.01 mm/s so that further separation of various contributions is not necessary. It was realized that the deviation between the observed shift, δ_{obs} and the δ_{sod} for $T < T_c$ appears to be a chemically induced effect due to onset of superconductivity. Actually the deviation is caused by the appearance of a gap in the dispersion relation of the superconductor. In $EuBa_2Cu_{2.98}Sn_{0.02}O_{7-\delta}$ there are two types of sites for the ^{119}Sn nuclei. A type in which the nearest neighbour oxygen sites are occupied and A' type in which the nearest neighbour oxygen sites are vacant. At the temperature of 30 K, the isomershift of ^{119}Sn at A site is 0.15 (5) mm/s and at A' site it is 0.28 (5) mm/s. From the expression (7.23) for θ = 320K, the first term is calculated to be about 0.046 mm/s and the second term is also of about the same value. Thus about 0.09 mm/s may be assigned to the second-order Doppler shift and the remaining to the isomer shift. Thus A sites the isomer shift is about 0.06 mm/s and for A' sites, it is 0.19 mm/s. Detailed measurements of the temperature dependence of the shift are not available for any further comparison between the theory and the epxeriments. The temperature dependence of the shift of the centre of the ^{151}Eu Mössbauer nuclei in $EuBa_2Cu_3O_{7-\delta}$ is given by Boolchand et al. The shift at 300 K is 0.825 mm/s, out of which 0.032 mm/s comes from the first term of (7.23) and including the second term of (7.23), $\delta_{sod} \simeq 0.064$ mm/s so that the isomer shift at room temperature is 0.761 mm/s. The superconducting contribution is about 0.025 mm/s. In view of the accuracy of the measurement, ± 0.025 mm/s, further analysis is not necessary. The shift of ^{57}Fe in $YBa_2(Fe_xCu_{1-x})_3O_y$ has been reported for several different values of the concentration of iron atoms at room temperature. For $x = 0.001$, the shift at room temperature is about 0.176 ± 0.006 mm/s. Although the spectra at 4.2 K are given, detailed measurements of the shift as a function of temperature have not been performed. Bottyan et al have measured the shift of 0.16 mm/s at 85 K and 0.07 mm/s at 295 K which shows that there is considerable amount of temperature dependence. Saitovitch et al have measured the linear dependence of the shift on temperature but pronounced effect at the transition temperature has not been seen. Pankhurst et al have suggested the use of relaxation model to understand the Mössbauer spectra at 1.3 K but the detailed temperature dependence of the centre shift has not been published. The Mössbauer spectra of ^{57}Fe in $Tl_2Ca_{1.5}Ba_{1.7}(Cu_{1-x}Fe_x)_3O_9$ have been reported at room temperature as well as at 77 K. The spectra have been analysed by using three doublets. The shift at room temperature is about 0.12 mm/s and 0.16 mm/s at 77 K for one of the three sites. Thus only 0.06 mm/s shift occurs in the range of temperatures studied part of which is due to superconductivity.

7.7 Goldanski-Karyagin effect

The absorption lines of the Moššbauer spectra are usually split by hyperfine interaction or by quadrupole interaction of the electronic field gradient. Usually several lines are found. In the case of ^{57}Fe the two lines split by the quadrupole interaction are often seen. The ratio of intensities of these lines is affected by the anisotropy of the Debye-Waller factor which determines the anisotropy of $< x^2 >$. Due to normal modes, the value of $< x^2 >$ is different in different directions. This observation was first made by Goldanski and Karyagin and hence is named after them with reference to Mössbauer effect in a normal material.

The value of $< x^2 >$ changes in going from the normal- to the superconducting state and hence the ratio of intensities of the Moššbauer lines changes in going from the normal- to the superconducting state. Thus there must be a modification of the Goldanski-Karyagin effect in a superconductor. We define x as the displacement due to a nuclear vibration. The displacement of an atom in a molecule is x_M and x_c in a crystal and $\hat{\imath}$ is the unit vector in the direction of the incident γ ray. The probability of a Mössbauer transition may be written as,

$$\begin{aligned} f' = f'_M f_c &= \exp\left\{-\left(\frac{1}{\lambda^2}\right)\left\langle [\hat{\imath}(x_M + x_c)]^2 \right\rangle\right\} \\ &= \exp\left\{-\left(\frac{1}{\lambda^2}\right) < (ix)^2 > \right\} . \end{aligned} \tag{7.80}$$

The electric field gradient has the symmetry of $3z^2 - r^2$ or Y_2^o, as an example, so that it has a principal axis along the z direction which is also the direction along which Y_2^o has a maximum. The angle which this direction makes with the direction of incident γ-ray is θ. The coordinates are chosen in such a way that they are the same as for the spherical harmonics Y_2^o. Then the angle θ is between $\hat{\imath}$ and $\hat{z}$. The components of $\hat{\imath}$ are as follows:

$$\begin{aligned} i_x &= \sin\theta\cos\varphi \\ i_y &= \sin\theta\sin\varphi \\ i_z &= \cos\theta \end{aligned} \tag{7.81}$$

and

$$(\vec{\imath}.\vec{x})^2 = [\langle x_x^2\rangle\cos^2\varphi + \langle x_y^2\rangle\sin^2\varphi]\sin^2\theta + \langle x_z^2\rangle\cos^2\theta \ . \tag{7.82}$$

In the case of axial symmetry of the Debye-Waller factor

$$\langle x_x^2\rangle = \langle x_y^2\rangle \tag{7.83}$$

so that

$$\langle(\vec{\imath}.\vec{x})^2\rangle = \langle x_\perp^2\rangle + \left[\langle x_\parallel^2\rangle - \langle x_\perp^2\rangle\cos\theta\right] \tag{7.84}$$

because

$$\begin{aligned} \langle x_\perp^2\rangle &= \langle x_x^2\rangle = \langle x_y^2\rangle && (7.85) \\ \langle x_\parallel^2\rangle &= \langle x_z^2\rangle \ . && (7.86) \end{aligned}$$

Substituting (7.84) into (7.80), we can write the recoilless fraction as,

$$f'(\theta) = \exp\left\{-\frac{\langle x_\perp^2\rangle}{\lambda^2}\right\}\exp\left\{-\frac{\langle x_\parallel^2\rangle - \langle x_\perp^2\rangle}{\lambda^2}\cos^2\theta\right\} \ . \tag{7.87}$$

The ratio of the intensities of the two quadrupole split components in the spectra of a polycrystalline sample, for the case of axial symmetry of the electric field gradient of ^{57}Fe is given by,

$$A = \frac{M_\pi}{M_\sigma} = \frac{\displaystyle\int_o^\pi f'(\theta)(1+\cos^2\theta)\sin\theta d\theta}{\displaystyle\int_o^\pi f'(\theta)(\frac{5}{3} - \cos^2\theta)\sin\theta d\theta} \tag{7.88}$$

which we express as,

$$A = \frac{\displaystyle\int_o^\pi \exp\{-(1/\lambda^2)[< x_\parallel^2 > - < x_\perp^2 >]\cos^2\theta\}(1+\cos^2\theta)\sin\theta d\theta}{\displaystyle\int_o^\pi \exp\{-(1/\lambda^2)[< x_\parallel^2 > - < x_\perp^2 >]\cos^2\theta\}(\frac{5}{3} - \cos^2\theta)\sin\theta d\theta} \ . \tag{7.89}$$

When Debye-Waller factor is isotropic, $A = 1$. In general for anisotropic cases, $A \neq 1$. In a polycrystalline sample,

$$f' = exp\left\{-(1/3\lambda^2)[< x_{\parallel}^2 > +2 < x_{\perp}^2 >]\right\} \quad . \tag{7.90}$$

We define,

$$N = (1/\lambda^2)[< x_{\parallel}^2 > - < x_{\perp}^2 >] \tag{7.91a}$$
$$-\ln f' = (1/3\lambda^2)[< x_{\parallel}^2 > +2 < x_{\perp}^2 >] \tag{7.91b}$$

which can be determined from experimental measurements,

$$\begin{aligned} < x_{\perp}^2 > &= 0.5 \times 10^{-18}\text{cm}^2 \\ < x_{\parallel}^2 > &= 1.7 \times 10^{-18}\text{cm}^2 \end{aligned} \tag{7.92}$$

for Sn in SnO. The intensities of the transitions are proportional to,

$$\begin{aligned} M_{\pi}(\theta) &= \text{const.} \left[2\sqrt{5}\bar{P}_o(\theta) + \bar{P}_2(\theta)\right] \\ M_{\sigma}(\theta) &= \text{const.} \left[2\sqrt{5}\bar{P}_o(\theta) - \bar{P}_2(\theta)\right] \end{aligned} \tag{7.93}$$

where $\bar{P}_o(\theta)$ and $\bar{P}_x(\theta)$ are Legendre's polynomials. The fraction f' may also be written as

$$f'(\theta) = \sum_k a_k \bar{P}_k(\theta) \quad . \tag{7.94}$$

The ratio of intensities of the Mössbauer quadrupole splitting components will be

$$A = \frac{1 + (1/2\sqrt{5})(a_2/a_o)}{1 - (1/2\sqrt{5})(a_2/a_o)} \quad . \tag{7.95}$$

According to the Zeeman splitting the ratio of intensities for a $\frac{3}{2} \rightarrow \frac{1}{2}$ transition is expected to be 3:2:1:1:2:3 for six lines. When anisotropy of the vibrations is taken into account it will change to,

$$\begin{aligned} M_{\pm 3/2 \rightarrow \pm 1/2} &= \frac{3(120 + 48a)}{120 + 40a} \\ M_{\pm 1/2 \rightarrow \pm 1/2} &= \frac{2(120 + 24a)}{120 + 40a} \\ M_{\pm 1/2 \rightarrow \mp 1/2} &= \frac{(120 + 48a)}{120 + 40a} \end{aligned} \tag{7.96}$$

where

$$\begin{aligned} a &= -(1/\lambda^2)[< x_{\|}^2 > - < x_{\perp}^2 >] - 1 \\ &\equiv f'_{\|}/f'_{\perp} - 1. \end{aligned} \tag{7.97}$$

In (7.97) the sum of the intensities is normalized over a value characteristic of a polycrystalline isotropic samples in the absence of f' asymmetry in the single crystal depending on the angular momentum change in the transition. The intenstities are affected by the anisotropy of lattice vibrations. Since there is pairing of electrons in superconductors, the intensities are changed in going from normal-to the superconducting state.

Suggested References

7.1. V.I. Goldanski and R.H. Herber, Chemical applications of Mössbauer spectroscopy, Academic Press, New York 1968.

7.2. N.N. Greenwood and T.C. Gibb, Moššbauer spectroscopy, Chapman and Hall, Ltd., London, 1971.

7.3. B.D. Josephson, Phys. Rev. Lett. **4**, 341 (1960).

7.4. R.V. Pound and G.A. Rebka, Phys. Rev. Lett. **4**, 274 (1960).

7.5. K.N. Shrivastava, Hyperfine Interact. **24-26**, 817 (1986).

7.6. M. Cardona, C. Thomsen, R. Liu, H.G. van Schnering, M. Hartweg, Y.F. Yan and Z.X. Zhao, Solid State Commun. **66**, 1225 (1988).

7.7. C. Thomsen, M. Cardona, B. Gegenheimer, R. Liu and A. Simon, Phys. Rev. B**37**, 9860 (1988).

7.8. C.M. Lin and S.T. Lin, J. Phys.: condens Matter **5**, L247 (1993).

7.9. Y. Wu, S. Pradhan and P. Boolchand, Phys. Rev. Lett. **67**, 3184 (1991).

7.10. K.N. Shrivastava, Phys. Rev. B**1**, 955 (1970); **7**, 921 (1973).

7.11. P. Boolchand, R.N. Enzweiler, I. Zitkovsky, J. Wells, W. Bresser, D. McDaniel, R.L. Meng, P.H. Hor, C.W. Chu and C.Y. Huang, Phys. Rev. B**37**, 3766 (1988).

7.12. P. Boolchand, R.N. Enzweiler, I. Zitkovsky, R.L. Meng, P.H. Hor, C.W. Chu and C.Y. Huang, Solid State Commun. **63**, 521 (1987).

7.13. E.B. Saitovitch, I.S. Azevedo, R.B. Scorzelli, H. Saitovitch, S.F. da Cunha, A.P. Guimaraes, P.R. Silva and A.Y. Takeuchi, Phys. Rev. B**37**, 7967 (1988).

7.14. L. Bottyan, B. Molnar, D.L. Nagy, I.S. Szucs, J. Toth, J. Dengler, G. Ritter and J. Schober, Phys. Rev. B**38**, 11373 (1988).

7.15. E.B. Saitovitch, R.B. Scorzelli, I.S. Azevedo and H. Midelitz, Phys. Rev. B**41**, 2103 (1990).

7.16. R.K. Siddique, S. Jha, W.C.H. Joiner and G.M. Julian, J. Phys.: Condens. Matter **6**, 1019 (1994).

7.17. Q.A. Pankhurst, S. Suharan and M.F. Thomas, J.Phys.: Condens. Matter **4**, 3551 (1992).

7.18. Z. Yiwei, Y. Xiahua, L. Mingzhu, R. Yanru, M. Xianren, L. Yongning, T. Qingyun, W. Xiaoshan and S. Lihua, Phys. Rev. B**40**, 727 (1989).

7.19. K.N. Shrivastava, Phys. Rev. B**13**, 2782 (1976).

7.20. K.N. Shrivastava, Phys. Lett. A**70**, 483 (1979).

7.21. C. Bansal and K.N. Shrivastava, Chem. Phys. Lett. **64**, 388 (1979); **80**, 87 (1981); Phys. Stat. Solidi B**109**, 743 (1982).

7.22. R. Jagannathan and K.N. Shrivastava, Hyperfine Interactions **7**, 377 (1979).

7.23. K.N. Shrivastava, J. Phys.: Condens. Matter **6**, 11151 (1994).

7.24. K.N. Shrivastava, M. Reissner and W. Steiner, J. Mag. Mag. Mat. **163**, 150 (1996).

7.25. L.S. Lingam and K.N. Shrivastava, Mod. Phys. Lett. B**10**, 1491 (1996).

7.26. H.K. Perkins and Y. Hazony, Phys. Rev. B**5**, 7 (1972).

7.27. Y. Hazony, Phys. Rev. B**7**, 3309 (1973).

7.28. J.R. Regnard, Phys. Stat. Solidi B**56**, 281 (1973).

7.29. L. Cianchi, Nuovo Cim. Lett. **8**, 492 (1973).

7.30. G.M. Kalvius, J. Phys. (Paris) **35**, C6-139 (1973).

7.31. T. Butz, Phys. Lett. A**50**, 127 (1974).

7.32. B. Brunot, J. Chem. Phys. **61**, 2360 (1974).

7.33. G. Shenoy and Wagner, Isomer shifts in Mošsbauer Spectra, North Holland Pub. Co., Amsterdam 1975.

7.34. S.C. Bharagava, J.E. Knudsen and S. Morup, J. Phys. C**12**, 2879 (1979).

7.35. E. Hartmann, Chem. Phys. **58**, 283 (1981).

7.36. W. Potzel, A. Forster, J. Moser and G.M. Kalvius, Phys. Lett. A**88**, 307 (1982); A**126**, 439 (1988).

7.37. C. Schafer, Phys. Rev. B**37**, 7247 (1988).

7.38. M. Steiner, Phys. Rev. B**41**, 1750 (1990).

7.39. H. Karzel, Hyperfine Intact. **70**, 1067 (1992).

7.40. K. N. Shrivastava, J. Phys. C **19**,L 389 (1986).

7.41. L. S. Lingam and K. N. Shrivastava, J. Phys.: Condens. Matter 7, L 231 (1995).

Chapter 8

Levitation

A pair of a superconductor and a magnet appears to levitate which means that the force between them completely cancells the gravitational force on the magnet as well as on the superconductor. The magnet as well as the superconductor freely float in air and do not fall under gravity. In the actual experiment the superconductor is seen suspended in liquid nitrogen. In the Meissner effect, field of a magnet at the site of the superconductor is zero. However, due to the London penetration depth there is a field at the superconductor which along with the magnet has considerable energy. The distance derivative of this energy produces a force which completely balances the gravitational force and the system levitates. In the type-II superconductors, the field of the magnet can be larger than the lower critical field so that a large magnet and a superconductor can levitate. The force between the superconductor and the magnet can be much larger than the gravitational force so that it is possible to load the magnet. This means that the force between the superconductor and the magnet becomes negative so that they can fly off the container. The load can be so large that the levitating system can be used to make a levitating train or ball bearings can be designed for making a fast rotating system. The force can be plotted to make a microscope.

8.1 Dipole-dipole interaction

We consider a spherical magnet levitating over a superconductor. There are two configurations for the magnetic moment which can be vertical or horizontal. In the vertical configuration, the magnetic moment is normal to the plane of the superconductor and in the horizontal configuration, it is parallel to the plane. The earth's field has been neglected here. Otherwise, the magnet will align along the earth's field which is at an angle from the horizontal axis. The magnetic moment of the magnet is very large such as several hundred gauss compared with the earth's magnetic field which is very small, $\simeq 0.3$ G. Therefore, if some rotations are setup by an external force, the magnet can freely rotate. We assume that the magnet is horizontal, parallel to the plane of the superconductor as shown in Fig. 8.1. The interaction between the pair of dipoles is given by

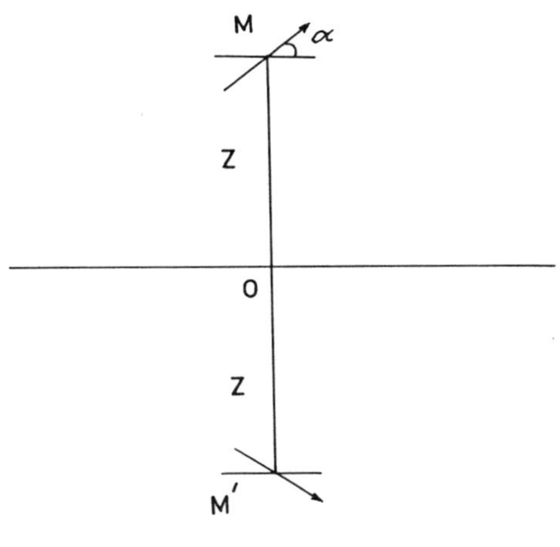

Fig.8.1 A magnetic point dipole with a moment of m placed above a superconducting plane.

$$I_{dd} = \frac{\mu_o}{8\pi}\left[\frac{M'\cdot M}{(2z)^3} - \frac{3(M'\cdot R)(M\cdot R)}{(2z)^5}\right]$$

$$= \frac{\mu_o}{8\pi}\frac{M^2}{(2z)^3}(\cos 2\alpha + 3\sin^2\alpha) \ . \quad (8.1)$$

The kinetic energy is

$$L = \frac{1}{2}m\dot{z}^2 + \frac{1}{2}I\dot{\alpha}^2 \quad (8.2)$$

and the potential energy is

$$V_{eff} = mgz + \frac{\mu_o}{8\pi}\frac{M^2}{8z^3}(\cos 2\alpha + 3\sin^2\alpha) \quad (8.3)$$

where m and I are the mass and moment of inertia of the magnetic sphere, respectively and g is the acceleration due to gravity. We minimize the potential energy as

$$\frac{\partial V}{\partial z} = 0 \ ,$$

and

$$\frac{\partial V}{\partial \alpha} = 0 \quad (8.4)$$

so that

$$mg - \frac{3\mu_o M^2}{(8\pi)8z^4}(\cos 2\alpha + 3\sin^3\alpha) = 0 \quad (8.5)$$

and

$$\frac{d}{d\alpha}(\cos 2\alpha + 3\sin^2\alpha) = 0$$

or

$$\alpha = 0, \quad (8.6)$$

so that,

$$z = \left(\frac{3\mu_o M^2}{64\pi mg}\right)^{1/4} \ . \quad (8.7)$$

The equilibrium distance depends inversely on the one-fourth power of the mass when the magnet levitates over the superconductor. Actually the interaction given by (8.1) treats the problem as if the superconductor and the magnet are separated by a distance. Calculation of the field inside the superconductor may be performed in a more realistic form in which the projection of the field at the site of the superconductor is found properly.

8.2 Magnetic energy

We consider a magnetic point dipole with moment m placed at a distance a above an infinite superconducting plane as shown in Fig. 8.2. From Maxwell and London equations, the vector potential can be written as

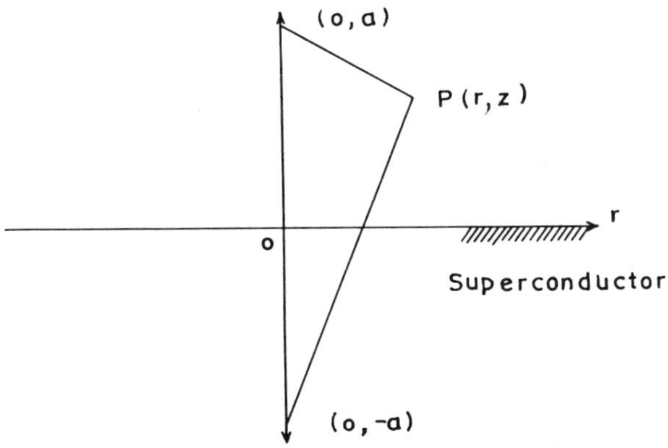

Fig.8.2 A magnetic moment on a conducting plane.

$$\nabla^2 A(x,y,z) = \mu_o m\delta(x-a)[-\hat{\imath}\delta(x)\delta'(y) + \hat{\jmath}\delta'(x)\delta(y)] \qquad (z>0)(8.8)$$

$$\nabla^2 A(x,y,z) - K^2 A(x,y,z) = 0 \qquad (z<0)(8.9)$$

where μ_o is the vacuum permeability,

$$\nabla^2 = \frac{\partial^2}{\partial x^2} + \frac{\partial^2}{\partial y^2} + \frac{\partial^2}{\partial z^2}$$

is the Laplace operator, $m = |\vec{m}|$ and $\hat{\imath}, \hat{\jmath}$ and $\hat{k}$ are the unit vectors along the three axes of the Cartesian coordinates, $\delta(x)$ is the Dirac δ function and $K = 1/\lambda$ and λ is the penetration depth.

A solution of (8.8) is given by,

$$A_1(x,y,z) = \frac{\mu_o m}{4\pi} \frac{-\hat{i}y + \hat{j}x}{[x^2+y^2+(z-a)^2]^{3/2}}. \tag{8.10}$$

The vector potential can be written in the form

$$\begin{aligned} A(x,y,z) &= A_1(x,y,z) + A_2(x,y,z) \quad (z>0) &(8.11a)\\ &= A_3(x,y,z) \quad\quad (z<0) &(8.11b) \end{aligned}$$

where $A_1(x,y,z)$ is the solution of (8.8), A_2 is the induced vector potential due to the superconductor and A_3 is the vector potential in the superconductor due to penetration depth. The magnetic fields can be calculated by taking the derivatives of $A(x,y,z)$ as,

$$B(x,y,z) = \vec{\nabla} \times \vec{A}(x,y,z) \tag{8.12}$$

and hence can be written as

$$\begin{aligned} B(x,y,z) &= B_1(x,y,z) + B_2(x,y,z) \quad z>0 &(8.13a)\\ &= B_3 \quad\quad z<0 &(8.13b) \end{aligned}$$

where B_1 is the direct contribution from the point dipole interaction, B_2 is the induced field due to the presence of the superconductor and B_3 is the magnetic field due to penetration depth inside the superconductor. These fields satisfy the following boundary condition,

$$B_1(x,y,0) + B_2(x,y,0) = B_3(x,y,0) \ . \tag{8.14}$$

By the use of cylindrical coordinates, it is easy to show that A has only the θ component $A_\theta(\rho,z)$ due to the symmetry of the problem. The solution is just written down,

$$A_{1\,\theta}(\rho,z) = \frac{\mu_o m}{4\pi} \frac{\rho}{[\rho^2+(z-a)^2]^{3/2}} \tag{8.15}$$

and $A_{2\theta}(\rho, z)$ and $A_{3\theta}(\rho, z)$ satisfy the following equations

$$\frac{1}{\rho}\frac{\partial}{\partial\rho}\left[\rho\frac{\partial A_{2\theta}}{\partial\rho}\right] + \frac{\partial^2 A_{2\theta}}{\partial z^2} - \frac{1}{\rho^2}A_{2\theta} = 0 \tag{8.16}$$

$$\frac{1}{\rho}\frac{\partial}{\partial\rho}\left[\rho\frac{\partial A_{3\theta}}{\partial\rho}\right] + \frac{\partial^2 A_{3\theta}}{\partial z^2} - \frac{1}{\rho^2}A_{3\theta} - \frac{1}{\lambda^2(T)}A_{3\theta} = 0\,. \tag{8.17}$$

The solutions $A_{2\theta}$ and $A_{3\theta}$ have the general forms,

$$A_{2\theta} = \int_o^\infty dk c_2(k) e^{-kz} J_1(k\rho) \tag{8.18}$$

$$A_{3\theta} = \int_o^\infty dk c_3(k) e^{(\sqrt{k^2+1/\lambda^2})z} J_1(k\rho)\ . \tag{8.19}$$

where $J_n(x)$ is the nth order Bessel function; c_2 and c_3 are determined from the boundary conditions (8.14),

$$c_2(k) = \frac{[1+k^2\lambda^2(T)]^{1/2} - k\lambda}{[1+k^2\lambda^2(T)]^{1/2} + k\lambda} e^{-ka} \tag{8.20}$$

$$c_3(k) = \frac{2k\lambda}{[1+k^2\lambda^2]^{1/2} + k\lambda} e^{-ka} \tag{8.21}$$

The z component of the induced magnetic field is calculated from

$$B_{2z} = \frac{1}{\rho}A_{2\theta} + \partial A_{2\theta}/\partial\rho \tag{8.22}$$

which upon substituting different quantities becomes,

$$B_{2z}(\rho, z) = -\frac{\mu_o m}{4\pi}\int_o^\infty k^2 dk \frac{[1+k^2\lambda^2]^{1/2} - k\lambda}{[1+k^2\lambda^2]^{1/2} + k\lambda}\ e^{-k(z+a)}\ J_o(k\rho)\ . \tag{8.23}$$

The self-interaction energy can be written in the form,

$$U = -\frac{1}{2}mB_2(0, a) \tag{8.24}$$

from which the levitation force acting on the magnetic dipole can be obtained by using the relation,

$$F_d = -\frac{\partial U}{\partial a}\ . \tag{8.25}$$

Making the change of variables and substituing the value of the energy from (8.24), we find,

$$F_d = \frac{\mu_o m^2}{64\pi a^4}\int_o^\infty dt t^3 e^{-t}\left[1 + \frac{1}{2}\left(\frac{\lambda}{a}\right)^2 t^2 - \frac{\lambda t}{a}\left\{1 + 1/4(\lambda/a)^2 t^3\right\}^{1/3}\right]\ . \tag{8.26}$$

This is the force which causes the levitation of the superconductor and the magnet when the magnetic field penetrates into the superconductor according to London penetration depth. The integral is approximately integrated to find,

$$F = \frac{3\mu_o m^2}{32\pi[a+\lambda]^4} \tag{8.27}$$

which shows that the levitation force depends on a^4 for $a/\lambda >> 1$. Expanding the denominator in (8.27) we find,

$$F = \alpha - \beta\frac{\lambda(T)}{a} \tag{8.28}$$

where $\alpha = 3\mu_o m^2/32\pi a^4$ and $\beta = 4\alpha$. The levitation force is thus linearly proportional to the London penetration depth, $\lambda(T)$.

In the type-II superconductors, the field penetrates the superconductor and forms vortices. Therefore, the right hand side of (8.9) is not zero. For the applied field near the H_{c1}, only one flux line exists in the superconductor. The force acting on the magnet contains two parts, one is due to the shielding field caused by the magnet itself and the other is due to the presence of the flux line. We assume that the radius of the vortex is of the order of coherence length, ξ. For a point dipole placed inside the superconductor, the London equation (8.9) is changed to,

$$\nabla^2 B - \frac{1}{\lambda^2}B = -\frac{\phi_o}{\lambda^2}\hat{z}\delta(x)\delta(y) \qquad z < 0 \ . \tag{8.29}$$

Here ϕ_o is the flux quantum. Above the superconductor (8.8) is valid for the type-II superconductors. In order to solve (8.29) and (8.8) under the boundary conditions (8.13) we write B in the form,

$$\begin{aligned} B(r) &= B_1(r) + B_2(r) \qquad && z > 0 \\ &= B_3(r) + B_4(r) && z < 0 \end{aligned} \tag{8.30}$$

where B_1 and B_3 are the solutions of (8.8) and (8.29) as,

$$B_1(r) = \frac{\mu_o m}{4\pi}\left\{\frac{3\rho(z-a)}{[\rho^2+(z-a)^2]^{5/2}}\,\hat{\rho} + \frac{2(z-a)^2-\rho^2}{[\rho^2+(z-a)^2]^{5/2}}\,\hat{z}\right\} \tag{8.31}$$

and

$$B_3(r) = \frac{\phi_o}{2\pi\lambda}\,K_o\left(\frac{\rho}{\lambda}\right)\hat{z} \tag{8.32}$$

where K_o is the zeroth-order Bessel function of the imaginary argument. The general solution of (8.8) and (8.29) is given by

$$B_2(r) = \int_o^\infty kdkD_2(k)e^{-kz}\left[J_1(k\rho)\hat{\rho} + J_o(k\rho)\hat{z}\right] \tag{8.33}$$

$$B_4(r) - \int_o^\infty kdkD_4(k)e^{(k^2+\frac{1}{\lambda^2})^{\frac{1}{2}}z}\left[-\left\{1+\frac{1}{k^2\lambda^2}\right\}^{\frac{1}{2}} J_1(k\rho)\hat{\rho} + J_o(k\rho)\hat{z}\right]. \tag{8.34}$$

From the boundary condition (8.14) we find,

$$D_2(k) = -\frac{1}{2\pi}\frac{k\lambda}{(1+k^2\lambda^2)^{1/2}+k\lambda}\left[\frac{\mu_o m}{2}e^{-ka}\left\{(k^2+1/\lambda^2)^{1/2}-k\right\}\right.$$
$$\left. - \frac{\phi_o}{k\lambda(1+k^2\lambda^2)^{\frac{1}{2}}}\right] \tag{8.35}$$

$$D_4(k) = \frac{1}{2\pi}\frac{k\lambda}{(1+k^2\lambda^2)^{\frac{1}{2}}+k\lambda}\left[\mu_o m e^{-ka}\left\{(k^2+\frac{1}{\lambda^2})^{\frac{1}{2}}-k\right\}\right.$$
$$\left. - \frac{\phi_o}{1+k^2\lambda^2}\right]. \tag{8.36}$$

The levitation force acting on the magnetic dipole is,

$$F = F_1 + F_2 = \frac{1}{2}\frac{\partial}{\partial a}[m.B_2(0,a)] \tag{8.37}$$

where F_1 is same as (8.26) caused by the magnet itself while F_2 is caused by the flux line inside the superconductor.

$$F_2 = -\frac{\phi_o m}{2\pi}\int_o^\infty dk\frac{k^2e^{-ka}}{1+k^2\lambda^2+k\lambda(1+k^2\lambda^2)^{1/2}}. \tag{8.38}$$

In the limit of $a/\lambda >> 1$, this force is also proportional to the London penetration depth. Thus the vortices generate a force which balances the gravitational force.

8.3 Quantized levitation

A spherical magnet levitates on a very small ring of a superconductor at a distance which is quantized due to the flux quantization. A cluster of magnetic atoms levitates over a small ring of a superconductor with several different values of the distance between them. Where there are current loops and weak links in the vicinity of a magnetic atom, the interaction between the magnetic atom and the current loop occurs in such a way that the distance between the two objects is quantized.

We consider a particle of magnetic moment $M = M_o z$, weight W and radius a as shown in Fig. 8.3. the superconducting loop or ring has a radius b and carries an induced current I, and self inductance L. The magnetic penetration depth of the superconductor is λ and the coherence length is ξ which diverges at T_c,

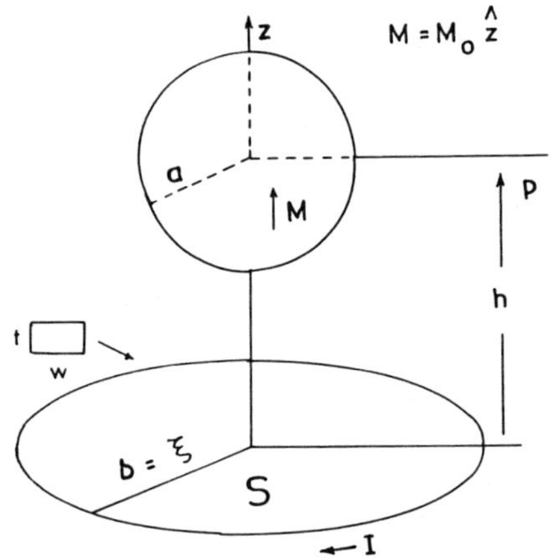

Fig.8.3 Spherical magnet above a very small superconducting ring.

$$\xi = \frac{\xi_o}{\left(1 - \frac{T}{T_c}\right)^{1/2}} \tag{8.39}$$

with the exponent $\nu = 1/2$. When the magnet levitates over the superconductor, the distance, h, between the two is obtained by minimizing the free energy. We suppose that the order parameter of the Ginzburg-Landau free energy is ψ with the square term, $N = |\psi|^2$ with,

$$\psi = \psi/\psi_{bulk} = \sqrt{N}\exp(i\theta) \tag{8.40}$$

and the kinetic energy density $\xi^2|p\psi|^2$. The free energy contains the gravitational potential, $mgh = Wh$, the magnetic field energy $\phi_a I + 0.5LI^2$, with ϕ_a as the applied magnetic flux in the superconducting ring due to M_o. The relevant terms of the free energy are then given by

$$\delta F = \frac{\Lambda}{V_s}\int dv\left\{-N + \frac{1}{2}N^2 + \frac{1}{N}\left[J_\psi^2 + \frac{1}{4}(\xi\nabla N)^2\right]\right\} + \phi_a I + \frac{1}{2}LI^2 + Wh\ , \tag{8.41}$$

where $\Lambda = \mu_o V_s H_c^2$ with V_s the volume of the superconductor and the quantum current density is defined by,

$$J_\psi = \xi Re(\psi^* p\psi) = NQ \tag{8.42}$$

where

$$Q = \xi\left[\nabla\theta + (2\pi/\phi_o)A\right] \tag{8.43}$$

is the normalized superfluid velocity. A contour integral of Q around the superconducting ring gives the fluxoid quantization as,

$$\oint d\vec{l}.\vec{Q} = \frac{2\pi\xi}{\phi_o}(\phi_a - LI + n\phi_o) \tag{8.44}$$

where n in the last term on the right hand side is zero or integer. The force on the magnet is

$$F = \mu_o\int dvj(\hat{\rho}H_z - \hat{z}H_\rho) - \hat{z}W \tag{8.45}$$

where j is the superconducting current density and H_z and H_ρ are the components of the field due to the magnet at the site of the superconducting ring. At the site of the magnet the flux and the components of the field are,

$$\phi_a = \frac{\mu_o M_o}{2b}\frac{1}{(1+x^2)^{3/2}} \tag{8.46}$$

$$H_\rho = -\frac{3M_o}{4\pi b^3}\frac{x}{(1+x^2)^{5/2}} \tag{8.47}$$

$$H_z = \frac{M_o(2x^2-1)}{4\pi b^3(1+x^2)^{5/2}} \tag{8.48}$$

with $x = h/b$. In equilibrium $F = 0$ and $\partial F_z/\partial h = 0$. Minimizing δF with respect to h, we find,

$$N = 1 - Q^2 \tag{8.49}$$

$$NQ = J(1 - \Omega) \tag{8.50}$$

$$\Omega = H_\rho \frac{\partial}{\partial h}\left(\frac{\phi_a}{H_\rho}\right)\left[\frac{\partial}{\partial h}\left(\phi_a + \frac{WL'}{2\pi H_\rho}\right)\right]^{-1} \tag{8.51}$$

where $L' = L/\mu_o b \approx \ln(16b/\omega) - 1.75$. The area of cross section of the wire of the superconducting loop is $s_o = \omega t$ with $t = T/T_c$. The normalized superconducting current density is,

$$J = \lambda I/\sqrt{2} s_o H_c \ . \tag{8.52}$$

Combining (8.49) and (8.51) we find the equilibrium height function, y,

$$y(x) = \frac{J}{J_c}(1 - \Omega) + \cos(Q/J_c) = 0 \tag{8.53}$$

with $J_c = 2/\sqrt{27}$ which is the normalized critical current density of the ring. The remaining variables are given by

$$Q = \frac{\xi}{b\phi_o}\left(\phi_a + \frac{WL'}{2\pi H_\rho} + n\phi_o\right) \tag{8.54}$$

$$J = -\frac{\xi^3 \kappa^2 W}{\phi_o s_o b}\frac{1}{H_\rho} \tag{8.55}$$

where $\kappa = \lambda/\xi$. The function $y(x)$ gives the normalized distance

$$x = h(t, n)/b \tag{8.56}$$

between the magnet and the superconductor. The value $y = x_o$ corresponds to minimum in the free energy. There is a minimum and a maximum value of n for which the root x_o exists, i.e., $n_{min} \leq n \leq n_{max}$. For each n as the temperature t increases x_o disappears at $t = t_{cutoff} \leq t_c$. The equation (8.53) has a discrete set of temperature independent levitation and suspension solutions arising from exact flux quantization, $Q = 0, \Omega = 1$.

8.4 Measurement of force between a magnet and a superconductor

A simple experiment has been setup to measure the weight of a magnet as a superconductor is brought near it from a large distance or taken away from it from a close contact. The initial weight of the magnet made of $Nd_2Fe_{14}B$ is 8.9g. As the superconductor is brought near the magnet from a large distance, the weight is measured as a function of distance between the magnet and the superconductor. As the distance, d, decreases from 0.9 cm to 0.3 cm the weight increases from 8.9 g to 24 g. The increase in weight is qualitatively consistent with a $mg \propto 1/d^n$ type behaviour expected from (8.7) and (8.27), which predict $mg \propto 1/z^4$. As d becomes very small the magnet becomes very heavy.

In another experiment, the starting position is with $d = 0.15cm$. As d is increased, the weight of the magnet reduces till at a particular value of d, the weight becomes zero and at this position the magnet freely levitates in the air and the superconductor in liquid nitrogen. The gravitational force mg is completely compensated by the dipolar force between the magnet and the superconductor. The initial weight of the superconductor made of $YBa_2Cu_3O_7$ is about 10 g. It is kept in a glass beaker containing liquid nitrogen. It is found that not only the magnet but also the superconductor levitates. It does not sit in the bottom of the beaker but it freely suspends itself in the liquid nitrogen. The actual graph of a laboratory measurement of the levitation is shown in Fig. 8.4. It shows the divergence in the weight of the magnet as a the superconductor is brought near the magnet. In another experiment the weight of the magnet is shown to reduce to zero when the starting distance between the magnet and the supercondcutor is very small and is increased in small steps till the magnet levitates. In this experiment the superconductor also levitates. In some other experiments the mass can become negative so that the magnet as well as the superconductor can fly off. Therefore, it is possible to load the system. The freely levitating magnet or a superconductor is subject to any kinetic energy which may be imparted externally. There are viscous forces acting on the superconductor. The levitating system can be made to oscillate by external forces. These oscillations are not connected with the actual force between the magnet and the superconductor. The dynamics of the system may be described by the dimensionless equation, .

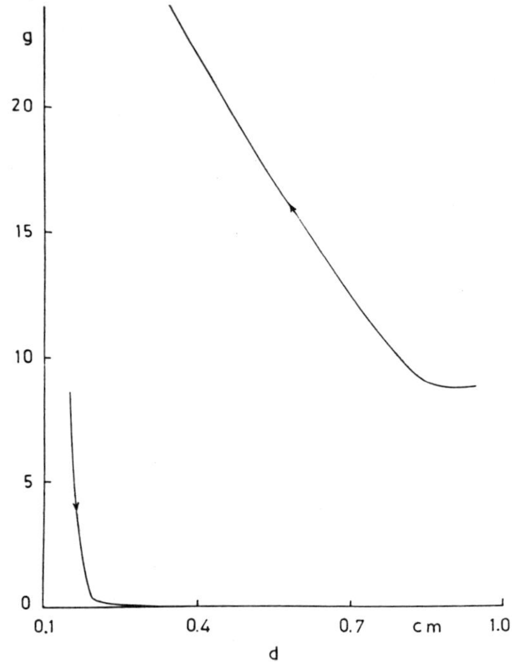

Fig.8.4 Weight of a magnet as a function of its distance from the superconductor. The straight line gives the normal weight. As the distance is reduced starting from a large distance, the weight increase. With increasing distance starting from a small distance, in field cooled samples, free levitation occurs.

$$\ddot{x} + \delta\dot{x} + x - \theta = A_o + A_1 \cos\omega t \tag{8.57}$$

where x is the displacement of the magnet from the surface of the superconductor, θ is the force between the superconductor and the magnet, δ represents a mechanical damping coefficient of the system and γ a relaxation coefficient between the superconductor and the magnet. A_o denotes the initial position of the magnet and A_1 the amplitude of the external force with the driving frequency ω. The shape of the hysteresis depends on the function

$$f(x,\dot{x}) = f_1(x)\{1+f_2(\dot{x})\} \tag{8.58}$$

$$f_1(x) = Fe^{-\beta x} \tag{8.59}$$

$$f_2(\dot{x}) = -\mu_1 - \alpha_1\dot{x} \qquad \epsilon \le \dot{x} \tag{8.60}$$

$$= -\dot{x}(\mu_1+\mu_2)/2\epsilon \qquad -\epsilon \le \dot{x} < \epsilon \tag{8.61}$$

$$= \mu_2 - \alpha_2\dot{x} \qquad \dot{x} < -\epsilon. \tag{8.62}$$

The equation is a modified hypo-elasticity relation which has been used in solid mechanics. The exponential function $f_1(x)$ shows the force and displacement relation without hysteresis which was motivated by the experimental work. F describes the force f_1 at $x = 0$. The integration of the force equation for $\dot{\theta}$ with cyclical variation in the position $x(t)$ and a small value of the velocity $\dot{t}$ yields a quasi-static force displacement relation,

$$\dot{\theta} = -\gamma\{\theta - f(x,\dot{x})\}\ . \tag{8.63}$$

There is a major hysteresis loop in the force versus displacement which drifts due to external vibrations.

A magnet and a superconductor with levitating distance can be used as a sensor in air borne and under sea moving platforms. It can work in place of a quartz transducer to detect shock waves and even forecast earth quakes. The levitation by superconducting magnetic bearings can be utilized in flywheels for energy storage. The high levitation force and low friction can be used to design ball bearings.

Suggested References

8.1 N. M. Krishna, L.S. Lingam and K.N. Shrivastava, Phys. Educ. (India) **12**, 214 (1995).

8.2 Z. J. Yang, J. Superconduc. **5**, 259 (1992).

8.3 T. Hikihara and F.C. Moon, Physica C**250**, 121 (1995).

8.4 J. H. Xu, J.H. Miller, Jr., and C.S. Ting, Phys. Rev. B**51**, 424 (1995).

8.5 M. W. Coffey, Phys. Rev. B**52**, 9851 (1995).

8.6 S. B. Haley, Phys. Rev. Lett. **74**, 3261 (1995).

8.7 S. B. Haley and H. J. Fink, Phys. Rev. B **53**, 3497 (1996).

8.8 S. B. Haley and H. J. Fink, Phys. Rev. B **53**, 3506 (1996).

Chapter 9

Fractals

We introduce the critical exponents and link them with dimensionalities. Several normal state properties such as Debye-Waller factor, Raman scattering of light and spin-lattice relaxation times are calculated. The flux penetration in superconductors, within a certain range of fields, is known to be represented by a fractal dimensionality. The change in the transition temperature upon the application of a magnetic field also depends on the diffusion exponent and hence on the dimensionalities. Within certain range of magnetic fields the superconductor shows a phase boundary which requires the use of fractal dimensions. The need of fractional dimensionality arises for the understanding of critical exponents associated with phase transitions.

9.1 Critical exponents

For a lattice with sites r_i, spins S_{r_i}, and spin components S_{xi}, etc, the susceptibility diverges as,

$$\chi_o = \frac{c}{t^\gamma} \tag{9.1}$$

$$t = \frac{T - T_c}{T_c} \tag{9.2}$$

where the critical exponent γ is observed to have values in the vicinity of 1.36 for ferromagnets like Ni and Fe but near 1.22 for anisotropic paramagnetic materials such as $CrBr_3$. The mean field or classical predictions away from T_c give the Curie-Weiss law which has $\gamma = 1$. The specific heat in zero-field displays a critical anomaly,

$$C_{H=0}(T) = \frac{A}{t^{\alpha}} \tag{9.3}$$

or

$$C_{H=0}(T) = \tilde{A}(t^{-\alpha} - 1)/\alpha \tag{9.4}$$

where $\alpha = 0.1$ for anisotropic magnets and ~ 0 for the λ-point anomaly of ^{4}He. The value of $\alpha = 0$ gives the logarithmic divergence whereas $\alpha < 0$ describes finite cusp at which $\alpha > -1$ and dC_H/dt diverges. The classical theory predicts only a jump discontinuity in the specific heat as a function of temperature. The singular part of the free energy varies as,

$$\begin{aligned} f(T,H) &= -\frac{1}{k_B T} F_{\text{singular}}(T,H) \\ &\approx t^{2-\alpha} Y(H/t^{\Delta}) \end{aligned} \tag{9.5}$$

where the gap exponent Δ is determined in terms of α and by

$$\Delta = \frac{1}{2}(2 - \alpha + \gamma)\ . \tag{9.6}$$

The spontaneous magnetization vanishes as $t \to 0$,

$$M_o(T) \approx B|t|^{\beta} \tag{9.7a}$$

$$\beta = \frac{1}{2}(2 - \alpha - \gamma) \tag{9.7b}$$

where $\beta \simeq 0.31$ for alloys and 0.36 for magnetic materials. The variation of the intensity of scattering of neutrons with wave vector q, near the critical region is given by Fourier transform of the correlation function,

$$G(q,T) = \sum_j \exp(iq.r_j) G(x_j, T) \tag{9.8}$$

where

$$G(x,T) = < S_o.S_x >\ . \tag{9.9}$$

At the critical point,

$$G_c(x) = D_c/x^{d-2+\eta} \qquad (x \to \infty) \tag{9.10}$$

$$G_c(x) = \tilde{D}_c/q^{2-\eta} \qquad (q \to 0) \tag{9.11}$$

where η varies from 0.03 to about 0.1. As

$$G(x,T) \approx x^{-d+2-\eta} D(x/\xi) \tag{9.12}$$

$$\xi \sim t^{-\nu} \tag{9.13}$$

where ν is the critical exponent of the correlation length ξ. The wave vector dependent Green function which describes the scattering diverges as,

$$\hat{G}(q,T) \approx ct^{-\gamma} \hat{D}(q^2/t^{2\nu}) \ . \tag{9.14}$$

The exponent of the correlation length is given by

$$\nu = \frac{\gamma}{2-\eta} \ . \tag{9.15}$$

We define the length, area and volume as $\int dx$, $\int dxdy$ and $\int\int\int dxdydz$, respectively, in Euclidean geometry. Therefore, the spatial dimension d in $\int d^d x$ gives the length, area or volume for $d = 1, 2$ or 3. We may extend the definition of dimensionality to continuous values of d and define a small expansion parameter, ϵ as, for example,

$$\epsilon = 4 - d \ . \tag{9.16}$$

We can also define the number of spin components in the hamiltonian as n. For example, s_z only $n = 1$; for s_x and s_z in the hamiltonian, we have $n = 2$; for s_x, s_y and $s_z, n = 3$; $s_x, s_y, s_z, s_x s_y, n = 4$; for s_x, s_y, s_x^2, s_y^2 and $s_z^2, n = 5$; etc. Therefore n gives the symmetry index of the hamiltonian and $1/n$ may be a small number. We thus have the definition of the critical exponents, $\alpha, \beta, \gamma, \Delta, \eta$ and ν and small parameters ϵ and $1/n$. For $n \to \infty$, we have the spherical model for which

$$\alpha = \frac{4-d}{d-x} \ , \quad \beta = \frac{1}{2} \ , \quad \gamma = \frac{2}{d-2} \tag{9.17}$$

and $\eta = 0$ for $2 < d \leq 4$. For $n \to \infty$, $d \geq 4$, we have $\alpha = 0, \beta = \frac{1}{2}, \gamma = 1$ and $\eta = 0$. The values of the critical point exponents follow the eigenvalues of the linearized renormalization group for,

$$2 - \alpha = d\nu \tag{9.18}$$

where α depends on the dimensionality d and on the symmetry index of the hamiltonian. Otherwise α is independent of the system and hence there is universality of the critical exponents. The values of the critical exponents and the character of the scaling functions do not depend on details of the hamiltonian. The naturally occuring exponents in real materials will occur in three groups, one value for fibers, another for sheet-like systems and yet another for three dimensional solids.

In the percolation problem, if p is the probability of occupation of a site, then minimum value of p for the formation of a connected s-cluster is ϕ_c. The ratio of the number of s-clusters to the total number of lattice sites is n_s, which summed over all s-clusters diverges as

$$\sum_n n_s(p) \sim |p - p_c|^{2-\alpha} \ . \tag{9.19}$$

The correlation length now varies as,

$$\xi(p) \sim |p - p_c|^{-\nu} \ . \tag{9.20}$$

In a d-dimensional system $2 - \alpha = d\nu$ and

$$\sum_p pn_s(p) \sim |p - p_c|^{\beta} \ . \tag{9.21}$$

The other percolation scaling laws are given by

$$2 - \alpha = \gamma + 2\beta = d\nu \tag{9.22}$$

and the effective dimensionality at $p = p_c$ as determined from the density profile is,

$$D = d - \frac{\beta}{\nu} \ . \tag{9.23}$$

Thus the dimensionality is reduced and the exponents are general properties of statistical systems.

9.2 Dimensionality of a fractal

We define a chemical lentgh to grow a fractal on a random lattice using at each time step, the solution of the Laplace equation,

$$\nabla^2\phi = 0 \ . \tag{9.24}$$

The discrete form of the potential on the two-dimensional alttice can be written as,

$$\phi_{ik} = \frac{1}{2}(\phi_{i+1,k} + \phi_{i-1,k} + \phi_{i,k+1} + \phi_{i,k-1}) \ . \tag{9.25}$$

The fractal so generated is shown in Fig.9.1 using small number of steps. Here the number of mass points as a function of distance defines the Hausdorff dimensionality, $N(r) \sim r^D$. The Hausdorff dimensionality is found to be $D = 1.7$ for a computer produced nondiffusing fractal dimensionality,

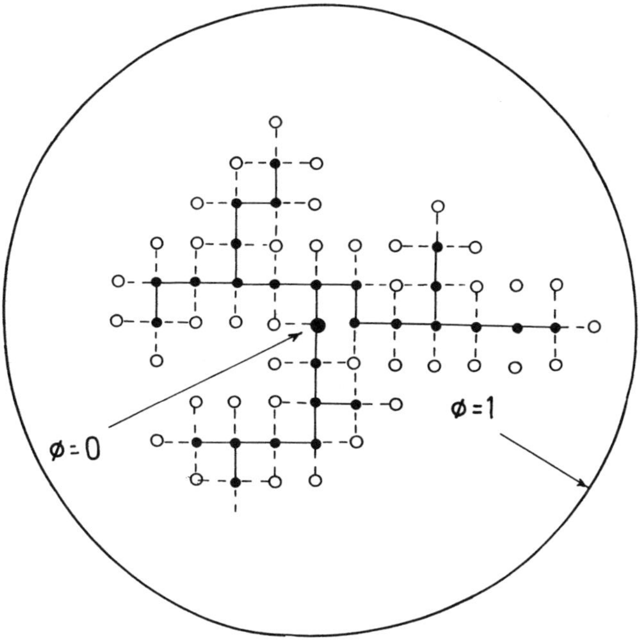

Fig.9.1: The thick lines connecting dots describe the fractal whereas the dashed bonds indicate all the possible growth processes. The probability for each of these processes is proportional to the local field.

$$d_R = \frac{2D}{2+\theta} \tag{9.26}$$

where the diffusion coefficient, $D(r) \simeq r^{-\theta}$. The elastic modulus of a bending mode in related to the correlation length as,

$$K \sim \xi^{-(t/\nu-d+2)} \tag{9.27}$$

where $M \sim \xi^D$, so that the frequency of a normal mode varies as,

$$\omega \sim \xi^{-(t/\nu-t+D+2)/2} \ , \tag{9.28}$$

along with the definition,

$$\theta = \frac{t}{\nu} - d + D = \frac{t-\beta}{\nu} \; . \tag{9.29}$$

The number of phonons as a function of frequency for three-dimensional systems is usually written as proportional to ω^2. However, for a general dimensionality such an expression requires to be generalized. Therefore the effective dimensionality in the density of states may be written as,

$$N(\omega) \sim \omega^{d_{eff}-1} \tag{9.30}$$

where d_{eff} may be the same as d_R given by (9.26). At high frequencies, the compressional modes are important for which $K \sim \xi^{-1}$ so that

$$d_{eff} \simeq \frac{2D}{D+1} \; . \tag{9.31}$$

For $d \geq 6$, the compressional spectral dimension approaches the limit 4/3. The Euclidean limit of D is d. Then (9.31) for $d = 2$ gives $d_{eff} = 4/3$. The density of vibrational states in a fractal is thus given by

$$N_{fr} \sim \omega^{d_R - 1} \tag{9.32}$$

which for $d_R \simeq 1.33$ gives $N_{fr}(\omega) \sim \omega^{0.33}$. This value is to be compared with that of a Debye solid, i.e.,

$$\begin{aligned} N_{ph}(\omega) &\sim \omega^2 && (d=3) \\ &\sim \omega && (d=2) \\ &\sim \omega^o = 1 \; . && (d=1) \; . \end{aligned} \tag{9.33}$$

The density of states for fractals, $N_{fr}(\omega)$, and for phonons, $N_{ph}(\omega)$ appears to cross at a certain frequency ω_{co} which we call as the crossover frequency. For $d = 3$,

$$N(\omega) \simeq 12\pi L^3 \omega^2 / (2\pi)^3 v^3 \tag{9.34a}$$

where L is the length of the lattice of volume $V = L^3$ and v is the sound velocity. For a general dimensionality the phonon density of states is given by

$$N(\omega) = 12\pi L^d \omega^{d-1} / (2\pi)^d v^d \; . \tag{9.34b}$$

When $d = 3$, the above gives the ω^2 dependence. For $d_R = 4/3$ for $d = 3$, $d_R - 1 = 1/3$ so that we find $\omega^{1/3}$ dependence in the density of states which is quite different from that of a Debye solid. We may define a composite dimensionality, $p = d_R d_\phi / D$ with $1 \leq q \leq d$, so that the density of states becomes,

$$N_{fr}(\omega) = 4\pi \left(\frac{L'}{2\pi}\right)^q 3\omega^{q-1}/v^q \tag{9.35}$$

where L' is the length of the fractal and $q = 1$ for $d = 3$. As $N_{ph}(\omega) \sim \omega^2$, $N(\omega)$ crosses $N_{fr}(\omega)$ at ω_{co},

$$N_{ph}(\omega_{co}) = N_{fr}(\omega_{co}) \tag{9.36}$$

so that

$$\omega_{co} = 2\pi v L'^{q/(d-q)} L^{-d'/(d-q)} \ . \tag{9.37}$$

For $L' = L$, we define $\omega_{co} = 2\pi v/L$. The resultant density of states thus starts as ω^2 for $d = 3$. The localization length than becomes,

$$l_{\omega_\alpha} = a(\omega_\alpha/\omega_{fD})^{-d_R/D} \tag{9.38}$$

where ω_{fD} is the fraction Debye frequency,

$$\omega_{fD} = \omega_D (L/a)^{\theta/2} \tag{9.39}$$

where a is the lattice distance. Thus d is the Euclidean dimensionality, D is the fractal dimensionality due to distribution of mass and d_R due to distribution of frequencies in the density of states. Some times in the literature a different notation is used for the same quantities, $\bar{d} = D$ and $\bar{\bar{d}} = d_R$.

9.3 Debye-Waller factor

We calculate the displacement of an atom in a fractal which exhibits a scale invariance for a noninteger fracton dimensionality. The Debye-Waller factor, also called as the Lamb-Mössbauer factor is given by

$$f' = \exp(-4\pi^2 < u^2 > /\lambda^2) \tag{9.40}$$

in which $< u^2 >$ is the mean square displacement of an atom. The above expression can be written as,

$$F = -\ln f' = \frac{4\pi^2}{\lambda^2} < u^2 > \tag{9.41}$$

where λ is the wave length of the γ-ray. The Debye-Waller factor which determines the intensity of the scattered X-ray of wave vector K is given by

$$u = \left(\frac{\hbar}{2M\omega_k}\right)^{1/2} (a_{k,\alpha} + a^{\dagger}{}_{-k,\alpha}) \ . \tag{9.42}$$

The square of the displacement of one atom in a lattice is given by

$$< u^2 >= \sum_k \left(\frac{\hbar}{2M\omega_k}\right) (2n_k + 1) \tag{9.43}$$

where the correlation function is,

$$(a_{k,\alpha} + a^{\dagger}{}_{-k,\alpha})(a_{k',\alpha} + a^{\dagger}{}_{-k',\alpha})\delta_{kk'} = 2n_k + 1 \tag{9.44}$$

and the number density of phonons is,

$$< a^{\dagger}{}_{k\alpha} a_{k\alpha} >= n_k \ . \tag{9.45}$$

Instead of adding contributions from three separate branches, two transverse and one longitudinal, we introduce a factor of 3. For the angular factors we substitute the values $\int d(\cos\theta) = 2$ and $\int d\phi = 2\pi$. The sume over k on the right hand side of (9.43) can be then be written as,

$$\sum_k = \frac{4\pi L^d}{(2\pi)^d} 3k^{d-1} dk \ . \tag{9.46}$$

We define the density of the system as $\rho = M/L^d$ so that,

$$< u^2 >= \frac{3(2\pi)^{1-d}\hbar}{\rho} \int \frac{(2n_k + 1)k^{d-1}dk}{\omega_k} \ . \tag{9.47}$$

In the case of linear dispersion, $\omega = vk$, where v is the sound velocity, we write the above as,

$$< u^2 >= \frac{6\hbar(2\pi)^{1-d}}{\rho v^d} \int (n + \frac{1}{2})\omega^{d-2} d\omega \ . \tag{9.48}$$

We make the change of variables to $x = \hbar\omega/k_BT$ so that

$$< u^2 >= \frac{6\hbar(2\pi)^{1-d}}{\rho v^d}\left(\frac{k_BT}{\hbar}\right)^{d-1}\int x^{d-2}\left(\frac{1}{e^x-1}+\frac{1}{2}\right)dx \tag{9.49}$$

which at $T = 0$ gives the zero-point vibrational contribution of

$$< u^2(T=0) >= \frac{3\hbar(2\pi)^{1-d}}{\rho v^d}\left(\frac{k_B}{\hbar}\right)^{d-1}\frac{\theta^{d-1}}{(d-1)} \tag{9.50}$$

and

$$< u^2 >= u^2(T=0) > +\frac{6\hbar(2\pi)^{1-d}}{\rho v^d}\left(\frac{k_B}{\hbar}\right)^{d-1}T^{d-1}\int_o^{\theta/T}\frac{x^{d-2}dx}{e^x-1} \tag{9.51}$$

which is the low-temperature limit varies as T^{d-1} becoming T^2 for $d = 3$, T for $d = 2$ and a constant for $d = 1$.

We now consider a simple model of a system in which the phonons do not experience the same dimensionality as the mass, which in turn does not experience the Euclidean dimensionality. In a fractal system, the mass is not arranged in a periodic array of atoms so that it experiences the Hausdorff dimensionality, D. The displacement of a mass point may be written as,

$$u = (\hbar/2M\omega_k)^{p/2}a^{1-p}(b_{k,\alpha} + b^{\dagger}{}_{-k,\alpha})\varphi_{k,\alpha}(\omega_{k,\alpha}) \ . \tag{9.52}$$

In this expression p need not equal q which obeys the percolation inequality $1 \leq q \leq d$. However, we may set $p = q$ when necessary. The square of the displacement is then found to be

$$< u^2 >= \sum_k \left(\frac{\hbar}{2M\omega_k}\right)^p a^{2-2p}(2n_{k,\alpha}+1)\varphi^2_{k,\alpha} \ . \tag{9.53}$$

In the first instance, we use the density of states as,

$$\sum_k = 6(2\pi)^{1-d_R}L^{d_R}\omega^{d_R-1}d\omega/v^{d_R} \tag{9.54}$$

where d_R is the Raman (spectral) dimensionality corresponding to the propagation of fractons in a fractal, so that (9.53) becomes

$$< u^2 >= 12(2\pi)^{1-d_R}a^{2-2p}L^{d_R}\left(\frac{\hbar}{2M}\right)^p\int(n_k+\frac{1}{2})\varphi^2_{k,\alpha}\omega^{d_R-p-1}d\omega/v^{d_R} \tag{9.55}$$

in which the zero-point vibrational contribution is,

$$< u^2(T=0) >= \frac{6(2\pi)^{1-d_R} a^{2-2p} L^{d_R} \hbar^p}{(2M)^p} \int \psi^2_{k,\alpha} \frac{\omega^{d_R-p-1} d\omega}{v^{d_R}} \ . \tag{9.56}$$

If the damping factor $\varphi_{k,\alpha}$ and the sound velocity v are independent of frequency, $\varphi_{k,\alpha} \simeq \varphi(o)$ and $v = v(o)$ then the above becomes,

$$< u^2(T=0) >= \frac{6(2\pi)^{1-d_R} a^{2-2p} L^{d_R} \hbar^p \psi^2(o) \omega_{fD}^{d_R-p}}{(2M)^p v^{d_R}(o)(d_R-p-1)} \tag{9.57}$$

where ω_{fD} is the fraction-Debye cutoff frequency. For $p = 1$ the above assumes a particularly simple form. The temperature dependent part of (9.55) after making the change of variables becomes,

$$\begin{aligned} < u^2 > &= < u^2(T=0) > \\ &+ \frac{12(2\pi)^{1-d_R} a^{2-2p} L^{d_R} \hbar^p}{(2M)^p} \int \frac{\varphi^2_{k,\alpha}}{v^{d_R}} \left(\frac{k_B T}{\hbar}\right)^{d_R-p} \frac{x^{d_R-p-1} dx}{(e^\alpha - 1)} \end{aligned} \tag{9.58}$$

which for $p = 1$ has a prefactor of T^{d_R-1} when $\varphi_{k,\alpha}$ is only a slowly varying function of frequency. The form of the decaying function may be chosen to be

$$\varphi_{k,\alpha} = \exp\left[-\frac{1}{2}\left(\frac{L}{l_{\omega_\alpha}}\right)^{d_\phi}\right] \ . \tag{9.59}$$

If $d_R = 4/3$, then the temperature dependent prefactor in (9.58) is predicted to be $T^{0.33}$. This calculated result is of importance for the understanding of width of the x-ray scattering. It also appears in the intensity of the Mössbauer γ-ray emission. The result for the fractals is very different from that of the crystals.

9.4 Light scattering

The exponent of the normalized intensity of the Raman scattered lines is found to vary with fractional dimensionalities, the fracton dimensionality, the range dimensionality and the Hausdorff dimensionality. The calculation is performed for the density of states for decaying as well as nondecaying fractons. It is found that the phonon and the fracton branches, may cross over or avoid crossing giving rise to new modes which we call as the fractons.

The intensity of the Raman line is given by

$$I_{ij}(\omega) = c_{ij}(\omega) g(\omega)(n+1)/\omega \tag{9.60}$$

where the Bose-Einstein factor, $(n+1)$, which occurs in the intensity of the Stokes line becomes n for the anti-Stokes line. Here $g(\omega)$ is the phonon density of states and $c_{ij}(\omega)$ describes the coupling of the vibrational modes to light for a particular polarization,

$$c_{ij}(\omega) \sim \omega^2 \ . \tag{9.61}$$

The phonon density of states is also proportional to the square of the frequency,

$$g(\omega) \sim \omega^2 \ . \tag{9.62}$$

Therefore, the intensity of the scattered light

$$I(\omega) \sim \omega^3 (n+1) \tag{9.63}$$

and hence the normalized intensity varies as,

$$\omega I(\omega) \sim \omega^4 (n+1) \ . \tag{9.64}$$

In this result, Euclidean dimensionality for the isotropic space $d = 3$ has been implied. We write a quantum mechanical hamiltonian applicable for fractional dimensionalities. We first consider the case in which the hamiltonian is independent of the dimensionality of the system in as much as the operator form is concerned. For the interaction linearly dependent on the wave vector, at long wave lengths (9.61) is independent of the dimensionality of the system but (9.62) may be written as,

$$g(\omega) \sim \omega^{d-1} \ . \tag{9.65}$$

For $d = 3$, the Debye ω^2 law may be recovered from (9.65) otherwise d defines the Euclidean dimensionality, $d = 1$ for a chain of atoms and $d = 2$ for a plane. The intensity of the scattered light is then

$$I(\omega) \sim \omega^d (n+1) \tag{9.66}$$

which agrees with (9.64) for $d = 3$. The normalized intensity is given by

$$\omega I(\omega) = \omega^{d+1}(n+1) \tag{9.67}$$

which gives ω^4 dependence for a three dimensional Euclidean solid.

We now assume that the wave vector dependence of the interaction is not linear but depends on an effective dimensionality $q = d_R d_\phi / D$ where the vibrational excitations experience a dimensionality of d_R. These excitations which we call as fractons decay with propgation so that there is a range dimensionality, d_ϕ. For nondecaying fractons, the range is infinity so that they have infinite life time. However, for finite life time of the fractons, we have the range limitation. The equation (9.61) now becomes,

$$c_{ij}(\omega) \simeq \omega^{2q} \tag{9.68}$$

so that (9.60) leads to

$$I(\omega) \sim \omega^{2q-1} g(\omega)(n+1) \ . \tag{9.69}$$

For the fractons, in analogy with (9.65), the density of states is given by

$$g(\omega) \sim \omega^{d_R - 1} \tag{9.70}$$

so that from (9.69) we conclude that,

$$\omega I(\omega)/(n+1) \sim \omega^{2q+d_R-1} \tag{9.71}$$

which gives the exponent as $2q + d_R - 1$. In the case of decaying fractons, the density of states is given by

$$g(\omega) \sim \omega^{q-1} \tag{9.72}$$

so that the scattered intensity from (9.69) and (9.72) becomes,

$$I(\omega) \sim \omega^{3q-2}(n+1) \tag{9.73}$$

and the normalized intensity is found to be

$$\omega I(\omega)/(n+1) \sim \omega^{3q-1} \ . \tag{9.74}$$

The index d_ϕ can be identified with the exponent which gives the bond resistance $R(x)$ between points at an Euclidean distance x from each other,

$$R(x) \sim x_\phi^d \ . \tag{9.75}$$

The range dimensionality is related to other dimensionalities as,

$$d_\phi \sim \frac{D(2-d_R)}{d_R} \ . \tag{9.76}$$

We assume that the derivative of the polarizability of the medium decays with distance in a fractal by the factor,

$$\phi_{k,\alpha}(\omega_{k,\alpha}) = \left(\frac{l_{\omega_\alpha}}{L}\right)^{-D/2} \exp\left[-\frac{1}{2}\left(\frac{L}{l_{\omega_\alpha}}\right)^{d_\phi}\right] \tag{9.77}$$

where l_{ω_α} is the localization length and L is the selfsimilarity length. The localization length is a function of the fracton frequency in the units of the fracton-Debye frequency,

$$l_{\omega_\alpha} = a\left[\frac{\omega_\alpha}{\omega_{fD}}\right]^{-d_R/D} \ . \tag{9.78}$$

Then (9.61) becomes,

$$c_{ij} = \omega^{2+d_R} \exp\left[-\left(\frac{L}{l_{\omega_\alpha}}\right)^{d_\phi}\right] \ . \tag{9.79}$$

The fracton density of states may be taken to be

$$\rho(\omega) \sim \omega^{p-1} \tag{9.80}$$

where the dimensionality p can give either of the values, the Euclidean dimensionality d as in (9.65), the fracton dimensionality d_R as in (9.70) or the dimensionality of the decaying fractons, q as in (9.72) which ever we like, depending upon the material. The intensity of the Raman line is then,

$$I(\omega) \simeq \omega^{p+d_R} \exp\left[-b\omega^2\right](n+1) \tag{9.81}$$

where

$$b = \left(\frac{L}{a}\right)^{d_\phi} \omega_{fD}^{-q} \tag{9.82}$$

and the normalized intensity varies as,

$$\frac{\omega I(\omega)}{(n+1)} = \omega^{p+d_R+1} \exp[-\omega^q] \ . \tag{9.83}$$

For $d_R = 0$ and the Euclidean dimensionality, $p = 3$ the above expression gives ω^4 dependence for the normalized intensity of the scattered light as in (9.64). We consider the wave vector dependence of the form of the change in polarizability,

$$c_{ij}(\omega) \sim \omega^{2q} \phi^2(\omega) \tag{9.84}$$

with the density of states

$$\rho(\omega) \sim \omega^{p-1} \tag{9.85}$$

and

$$\phi^2(\omega) \sim L^D a^{-D} (\omega/\omega_{fD})^{d_R} \exp[-b\omega^q] \tag{9.86}$$

so that the intensity of the Raman line becomes,

$$I(\omega) \simeq c_{ij}(\omega)\rho(\omega)(n+1)/\omega^q \tag{9.87}$$

and hence the normalized intensity

$$\frac{\omega I(\omega)}{n+1} \sim \omega^{q+p+d_R} \exp[-b\omega^q] \tag{9.88}$$

of the Raman line has a dimensionality dependent decay factor, apart from the power law. At low temperatures only the phonons are excited so that the light is scattered from a phonon branch only. However, at elevated temperatures fractons of comparatively large wave vectors are excited, so that the light is scattered from fractons. In anisotropic systems at some characteristic frequency, we expect a cross over from the phonon to the fracton behaviour. The phonon density of states, ω^{d-1}, then becomes $\rho(\omega) \sim \omega^{d-1}\omega_c^{d_R-d}$ which near ω_c has fractal behaviour. The phonon-fracton cross over may be represented as mixing of quasiparticles in the dispersion relation with a resonance. With the arbitrarily small interaction between phonons and fractons the crossing will be avoided as a result of nondiagonal matrix elements of the interaction. Then there must occur quasiparticles which are neither phonons nor fractons but have mixed character. At small wave vectors we expect a phonon density of states, ω^{d-1} which becomes fracton like at large wave vectors, ω^{d_R-1}, and has resonance character near the cross over frequency. We may call this region as the fracton region. Here the density of states may look like $\omega^{\frac{1}{2}(d+d_R)-1}$ or like ω^{f-1} with $f = dd_R/D$. For distances less than a certain length, the fractons are excited and for the larger distances, the phonons are exited. In the intermediate region, we refer to the quasi particles as fracnons.

9.5 Measurements of fractional dimensions

We determine the dimensionality of space corresponding to the scale invariance by calculating the displacement of atoms in a fractal. The calculations are applied to find fractal nature in frozen solutions of deoxygenated myoglobin using the recoilless fraction of $^{57}Fe^{3+}$ Mössbauer spectra. Since very accurate measurements of the root mean square of a distance as a function of temperature are obtained, we are able to get a good value of the dimensionality. The haeme group is a plane with ^{57}Fe atom in the centre so that Euclidean dimensionality is $d = 2$. The principal axis of the electric field gradient is in the haeme plane. Myoglobin contains two haemes within one unit cell. The one being transformed into the other by a rotation of 180° around the crystallographic b axis and translation. The orientation of the electric field gradient in deoxygenated myoglobin has been measured and the root mean square distance from the recoilless fraction as a function of temperature has been determined. It does not follow what is expected from a three dimensional solid. The atoms in a haeme group are arranged at random as compared with a two-dimensional square lattice with fully occupied sites. In a myoglobin structure we count the number of atoms $N(r)$ as a function of distance from a fixed point. It is found that $N(r) \simeq r^D$ with $D \simeq 1.7$. Therefore we assume that the haeme group can be treated as a fractal such that the length of the fractal is smaller than the length of the unit cell. We examine the recoilless fraction of ^{57}Fe in a frozen solution of deoxygenated myoglobin.

The values of $F = -lnf'$ as a function of temperature are available. We plot $\ln F$ as a function of $\ln T$ from the experimental measurements of the ^{57}Fe Mössbauer effect in the frozen solutions of deoxygenated myoglobin, the slope of which gives $T^{0.38}$. Accordingly, we estimate the experimental value of the dimensionality as 1.38. Upon comparing this value with the fracton dimensionality, we estimate, $d_R \simeq 1.38$. The measurement of the fracton dimensionality, $d_R \simeq 1.38$ is in accord with the expression (9.26) as it is smaller than $D = 1.7$ found by counting atoms as a function of distance taking into account only the atoms connected by a chemical length. For $d \geq 6$, the compressional spectral dimension approaches the limit 4/3 which is quite close to our value 1.38. The value of $d_R \simeq 1.38$ found from the recoilless fraction in the deoxygenated myoglobin is thus in accord with the value of the spectral dimension.

The normalized Raman scattering from silica gel, $(C_2H_5O)_4Si$ polymerized from 20% water and 55% ethyl alcohol, has been used to measure the exponent of frequency from (9.71) in the range from 20 to 180 cm^{-1}, which gave,

$$2(d_R d_\phi / D) + d_R - 1 = 1.39 \pm 0.02 \tag{9.89}$$

which gives $d_\phi = 1.49$ in reasonable agreement with $d_\phi = D/1.8 = 1.4$. Therefore, the density of states (9.72) is a closer representation to the density of states in silica gel than (9.70).

9.6 Fractal upper critical field

Some wire networks made from superconducting materials have fractal dimensionalities. Triangular arrays of Sierpinski gasket exhibit both fractal and two dimensional properties. When the probe length, ξ_s is varied from values smaller than the correlation length to larger than correlation length there is cross over from integer dimensions to fractal dimensions. The arrays of wires exhibit cross over phenomena when their length crosses over the correlation length. The percolation net works are formed on a lattice of bonds which are present with probability p and absent with probability $(1 - p)$. We consider the two dimensional superconductors.

The superconducting coherence length may be written as,

$$\xi_s^2(T) = D_1 \tau_{GL}(T) \tag{9.90}$$

where $\tau_{GL}(T) = (\pi\hbar/8k_B)(T_c - T)^{-1}$ is the Ginzburg-Landau life time for Cooper pairs and D_1 is the diffusivity. The perpendicular critical field is quantized as,

$$H_{c2} = \frac{\phi_o}{2\pi\xi_s^2} = \frac{\phi_o}{2\pi D\tau_{GL}} = \frac{4\phi_o k_B(T_c - T)}{\hbar\pi^2 D} \tag{9.91}$$

which is usually the case in two dimensions. In the one-dimensional case, where both the width ω and the thickness, t, of the superconductor are less than ξ_s,

$$\begin{aligned} W &< \xi_s < L \\ t &< \xi_s < L \end{aligned} \tag{9.92}$$

in the expression for H_{c2} in the denominator, ξ_s^2 may be replaced by $\xi_s W/\sqrt{12}$. According to de Almeida-Thouless exponent, the change in T_c upon the application of a magnetic field is given by,

$$T_c - T \sim H^{2/d} \tag{9.93}$$

where $d = 1, 2$ or 3 or a fraction. The homogeneous superconductor is obtained by $\xi_s > \xi_p$ where ξ_p is the correlation length or the size of the fractal region. In this regime, the superconducting order parameter varies only over distances greater than ξ_p and diffusion properties follow the usual laws,

$$\xi_s \propto \frac{1}{(T - T_c)^{1/2}} \; . \tag{9.94}$$

In the case of Sierpinski's gasket, the correlation length is identified with the size of the third-order gasket or $L_3 = 2^3 a = 8a$, while for the percolation model the correlation length is,

$$\xi_p = a(p - p_c)^{-\nu} \tag{9.95}$$

where $\nu = 4/3$ is the correlation length exponent. Below T_c the coherence length decreases as

$$\xi \sim (T_c - T)^{-1/2} \sim (1 - T/T_c)^{-1/2} \tag{9.96}$$

and at sufficiently low temperatures, there is the inhomogeneous region,

$$a < \xi_s(T) < \xi_p \tag{9.97}$$

where diffusion with fractal dimensionalities occurs

$$< x^2 > \sim t^{D/d_R} \sim t^{1/(1+\theta/2)} \tag{9.98}$$

where D is the fractal dimension and d_R is the fracton or spectral (Raman) dimension. The ratio of the two dimensionalities is related to the diffusion exponent θ where $D_1 \sim L^\theta$. In (9.90) the square of the coherence length is shown to be linearly related to the GL impurity scattering life time with the exponent of unity, whereas due to diffusion,

$$\xi_s^2 \simeq \tau_{GL}^{1/(1+\theta/2)} \tag{9.99}$$

which modifies (9.90). Similarly, the (9.93) is changed to

$$T_c - T = H^{1+\theta/2} \quad . \tag{9.100}$$

At very small values of $T_c - T$ and H, the field temperature relationship becomes linear. A cross over is expected at the point,

$$\xi_s(T) \simeq L(H) \simeq \xi_p. \tag{9.101}$$

The value of the diffusion exponent is $\theta = 0.32$ for the Sierpinski gasket and $\theta = 0.8$ for an infinite cluster. The H_{c2} is enhanced as p approaches p_c as in (9.95) where ξ_o varies with varying p. At temperatures where $\xi_s \simeq \xi_p$, in homogeneous regime, the slope of the critical field depends on $\delta p = p - p_c$ as,

$$\left. \frac{d}{dT} H_{c2} \right|_{T=T_c} \simeq \delta p^{-k} \tag{9.102}$$

where $k = \nu\theta$. Taking $\nu = 4/3$ and $\theta \simeq 0.8$ in the percolation limit, we find $k = 1.06$. When $\xi_p > \xi_s$ the upper critical field does not depend on p. We define a parameter A which is proportional to the derivative dH_{c2}/dT,

$$A = \left\{\xi^2(o) T_c/\phi_o\right\} \left[dH_{c2}/dT\right]_{T=T_c}$$

so that

$$H_{c2} = A \left\{\phi_o/\xi^2(o)\right\} \frac{\delta T_c}{T_c} \tag{9.103}$$

for small variation near the transition temperature.

9.7 Fractal flux lattice melting

The penetration of the field in superconductors within a certain range of fields takes place through the growth of magnetic dendrites. The remnant magnetic field for a sample of Nb cooled from above T_c in the presence of a field of 135 Oe penetrates in the form of a fractal. The magnetic field in the dendrites penetrates the sample as vortices and that different field intesities correspond to different vortex densities as in usually the case for type-II superconductors. The dendrites look as though they are filled with a vortex liquid. The fluctuations from the mean field theory in a type-II superconductor have been used to determine the field versus temperature phase boundary which under special circumstances becomes analogous to the de Almeida-Thouless line. We have discussed that quantum effects occur in the flux-lattice melting line. This calculation can be extended to take into account the effect of the fractal dimensionality. The physical properties of fractals have been described by an effective dimensionality, D, which is different from the usual Euclidean dimensionality, d. We have introduced the spectral dimensionality, $d_R(d_R = \bar{d})$, to describe the density of states in a fractal, $N(\omega) \propto \omega^{\bar{d}-1}$ which is similar to that of the phonon density of states proportional to the square of the frequency, $N_{ph}(\omega) \propto \omega^2$ for a three dimensional solid. The phonons in a solid are associated with the mass per atom. Although there is no analog of the mass in the vortex lattice of a type-II superconductor it is said to melt at a finite temperature. Thus there is a phase boundary within which the fractal phase is found.

The dendritic phase of a superconductor may be described by a model in which the superfluid density and the shear modulus vanish at the upper critical field with the exponent represented by a fractal spectral dimensionality. Usually the order parameter has the periodicity of a triangular lattice with the length of the sides of the triangle given by l. The co-ordinates of the three vortex points of the triangle are $(x_1, y_1) = (l, 0)$, $(x_2, y_2) = (\frac{1}{2}, \frac{\sqrt{3}}{2})l$ and $(0, 0)$ so that the area becomes $Q = 2\pi P = x_1 y_2 = l^2 \frac{\sqrt{3}}{2} = \phi_o / B$ where ϕ_o is the unit flux $hc/2e$ and B is the induction field inside the type-II superconductor. The superfluid energy density in the mean field theory is given by

$$\rho_s = \phi_o H^o_{c2} h_{c2} \left(1 - \frac{B}{H_{c2}}\right) \left[4\pi^2 \left\{1 + (2\kappa^2 - 1)\beta\right\}\right]^{-1} \tag{9.104}$$

where $h_{c2} = H_{c2}/H^2_{c2}$ is the upper critical field in dimensionless units with H^o_{c2} the upper critical field at $T = 0$, $\kappa = \lambda/\xi$ is the Ginzburg-Landau parameter and $\beta = 1.16$. Since the superfluid density is the analog of the elastic constants, spectral dimensionality, d_f may be used to describe it for a fractal as,

$$\rho_f = \phi_o H^o_{c2} h_{c2}^{(d_f-1)/2} \left(1 - \frac{B}{H_{c2}}\right)^{(d_f-1)/2} \left[4\pi^2 \left\{1 + (2\kappa^2 - 1)\beta\right\}\right]^{-1} \tag{9.105}$$

which for $d_f = 3$ reduces to (9.104) and the upper critical field, H_{c2}. The shear modulus of the flux lattice is given by

$$c_{66} = 0.48(H^o_{c2})^2 h^2_{c2}(2\kappa^2 - 1) \left(1 - \frac{B}{H_{c2}}\right)^2 \left[8\pi \left\{1 + (2\kappa^2 - 1)\beta\right\}^2\right]^{-1} \tag{9.106}$$

which for a fractal system becomes,

$$c_{66f} = 0.48(H^o_{c2})^2 h_{c2}^{d_f-1}(2\kappa^2 - 1) \left(1 - \frac{B}{H_{c2}}\right)^{d_f-1} \left[8\pi \left\{1 + (2\kappa^2 - 1)\beta\right\}^2\right]^{-1} \tag{9.107}$$

reducing to c_{66} for $d_f = 3$. The fluctuation in the order parameter, ψ of the Ginzburg-Landau phenomenological free-energy density functional is given by

$$D = \left\langle |\psi - \alpha\phi(r|0)|^2 \right\rangle / \alpha^2 \tag{9.108}$$

where ϕ is obtained by minimizing the functional F with respect to ψ. For cylindrical symmetry,

$$D = \int \frac{dq}{2\pi} \int_{Bz} \frac{d^2k}{(2\pi)^2} \frac{k_B T}{(\rho_f q^2 + c_{66} P^2 k^4)} \tag{9.109}$$

where q is the wave vector in the z direction. Using the method of contour integration, the q space integral is solved and only k space is left such that

$$D = \frac{k_B T}{4\pi P(\rho_f c_{66f})^{1/2}} \int \frac{dk}{k}. \tag{9.110}$$

The maximum value of k corresponds to the flux-lattice cell size so that $k_{max} = (2/P)^{1/2}$ whereas the minimum value of k is the inverse length, $1/L$, of the sample, so that,

$$D = \frac{k_B T}{4\pi P(\rho_f c_{66f})^{1/2}} \ln \frac{2^{1/2} L}{P^{1/2}} \ . \tag{9.111}$$

Substituting (9.105) and (9.107) in (9.111) we find,

$$D = \frac{2.05\pi^{1/2} k_B T \ln(\sqrt{2}L/\sqrt{P})\{1 + (2\kappa^2 - 1)\beta\}^{3/2}}{P\phi_o^{1/2}(H_{c2}^o)^{3/2}[h_{c2}(1 - \frac{B}{H_{c2}})]^{3(d_f-1)/4}} \ . \tag{9.112}$$

Since the upper critical field varies with temperature as,

$$H_{c2} = H_{c2}^o \left(1 - \frac{T}{T_c}\right) \tag{9.113}$$

it is found that the fluctuation diverges as,

$$D = \left(1 - \frac{T}{T_c}\right)^{-3(d_f-1)/4} \tag{9.114}$$

for $d_f = 3$, $D \propto (1 - T/Tc)^{-3/2}$. Assuming that there is a range of magnetic fields in which the magnetic field growth inside a superconductor is like a fractal and not like a vortex lattice we expect that the actual spectral dimensionality for the field in a superconductor as in the case of phonons, is about $d_f = 4/3$ so that the deviation (9.114) is expected to become $D \propto (1 - T/T_c)^{-1/4}$ in which the exponent of 1/4 is very different from 3/2. Fractals appearing in different situations clearly can have differnet spectral dimensions. However, the value of 4/3 may describe a resonable approximation. Thus the fluctuations in the dendritic region of the type-II superconductor are reasonably distinguished from those in the region of uniform vortices.

The fluctuation in the distance between two flux lines,

$$d^2(T) = \langle (u_x^2 + u_y^2) \rangle \tag{9.115}$$

determines the stability of the lattice thermal fluctuations. For cylindrical symmetry it can be written as,

$$d^2(T) = P^2 \int \frac{dq}{2\pi} \int \frac{d^2k}{(2\pi^2)} \frac{k_B T k^2}{(\rho_f q^2 + c_{66f} P^2 k^4)} \ . \tag{9.116}$$

In the q-space integral there are two poles out of which only one is inside the contour in the upper half plane so that this integral is solved. Then the k-space integral is solved leading to,

$$d^2(T) = \frac{k_B T}{4\pi(c_{66f}\rho_f)^{1/2}} \ . \tag{9.117}$$

It has been proposed that at a temperature T_{FM} below the transition temperature, the flux lattice melts leading to a flux liquid. We apply the Lindemann criterion of melting, $d^2 = c^2 l^2$ to find the flux-lattice melting line where c is the Lindemann's constant $c \simeq 0.1$ and l is the vortex-vortex separation such that $l^2 B = \phi_o$. Substituting ρ_f and c_{66f} from (9.105) and (9.107) into (9.117) we find that,

$$\frac{c^2\phi_o}{B} = \frac{3.63 k_B T_{FM}\{1 + (2\kappa^2 - 1)\beta\}^{3/2}}{\phi_o^{1/2}(H_{c2}^o)^{3/2}[h_{c2}(1 - \frac{B}{H_{c2}})]^{3(d_f-1)/4}} \tag{9.118}$$

which means that

$$B \propto (1 - T_F M/T_c)^{3(d_f-1)/4} \ . \tag{9.119}$$

Thus for $d_f = 3$, $B^{2/3} \propto (1 - \frac{T}{T_c})$ is predicted which is the analog of the deAlmeida-Thouless line. For a fractal structure, $d_f = 4/3$ so that at such fields near which the dendritic growth is observed, the de Almeida-Thouless line is predicted to be considerably modified to

$$B^4 \propto 1 - \frac{T_F M}{T_c} \ . \tag{9.120}$$

Thus the value of 2/3 for a perfect lattice is changed to 4 for a fractal lattice structure. The melting temperature of the flux-lattice in the fractal superconductor is given by

$$k_B T_{FM} = \frac{c^2\phi_o^{3/2}(H_{c2}^o)^{3/2}[h_{c2}(1 - \frac{B}{H_{c2}})]^{3(d_f-1)/4}}{3.63B\{1 + (2\kappa^2 - 1)\beta\}^{3/2}} \tag{9.121}$$

only as long an $T_{FM} << T_c$ or $h_{c2} \simeq 1$, otherwise, the flux lattice melting occurs at temperature determined by,

$$\frac{T_{FM}}{\left(1 - \frac{T_{FM}}{T_c}\right)^{3(d_f-1)/4}} = \text{constant} \ , \tag{9.122}$$

of which there are multiple solutions. For a given magnetic field several melting temperatures occur. For $d_f = 4/3$ upto four solutions of $T_{FM}^4(1 - \frac{T_{FM}}{T_c})^{-1}$ =constant are predictable so that the phase boundary appears to be diffused.

In the previous discussion we have considered that the wave vector spacc has cylindrical symmetry. However, the boundary of the growth region is a fractal so that the cylindrical symmetry is modified by the fractal (Hausdorff) dimensionality, D. The integral in (9.107) is then modified to,

$$D = \int \frac{dq}{2\pi} \int \frac{y^{D-1} dy k_B T}{(2\pi)^D L^2(\rho_f q^2 + c_{66f} P^2 L^{-4} y^4)} \tag{9.123}$$

where we have made the change to the dimensionless variables $y = kL$ when $k = \frac{1}{L}, y = 1$ and $y_{\max} = k_{\max} L$ with $k_{\max} = (4\pi B/\phi_o)^{1/2}$. When $D = 2$, the equation (9.109) is recovered. The spectral dimensionality is related with the Hausdorff dimensionality by the relation, $d_f = 2D/(2+\theta)$ where θ is the exponent of distance in the diffusion constant, $D(r) \propto r^{-\theta}$. The minimum value of q is zero and the maximum value is $q_m = (2b)^{1/2}/\xi$ where $b = B/H_{c2}$ and ξ_z is the coherence length along the z-direction. For large values of the upper critical field, the superfluid density term in the denominator of (9.123) is much smaller than the second term containing c_{66f} so that (9.123) becomes,

$$D = \frac{(2b)^{1/2} k_B T L^2}{\xi (2\pi)^{D+1} c_{66f} P^2 (D-4)} \left[\left(\frac{4\pi B}{\phi_o} \right)^{(D-4)/2} L^{D-4} - 1 \right] \tag{9.124}$$

so that the fluctuation depends on $B^{(D-4)/2}(1 - \frac{T}{T_c})^{1-d_f}$. Using the Hausdorff dimensionality D, the square of the fluctuation in distance (9.116) between flux lines is given by

$$d^2(T) = \frac{P^2 k_B T (2b)^{1/2}}{(2\pi)^{D+1} L^4 \xi} \int \frac{y^{D+1} dy}{\rho_f q^2 + c_{66f} P^2 y^4 L^{-4}} \, . \tag{9.125}$$

Usually the upper critical field is so large that the second term in the denominator is much larger than the first term. Due to the Lindemann criterion, $d^2 = c^2 l^2 = c^2 \phi_o / B$. Therefore, at the melting temperature,

$$\frac{c^2 \phi_o}{B} = \frac{k_B T \sqrt{2}\, B^{1/2}}{(2\pi)^{D+1} \xi H_{c2}^{1/2} (D-2) c_{66f}} \left[\left(\frac{4\pi B L^2}{\phi_o} \right)^{(D-2)/2} - 1 \right] \tag{9.126}$$

which has one of the terms of the phase boundary described by

$$B^{(D+1)/2} \propto \left(1 - \frac{T}{T_c}\right)^{d_f - \frac{1}{2}} . \tag{9.127}$$

For $D = 2$ and $d_f = 3$ it reduces to $B^{3/5} \propto (1 - \frac{T}{T_c})$. However, for a $3-d$ fractal system we expect $d_f = 4/3$ and $D = 5/3$ so that the phase boundary appears as $B^{8/5} \propto 1 - \frac{T}{T_c}$. The second term on the right hand side of (9.126) bends the phase boundary by $B^{9/5} \propto 1 - \frac{T}{T_c}$. Thus the phase boundary of the dendritic phase is predicted to be considerably shifted from that of a type-II superconductor. Depending on the distortions and size of the system, the morphology of the growth of the dendrite is that of a fractal. Therefore, we conclude the a solid dendrite is grown from a flux liquid phase in which both the liquid as well as the solid phase are superconducting. These fractal phase boundaries in the magnetic field versus temperature diagram are hysteric and show dendritic penetration of magnetic field in a superconductor. The exponents depend on vortex spectral dimensionality as well as on the fractal (Hausdorff) dimensionality. Our theoretical predictions form a model of the experimental observations of dendritic growth of magnetic fields in superconductors.

9.8 Fractal growth of magnetic field

A detailed study of out of equilibrium patterns of flux penetration into thin superconducting niobium films has been reported by Duran et al. Pure niobium thin films grown by d.c. magnetron sputtering in argon from 8-inch diameter target with substrates positioned 230 nm above the target were used. The system was cryopumped to a base pressure of 5×10^{-9} Torr. The evaporations were done with an argon partial pressure of 3.5 Torr with a niobium deposition rate of 1.7 mm/s. These high quality films used to produce excellent strain-free Nb-Al_2O_3-Nb junctions with $T_c(o) = 9.1$K and a critical current of 2×10^6 A/cm^2 at 4.2 K. The remnant magnetic field distribution for a sample which was cooled from above T_c in the presence of a field of 135 Oe resembles a non-uniform distribution. At 7.42 K, the temperature was stabilized and the field was subsequently removed. The observed distribution is qualitatively what would be expected from conventional critical state model. The vortices leave the sample from edges giving a rooflike field distribution whose slope is given by the critical current density. There are some inhomogeneities in the distribution due to the intrinsically random nature of the pinning but in general the gross features agree with one's expectations. A more careful study of the flux front does show evidence of a coarsening of the front as it propagates. At 5.2 K, the pattern is much more complex with a critical state component near the edges of the sample but also with large dendritic structures with significant branching. These structures are not at all predicted in a critical state model and they are characteristic of a fractal model. The dendritic growth of magnetic field penetration in a type-II superconductor is shown in Fig. 9.2. The pattern is obtained experimentally by Duran et al and the fractal phase boundary is explained in ref. 9.7.

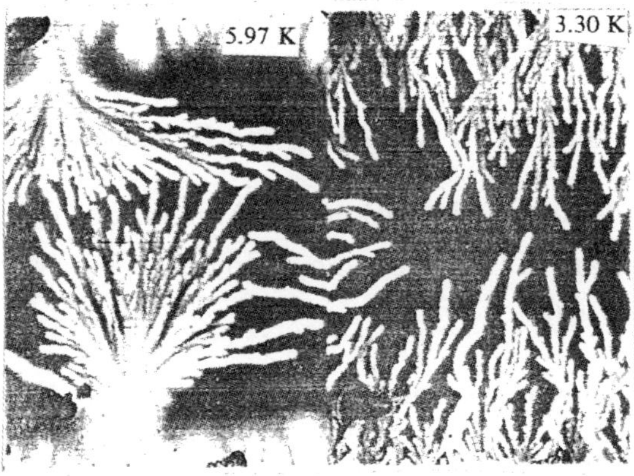

Fig. 9.2: Dendritic growth of magnetic field penetration in a type-II superconductor as experimentally observed by Duran et al and theoretically explained in ref. 9.7.

Suggested References

9.1 K.N. Shrivastava, J. Phys. C**19**, L389 (1986).

9.2 K.N. Shrivastava, Phys. Stat. Solidi B**138**, 699 (1986).

9.3 K.N. Shrivastava, J. Phys. C**20**, L 381 (1987).

9.4 K.N. Shrivastava and S.K. Misra, Solid State Commun.**65**, 597 (1988).

9.5 K.N. Shrivastava, Solid State Commun.**69**, 301 (1989).

9.6 K.N. Shrivastava Phys. State Solidi B**153**, 547 (1989).

9.7 L.S. Lingam and K.N. Shrivastava, Mod. Phys. Lett B**10**, 1447 (1996).

9.8 S.H. Liu, Solid State Physics, eds., H. Ehrenreich, Acad. Press., New York, **39**, 207 (1986).

9.9 J.L. Langer and H. Muller-Krumbhaar, Acta Metall. **26**, 1681, 1689, 1697 (1978).

9.10 C.A. Duran, P.L. Gammel, R.E. Miller and D.J. Bishop, Phys. Rev. B**52**, 75 (1996).

9.11 B. Lehndorff, H.G. Kürschner and B. Lücke, EUCAS 1995.

9.12 N. Vandewallee, M. Ausloos, N. Mineur and R. Cloots, Proc. 10th Anniversary HTSC Workshop, Houston, Tx., World Scientific, Singapore, 1996.

9.13 N. Vandewalle, M. Ausloos and R. Cloots, J. Cryst. Growth 1996.

9.14 H. Frahm, S. Ullah, A.T. Dorsey, Phys. Rev. Lett. **66**, 3067 (1991).

9.15 F. Liu, M. Mondello and N. Goldenfeld, Phys. Rev. Lett. **66**, 3071 (1991).

9.16 R.E. Goldstein, D.P. Jackson and A.T. Dorsey, Phys. Rev. Lett. **76**, 3818 (1996).

9.17 J.M. Gordon, A.M. Goldman and B. Whilehead, Phys. Rev. Lett. **59**, 2311 (1987).

9.18 L. Niemeyer, L. Pietronero and H.J. Weismann, Phys. Rev. Lett. **52**, 1033 (1984).

Chapter 10

Nuclear Magnetic Resonance

10.1 Nuclear magnetic resonance

The nuclear spin in a magnetic field shows energy levels corresponding to the z component of the angular momentum. For example, the nuclear spin of protons in 1/2 which in a magnetic field splits into two levels, one corresponding to $+1/2$ and the other $-1/2$. The energy is given by $-\mu_n.H$ where μ_n is the nuclear magnetic moment, and H is the magnetic field. The hamiltonian of non-interacting nuclear spins in $\mathcal{H} = g_n \mu_N H.I_z$ where I is the nuclear spin and I_z is its z component. The external magnetic field H is also along the z direction. The nuclear mangetic moment μ_N of the proton is inversely proportional to the mass of the proton and hence it is about 10^3 times smaller than the Bohr magneton for the electron, μ_B. Here g_n is the nuclear gyromagnetic ratio, about 1.38 for protons in water. A small modulation field is applied at an angle of 90° to the external magnetic field so that I_x and I_y components are introduced. These spin components excite transitions amongst the energy level corresponding to different values of I_z so that there is absorption of the radio frequency waves at,

$$\hbar\omega = g_n \mu_N H \ . \tag{10.1}$$

This phenomenon is called the nuclear magnetic resonance and in principle, it is analogous to that of the electron spin resonance discussed in Ch.5. There is the Heisenberg exchange interaction between the protons in the organic materials, such as CH_3-CH_2-OH, so that the resonance corresponding to differnetly bonded protons occurs at slightly different frequencies. Thus the protons bonded to CH_3 group can be distinguished from those of CH_2 or OH. In a solid the μ_N can be replaced by the magnetic moment per atom. The electrons of up spin in the 1s orbit interaction with the up spin of the unpaired valence electrons so that the wave function of the $1s \uparrow$ electrons at the site of the proton is different from that of the non-interacting electron. The down spin electron, $1s \downarrow$, also interacts with the up spin unpaired electron but in this case the interaction is different because both the electrons have different spins. Thus the square of the wave function of the $1s \uparrow$ electron evaluated at the site of the proton is not equal to that of the $1s \downarrow$ electron. Thus there is a spin dependent amplitude of the electron wave function which causes a field at the site of the proton. This field is called the hyperfine field, H_{hyp}. There is a finite energy associated with this magnetic field due to finite magnetic moment of the proton, μ_n,

$$\mathcal{H} = -\mu_n . H_{hyp} \quad . \tag{10.2}$$

The nuclear magnetic moment μ_n in terms of nuclear spin, I, is

$$\mu_n = g_n \mu_N I \tag{10.3}$$

where g_n is the nuclear gyromagnetic ratio and μ_N is the nuclear magneton. The hamiltonian of the hyperfine interaction then becomes,

$$\mathcal{H} = -g_n \mu_N I . H_{hyp} = AI.S \tag{10.4}$$

where s is the atomic spin so that

$$A = -g_N \mu_N \frac{I.H_{hyp}}{I.S.} \quad . \tag{10.5}$$

The field of the electrons at the site of the nucleus is thus,

$$H_{hyp} = \frac{AS_z}{g_N \mu_N} \quad . \tag{10.6}$$

Assuming that the wave function of the electron is $\psi(r)$, the isotropic part of the hyperfine constant can be written as

$$A = -\frac{8\pi}{3} g g_N \mu \mu_N |\psi(o)|^2 \tag{10.7}$$

where g is the Lande's splitting factor for the electron, g_N the nuclear gyromagnetic ratio for the proton, μ the elctronic magnetic moment and μ_N that of the proton. Since the amplitude of the $s \uparrow$ electrons at the site of the nucleus is not equal to those of $s \downarrow$, we can define the atomic susceptibility as,

$$\chi_a = \frac{4\pi}{S} \sum \left[|\psi^{+}_{ns}(o)|^2 - |\psi^{-}_{ns}(o)|^2 \right] \ . \tag{10.8}$$

The hamiltonian for the nuclear spin resonance along with the hyperfine interaction is thus

$$\mathcal{H}_{nmr} = g_n \mu_N H.I + AI.S \tag{10.9}$$

where $g\mu_B H.S$ term has not been added because it is far away from the nuclear magnetic resonance line. Thus we have the NMR line split by the hyperfine interaction.

10.2 Knight shift

Due to the hyperfine interaction there is a field at the site of the nucleus. Therefore, there is a small shift in the frequency of the nuclear magnetic resonance. From the hamiltonian appropriate to the nuclear spins, we can write,

$$\mathcal{H} = \left[\sum_i g^i_n \mu_N H_z + \sum_i A < s^i_z > \right] I^i_z \tag{10.10}$$

since the hyperfine interaction can be written as a hyperfine field, using (10.6) we write the above expression as,

$$\mathcal{H} = \left[\sum_i g^i_n \mu_N H_z + \sum_i < s^i_z > g_n \mu_N H_{hyp} / < S_z > \right] I_z \tag{10.11}$$

which gives the shifted field,

$$\mathcal{H} = \sum_i g_n \mu_N \left[H_z + H_{hyp} \right] I_z \ . \tag{10.12}$$

Therefore, the line shift is

$$\Delta H = \sum_i g_n \mu_N H_{hyp} = \sum_i A < S^i_z > \tag{10.13}$$

between two levels corresponding to $I_z = +1/2$ and $-1/2$. In a magnetic material

$$\sum_i < S_z^i >= NS_z = M \ , \tag{10.14}$$

where N is the number of magnetic atoms so that M is the magnetization. The parametric susceptibility is defined as,

$$\chi = \frac{M}{H} \ . \tag{10.15}$$

Therefore, the line shift (10.13) can be written as

$$\Delta H = A\chi H \tag{10.16}$$

or the relative shift becomes

$$\frac{\Delta H}{H} = A\chi \tag{10.17}$$

which is proportional to the susceptibility. It was found by Walter Knight that the nuclear magnetic resonance lines in the metals are shifted with respect to their position in an insulator. Therefore, the NMR line shift in a metal is called the Knight shift. In the metals we have the electronic wave function appropriate to that of the conduction band whereas in the insulators the conduction band is fully occupied so we do not have the concept of Fermi surface. The Knight shift therfore determines the wave function at the Fermi surface and hence plays an important role.

10.3 Nuclear relaxation

The nuclear spin states relax to lower energy states by making contact with the environment via the hyperfine interaction. In magnetically ordered materials, the nuclear spin operators can be represented by nuclear spin waves which are bosons and the electronic spin operators can also be expanded in the form of spin waves. The number density of excited bosons at low temperatures is small so that at very low temperatures another mechanism may be needed. At high temperatures phonons play an important role. Therefore three types of nuclear relaxation mechanisms are commonly found, the radiative relaxation, the relaxation via magnons and the relaxation via phonons. Near a phase transition, the short range forces become important and hence the exchange interaction depends on the wave vector. The wave vector dependent susceptibility plays the role of a response function and hence the relaxation rate may depend on the wave vector dependent susceptibility.

(a) Radiative relaxation

The power radiated classically by a precessing dipole moment, $\mu_\perp$, is

$$P = \frac{2}{3}ck^4\mu_\perp^2 \tag{10.18}$$

whereas the power absorbed is

$$P_a = \frac{1}{2}\omega\chi''Vh_1^2 \tag{10.19}$$

where k is the radiation wave vector, c is the velocity of hight and ω its frequency, V is the volume and h_1 is the amplitude of the magnetic vector field of the electromagnetic wave and χ'' is the susceptibility function,

$$\chi'' = \gamma^2 M\Delta H/[(\omega - \omega_o)^2 + (\Delta H)^2] \tag{10.20}$$

which at resonance, $\omega = \omega_o$, becomes,

$$\chi'' = M/\Delta H \ . \tag{10.21}$$

Here ΔH is the line width at the resonance. We define the dynamic susceptibility as,

$$\chi'' = \mu_\perp/Vh_1 \tag{10.22}$$

so that from (10.21) and (10.22),

$$\mu_\perp = Vh_1M/\Delta H \ . \tag{10.23}$$

If all the power is rediated at resonance, $P_r = P_a$, which solved for the width gives

$$\Delta H = \frac{4}{3}VM\omega^3/c^3 \ . \tag{10.24}$$

Thus the classical radiation emission from a precessing dipole in nuclear magnetic resonance leads to a damping constant,

$$\tau = (2\pi\gamma\eta MQ)^{-1} \tag{10.25}$$

where $\gamma = g_N\mu_N/\hbar$, η is a filling factor and Q the quality factor of the coil. For proton NMR in 1 g of water, the proton life time from this process is of the order of 60 s. The interaction of nuclear spins with the electromangetic field as appropriate to a nuclear magnetic resonance experiment can be written as,

$$\mathcal{H} = \mathcal{H}_o + \mathcal{H}' \tag{10.26}$$

$$\mathcal{H}_o = \sum_i g_n\mu_N H.I_z^i + \sum_q \hbar\omega_q b_q^\dagger b_q \tag{10.27}$$

$$\mathcal{H}' = \sum_{iq} A_{iq}(I_i^+ + I_i^-)(b_q - b_{-q}^\dagger) \tag{10.28}$$

with

$$A_{iq} = \frac{1}{2}\hbar\gamma(2\pi\hbar\omega_q/V)^{1/2}\exp(iq.\gamma_i) \tag{10.29}$$

and $b_q, b^\dagger{}_{-q}$ are r.f. photon annihilation and creation operators. From the above interaction of nuclear spins with photons, we calculate the relaxation time as,

$$\frac{1}{\tau_1} = \frac{\gamma^2\hbar\omega_o^3 N}{2c^3}(2n_o + 1)[I(I+1) - m(m-1)] \tag{10.30}$$

which is the quantum mechanical version of the radiative life time of a nuclear spin state. The relaxation rate is proportional to the number of atoms N and to the r.f. power through n_o, the number density of r.f. photons, and varies as the cube of the resonance frequency, ω_o^3. Under normal working conditions such as the quantity of the salt required, it is of the order of 100 s.

(b) Magnon scattering

Usually the magnon energies are much larger than the nuclear spin splitting energy so that direct nuclear relaxation by magnon emission or absorption is not possible. Due to the boson nature of the magnon statistics, magnons of small number density may be available but it will cause long relaxation times. The hamiltonian of unperturbed nuclear spins and magnons is,

$$\mathcal{H}_o = \sum_i g_n \mu_N H.I_z + \sum_k \hbar\omega_k b^\dagger{}_k b_k \ . \tag{10.31}$$

The hyperfine interaction,

$$\mathcal{H}' = \sum_j A I_j . S_j \tag{10.32}$$

can also be expanded using the Holstein-Primakopf transformation which converts spin operators into site variables which create or destroy bosons at the particular sites as,

$$\begin{aligned} S_i^+ &= (2S)^{1/2} f_i a_i \\ S_j^+ &= (2S)^{1/2} b^\dagger{}_j f_j \\ S_i^- &= (2S)^{1/2} a^\dagger{}_i f_i \\ S_j^- &= (2S)^{1/2} f_j a_j \\ S_{zi} &= S - a^\dagger{}_i a_i \\ S_{zj} &= -S + b^\dagger{}_j b_j \\ f_i &= (1 - a^\dagger{}_i a_i / 2S)^{1/2} \ . \end{aligned} \tag{10.33}$$

Here S_{zi} is directed in the positive z direction while S_{zj} is along the negative z direction as in a two sublattice model of an antiferromagnet. In the ferromagnets there is only one lattice so that sublattices are not necessary. The site variables are changed to Fourier transforms,

$$\begin{aligned} a_i &= \frac{1}{\sqrt{N}} \sum_k a_k \exp(i\vec{k}.\vec{r}_i) \\ b_j &= \frac{1}{\sqrt{N}} \sum_k b_k \exp(-i\vec{k}.\vec{r}_j) \end{aligned} \tag{10.34}$$

and their hermitian conjugates. A Bogoliubov transformations

$$\begin{aligned} c_k &= u_k \alpha_k + v_k \beta_k^\dagger \\ d_k &= u_k \beta_k + v_k \alpha_k^\dagger \end{aligned} \tag{10.35}$$

absorbs part of the potential into the unperturbed hamiltonian so that the resulting potential is small compared with the transformed unperturbed hamiltonian. Two relations are generated to determine u_k and v_k which are some times called the coherence factors. One of the relations is obtained by demanding that $c_k(c_k^\dagger)$ are bosons obeying commutators and the other relation is obtained by setting the coefficient of two operator terms such as $c_k^\dagger c_{k'}^\dagger$ to zero. The terms in the hamiltonian are the exchange energy $\hbar\omega_e = g\mu_B H_e - 2JzS$, the anisotropy energy, $\hbar\omega_A = g\mu_B H_A$ and the hyperfine interaction

$$AI_j.S_j = \frac{1}{2}A\left(I_j^+S_j^- + I_j^-S_j^+ + 2I_j^zS_j^z\right) \tag{10.36}$$

which can be written in terms of components of the nuclear spin

$$I_j^\pm = I_{jx} \pm iI_{jy} \tag{10.37}$$

and those of the electronic spin,

$$\begin{aligned} S_j^+ &= S_{jx} + iS_{jy} = \left(\frac{2S}{N}\right)^{1/2}\left\{\sum_k \exp(-ik.r_j)b_k \right. \\ &\quad \left. -\frac{1}{4SN}\sum_{k,k',k''}\exp\left[i(k-k'-k'').r_j\right]b_k^\dagger b_{k'}b_{k''} + \cdots\right\} \\ S_j^- &= S_{jx} + iS_{jy} = \left(\frac{2S}{N}\right)^{1/2}\left\{\sum_k \exp(-ik.r_j)b_k^\dagger \right. \\ &\quad \left. -\frac{1}{4SN}\sum_{k,k',k''}\exp\left[i(k+k'-k'').r_j\right]b_k^\dagger b_{k'}^\dagger b_{k''} + \cdots\right\} \\ S_{jz} &= S - \frac{1}{N}\sum_{k,k'}\exp[i(k-k').r_j]b_k^\dagger b_{k'}\ . \end{aligned} \tag{10.38}$$

Here N is the number of atoms in the solid, r_j is the position coordinate of the j-th atom of spin S_j and $b_k^\dagger$ and b_k are the boson creation and annihilation operators of spin waves of wave vector k. After substituting the transformation we find,

$$\begin{aligned} S_i^+ &= \left(\frac{2S}{N}\right)^{1/2}\left\{\sum_k \exp(-ik.r_i)c_k + \cdots\right\} \\ S_j^- &= \left(\frac{2S}{N}\right)^{1/2}\left\{\sum_k \exp(ik.r_i)c_k^\dagger + \cdots\right\} \end{aligned}$$

$$\begin{aligned}
S_i^z &= S - N^{-1}\sum_{k,k'}\exp[i(k-k').r_i]c_k^\dagger c_{k'} \\
S_j^+ &= \left(\frac{2S}{N}\right)^{1/2}\left\{\sum_k \exp(-ik.r_j)d_k^\dagger + \cdots\right\} \\
S_j^- &= \left(\frac{2S}{N}\right)^{1/2}\left\{\sum_k \exp(ik.r_j)d_k + \cdots\right\} \\
S_j^z &= -S + \frac{1}{N}\sum_{k,k'}\exp[-i(k-k').r_j]d_k^\dagger d_{k'} \ . \qquad (10.39)
\end{aligned}$$

The coherence factors are found from,

$$2u_k v_k = -(\omega_e \gamma_k/\omega_k) \qquad (10.40)$$

$$u_k^2 + v_k^2 = (\omega_e + \omega_A)/\omega_k \qquad (10.41)$$

with

$$\omega_k = [2\omega_e\omega_A + \omega_e^2 b^2 k^2 a^2] \ . \qquad (10.42)$$

Here a is the distance between atoms, z is the number of nearest neighbours and $b^2 = 2/z$. The nuclear relaxation time can be calculated by using the Fermi's golden rule. There are two probabilities, one for $|a>=|-1/2>$ to $|b>=|+1/2>$ and the other for $|+1/2>$ to $|-1/2>$. Due to quantum mechanics, one depends on n_k and the other on n_k+1. The sum of the two probabilities, $P_{ab}+P_{ba}=1/\tau$, determines the relaxation rate. the golden rule is,

$$P_{fi} = \frac{2\pi}{\hbar}\sum_j |<f|\mathcal{H}'|i>|^2\rho_f \qquad (10.43)$$

where the density of final states, $\rho_f = \delta(E_i - D_f)$, ensures that the energy is conserved. The matrix element between initial and final states is determined by the interaction. In the present case, the interaction of nuclear spins with magnons can be written as,

$$\mathcal{H}' = \frac{1}{2}AI^+\left(\frac{2S}{N}\right)^{1/2}\sum_k b^\dagger{}_k \ . \qquad (10.44)$$

The relaxation times from this interaction are very long so that this interaction does not form an effective process of relaxation. If the hyperfine interaction is anisotropic we can obtain,

$$\mathcal{H}' = -\frac{A}{2N} I^+ \sin\theta \sum_{k,k'} b^\dagger{}_k b_{k'} \ . \tag{10.45}$$

The difference between the energies of two magnons, ω_k and $\omega_{k'}$ can be equal to the hyperfine energy required for a transition induced by I^+. Therefore, the transition probabilities from this interaction can provide an effective relaxation mechanism,

$$\frac{1}{T_1} = \frac{4\pi}{\hbar}\left(\frac{A}{2N}\right)^2 \sin^2\theta \sum_{k,k'} n_{k'}(1-n_k)\delta(\hbar\omega_k - \hbar\omega_{k'} - AS) \tag{10.46}$$

where the Bose distribution for magnons is given by,

$$< n_k >= [\exp(\hbar\omega_k/k_BT) - 1]^{-1} \ . \tag{10.47}$$

The change from k space to frequency space is affected by the dispersion relation,

$$\hbar\omega_k = g\mu_B H + 2JSk^2a^2 \tag{10.48}$$

which shows that frequency of the antiferromagnetic magnon is proportional to the square of the wave vector. Integrating the wave vector space by multiplying by density of states and removing the sums, we find,

$$\begin{aligned}\frac{1}{T_1} &= \frac{4\pi}{\hbar}\left(\frac{A}{2N}\right)^2 \sin^2\theta \frac{V^2}{(2\pi)^6}(4\pi)^2 \\ &\times \int_o^{k_{max}} \int_o^{k'_{max}} \frac{k^2dk k'^2 dk' \exp(\hbar\omega_k/k_BT)\delta(\hbar\omega_k - \hbar\omega_{k'})}{[\exp(\hbar\omega_k/k_BT) - 1]^2} \ .\end{aligned} \tag{10.49}$$

For $k_BT << 2JS$

$$\frac{1}{T_1} = \frac{A^2\sin^2\theta}{\hbar 2(2\pi)^3(\hbar\omega_e)^3} \int_{g\mu_B H}^{\infty} \frac{dE(E - g\mu_B H)\exp(E/k_BT)}{[exp(E/k_BT) - 1]^2} \ . \tag{10.50}$$

We make the change of variables,

$$\hbar\omega_e = 2JS, \quad x = E/k_BT \quad \text{and} \quad x_o = g\mu_B H/k_BT \ ,$$

so that

$$\frac{1}{T_1} = \frac{A^2}{\hbar^2\omega_e^2}\frac{\sin^2\theta}{2(2\pi)^3}\left(\frac{k_BT}{\hbar\omega_e}\right)^2\int_{x_o}^{\infty}\frac{dx}{e^x-1} \tag{10.51}$$

which for $g\mu_BH << k_BT$ becomes,

$$\frac{1}{T_1} = \frac{A^2}{\hbar^2\omega_e}\frac{\sin^2\theta}{2(2\pi)^3}\left(\frac{k_BT}{\hbar\omega_e}\right)^2\ln\frac{k_BT}{g\mu_BH}\ . \tag{10.52}$$

For two sublattice antiferromagnets,

$$\frac{1}{T_1} = \frac{8A^2\sin^2\theta}{\hbar^2\omega_e(2\pi)^3b^6}\left(\frac{k_BT}{\hbar\omega_e}\right)^2\int_{T_{AE}/k_BT}^{\infty}\frac{xdx}{e^x-1} \tag{10.53}$$

which determines the temperature dependence of the nuclear relaxation rate. Here $k_BT_{AE} = \hbar\omega_{AE} = g\mu_B(2H_eH_A)^{1/2}$ with $\hbar\omega_e = 2JSz$ gives the anisotropy energy in terms of an average of exchange and anisotropy energies. For $T > T_{AE}$, the integral approaches a maximum value of $\pi^2/6$. For $T < T_{AE}$ the integral is $(T_{AE}/T)\exp(-T_{AE}/T)$. Thus the temperature dependence of the nuclear spin lattice relaxation is obtained from the magnon scattering. By further expansion of Holstein-Primakopf transformation multiple magnon scattering terms can be calculated. These multimagnon terms can easily carry the nuclear spin splitting energy and hence contribute effectively to the nuclear relaxation time.

(c) Wave vector dependence

Due to the short range forces, the susceptibility depends on the wave vector as distinguished from the Curie-Weiss law. A component of the Knight shift, K, depends on the imaginary part of the wave vector dependent susceptibility, $\chi''(q,\omega)$. It is found by using experimental data that

$$T_1TK^2 = \text{constant} \tag{10.54}$$

for many nuclei. The experimentalists are able to make good use of this relation, called the Korringa relation.

The relaxation rate depends on the imaginary part of the susceptibility so that,

$$\frac{1}{T_1} \propto T\sum_q \frac{\chi''(q,\omega_N)}{\omega_N}\ . \tag{10.55}$$

Therefore, the needed q dependence may be prescribed by a model which introduces the coherence length.

10.4 Coherence length and susceptibility

We assume that in the CuO_2 plane there is at most one $S = 1/2$ electronic spin degree of freedom per CuO_2 unit cell. This spin is centred on the Cu sites which are antiferromagnetically correlated. As frequency tends to zero, the spin dynamics are described by $\chi''(q,\omega)$,

$$\chi''(q, \omega \to 0) = \frac{\pi\chi_o\omega}{\Gamma}\left[1 + \beta\frac{(\xi/a)^4}{(1+\xi^2q^2)^2}\right] \tag{10.56}$$

where χ_o is the uniform susceptibility. However, the wave vector dependence of χ_o is small so that χ_o is also about equal to its value near $q = \pi/a$ where a is the distance between atoms. Here ξ is the spin correlation length, a is the lattice constant and ξ is the spin correlation length. It is assumed that $\chi''(q, \omega \to 0) \sim \omega$ at all values of q and the relaxation is determined by correlations. The value for the dynamic exponent is $z = 2$ and the exponent for susceptibility is $2 - \eta$ where η is defined by the neutron scattering cross section dependence on the wave vector, $\sigma \propto k^{-2+\eta}$ as in chapter 9. The mean field expression for χ of an interacting spin system in terms of the susceptibility, $\bar{\chi}$ of a noninteracting spin system is given by,

$$\chi(q,\omega) = \frac{\bar{\chi}(q,\omega)}{1 - J_q\bar{\chi}(q,\omega)}\,. \tag{10.57}$$

We assume that,

$$\lim_{\omega\to 0}\bar{\chi}''(q,\omega) = \frac{\pi\omega}{\Gamma_q}\bar{\chi}'(q,\omega\to 0) = \frac{\pi\omega}{\Gamma_q}\bar{\chi}_q \tag{10.58}$$

where Γ_q is a characteristic energy of spin fluctuations at wave vector q, then,

$$\lim_{\omega\to 0}\chi''(q,\omega) = \frac{\pi\omega}{\Gamma_q}\bar{\chi}_q\frac{1}{(1-F_q)^2 + [(\pi\omega/\Gamma_q)\Gamma_q]^2} \tag{10.59}$$

where $J_q\bar{\chi}_q = F_q$. We expand F_q about its value F_Q at the zone corner,

$$F_q = F_Q - q^2\xi_o^2 \tag{10.60}$$

and define ξ by the relation,

$$1 - F_Q = (\xi_o/\xi)^2 \ . \tag{10.61}$$

We set $F_q = 1$ in the second term of the denominator in (10.59) so that

$$\chi''(q,\omega) = \frac{\pi\omega\bar{\chi}_q}{\Gamma_q} \frac{(\xi/\xi_o)^4}{(1+q^2\xi^2)^2 + (\pi^2\omega^2/\Gamma_q^2)(\xi^4/\xi_o^4)} \ . \tag{10.62}$$

For large ξ, the q dependence of $\bar{\chi}_q$ and Γ_q gives only a small correction. Therefore, we ignore the q dependence, setting $\bar{\chi}_q = \bar{\chi}$ and $\Gamma_q = \bar{\Gamma}$. For $\omega \to 0$, the (10.62) gives approximately the first term of (10.56). For large q, $qa > 1$, (10.62) gives a much smaller value than the first term of (10.56). Therefore, we add a q- independent term in (10.62) by identifying β as,

$$\beta = \left(\frac{a}{\xi_o}\right)^4 \left(\frac{\bar{\chi}}{\chi}\right) \left(\frac{\Gamma}{\bar{\Gamma}}\right) \ . \tag{10.63}$$

From (10.62) the characteristic energy for a zone corner spin fluctuation in this mean field theory is,

$$\hbar\omega_{SF} = \hbar\Gamma\xi_o^2/\xi^2 \tag{10.64}$$

which is much smaller than Γ if ξ^2 is large. Here

$$\omega_{SF} \propto \frac{1}{\xi^2} \tag{10.65}$$

so that the dynamic exponent is $z = 2$. From (10.57) and (10.61), $\chi(o) \sim \xi^{-2}$ which agrees with the mean field value $\eta - 2$. For the planar systems, $d = 2$ and $z = 1$ so that $\omega_{SF} \propto \frac{1}{\xi}$. For the temperature dependence of spin correlation length,

$$\left(\frac{\xi}{a}\right)^2 = \left(\frac{\xi(T=0)}{a}\right)^2 \left(1 + \frac{T}{T_x}\right)^{-1} \qquad (T > T_c) \tag{10.66}$$

in the normal state, where T_x is the temperature at which the Cu relaxation rate crosses over from high temperature to its low temperature behaviour. At $T > T_x$, $\xi \sim q/T^{0.5}$. In the antiferromagnetic state,

$$\xi \sim \left(1 - \frac{T}{T_N}\right)^{-1/2} \tag{10.67}$$

in the mean field theory. However, -0.7 may come in the scaling theory instead of $-1/2$. The changes in the coherence length as a function of temperature changes the density of low-frequency spin excitations and hence the width as well as the shifts of the NMR line.

10.5 Structure factors

We assume that the hyperfine interaction of ^{63}Cu is anisotropic with one isotropic component and one anisotropic component with axial symmetry. The hyperfine interaction at ^{17}O and ^{89}Y atoms is also important in $YBa_2Cu_3O_7$ structure. The electron-nuclear hyperfine interaction may therefore be written as,

$$\chi_{hyp} =^{63} I_n A S_n + B\sum_{\delta} {}^{63}I_n.S_{n+\delta} + C\sum_{\delta'} {}^{17}I_n.S_{n+\delta'} + D\sum_{\delta''} {}^{89}I_{n'}.S_{n+\delta''}. \tag{10.68}$$

Here S labels the four unit cells nearest to Cu site n, δ' labels the two unit cells sharing O at site n and δ'' labels eight unit cells, four in one plane and four in another with nearest Y at site n. A is the isotropic component of ^{63}Cu hyperfine interaction, C is for ^{17}O nuclei and D for ^{89}Y atoms. When electronic spin and the nuclear spin belong to one and the same atom, the interaction is called the hyperfine interaction. In the case of second term in (10.68) the electronic spin belongs to $n+\delta$ th atom ($\delta \neq 0$) and the I_n belongs to Cu atoms. Therefore S and I do not belong to one and the same atom. Instead of belonging to one atom, I_n belongs to one and S to another. This type of interaction is called the transferred hyperfine interaction. In terms of the above hamiltonian, the Knight shifts are given as,

$$\begin{aligned} {}^{63}K_{\|} &= \frac{(A_{\|}+4B)\chi_o}{\gamma_e {}^{63}\gamma_n \hbar^2} \\ {}^{63}K_{\perp} &= \frac{(A_{\perp}+4B)\chi_o}{\gamma_e {}^{63}\gamma_n \hbar^2} \\ {}^{17}K_{iso} &= \frac{2C\chi_o}{\gamma_e {}^{17}\gamma_n \hbar^2} \end{aligned}$$

$$ {}^{89}Y_{iso} = \frac{8D\chi_o}{\gamma_e {}^{89}\gamma_n \hbar^2} \; . \tag{10.69} $$

Here $\gamma_e = g\mu_B/\hbar$. The ^{63}Cu nuclear spin lattice relaxation rate for field parallel to c axis is given by

$$ {}^{63}W_{\|} = \frac{3}{4\mu_B^2\hbar^2} \lim_{\omega\to 0} \sum_q \left[A_\perp - 2B(\cos q_x a + \cos q_y a)\right]^2 S(q,\omega) \tag{10.70} $$

where $S(q,\omega)$ is the spin dynamical structure factor which is related to the imaginary part of the spin-spin correlation fucntion, $\chi''(q,\omega)$ by,

$$ S(q,\omega) = \frac{e^{\beta\omega}}{e^{\beta\omega}-1}\chi''(q,\omega) \tag{10.71} $$

where $\beta = 1/k_B T$. At low frequencies, we can expand the exponential factors in the above and retain only the first term so that,

$$ S(q,\omega) = \frac{k_B T}{\hbar\omega}\chi''(q,\omega) \; . \tag{10.72} $$

The ^{63}Cu relaxation rate for field in the ab plane is

$$ \begin{aligned} {}^{63}W_\perp &= \frac{3}{8\mu_B^2\hbar}\sum_q \Big[\left\{A_\| - 2B(\cos q_x a + \cos q_y a)\right\}^2 \\ &+ \left\{A_\perp - 2B(\cos q_x a + \cos q_y a)^2\right\}^2\Big] S(q,\omega) \; . \end{aligned} \tag{10.73} $$

For field applied in the ab plane, for ^{17}O nuclei,

$$ {}^{17}W(2) = \frac{3}{4\mu_B^2\hbar}\sum_q [2C^2(1-\cos q_x a)] S(q,\omega) \tag{10.74} $$

and for ^{89}Y,

$$ {}^{89}W = \frac{3}{4\mu_B^2\hbar}\sum_q \left[16D^2\left\{\cos^2\frac{q_a a}{2}\right\}(1-\cos q_y a)\right] S(q,\omega) \; . \tag{10.75} $$

Four different moments of $S(q,\omega)$

$$ \begin{aligned} S_o &= \left(\frac{a}{2\pi}\right)^2 \int d^2q S(q,\omega) \\ S_1 &= \left(\frac{a}{2\pi}\right)^2 \int d^2q \left[1 - \frac{1}{2}(\cos q_x a + \cos q_y a)\right] S(q,\omega) \\ S_2 &= \left(\frac{a}{2\pi}\right)^2 \frac{4}{5} \int d^2q \left[1 - \frac{1}{2}(\cos q_x a + \cos q_y a)\right]^2 S(q,\omega) \\ S_3 &= \left(\frac{a}{2\pi}\right)^2 \int d^2q (1-\cos q_x a)(1-\cos q_y a) S(q,\omega) \; . \end{aligned} \tag{10.76} $$

Substituting (10.56) in the above, evaluating the integrals and dropping terms of $0(a/\xi)$ and higher, we obtain,

$$\begin{aligned} S_o &= \frac{\pi\chi_o k_B T}{\hbar\Gamma}[0.08\beta(\xi/a)^2 - 0.007\beta + 1] \\ S_1 &= \frac{\pi\chi_o k_B T}{\hbar\Gamma}[0.04\beta\ln(\xi/a) + 0.017\beta + 1] \\ S_2 &= \frac{\pi\chi_o k_B T}{\hbar\Gamma}[0.026\beta + 1] \\ S_3 &= \frac{\pi\chi_o k_B T}{\hbar\Gamma}[0.020\beta + 1] \ . \end{aligned} \tag{10.77}$$

Thus we see that the nuclear relaxation rate depends on the coherence length. Therefore, there is a change in line width in going from the normal to the superconducting state. Actually, the susceptibility shows a large change in going from the normal to the superconducting state. The nature of the spin of the ground state is identified from the susceptibility. The triplet state shows much larger value for χ_o than the singlet state. Hence, the nuclear relaxation rates and the line widths give useful information pertaining to the spin degeneracy and the coherence of the system.

10.6 Experimental relaxation rates

The experimentally measured values of the nuclear magnetic moments are given by

$$\begin{aligned} {}^{63}\gamma_n\hbar &= 7.5 \times 10^{-24} \text{ erg/G} \\ {}^{17}\gamma_n\hbar &= 3.8 \times 10^{-24} \text{ erg/G} \\ {}^{89}\gamma_n\hbar &= 1.3 \times 10^{-24} \text{ erg/G} \ . \end{aligned} \tag{10.78}$$

and the Knight shifts are

$$\begin{aligned} {}^{63}K_{\parallel}(2) &= -0.01 \pm 0.01\% \\ {}^{63}K_{\perp} &= 0.30 \pm 0.02\% \\ {}^{17}K_{iso} &= 0.21 \pm 0.02\% \\ {}^{89}K_{iso} &= 0.03 \pm 0.01\% \ . \end{aligned} \tag{10.79}$$

The hyperfine constants in the Knight shift expressions (10.69) are found to be,

$$\begin{aligned} A_{\|} &= -4B \ , \\ B+\frac{A_{\perp}}{4} &= B(1+\alpha) \ . \\ K &= (1+\alpha)B\chi_o/\mu_B^2 = 1.21\times 10^{-6} \ , \\ C &= (0.72\pm 0.05)B(1+\alpha) \ , \\ D &= (1\pm 0.3)\times 10^{-2}B(1+\alpha) \ . \end{aligned} \tag{10.80}$$

The susceptibility is

$$\chi_o = 3\mu_B^2 \tag{10.81}$$

in states/eV implies that,

$$B(1+\alpha) = 4\times 10^{-7}\ \text{eV} \ = \ 42.9\ \text{kOe}/\mu_{\text{B}} \ . \tag{10.82}$$

We define a dimensionless parameter Δ,

$$\Delta = \chi_o\hbar\Gamma/\mu_B^2 \tag{10.83}$$

so that the relaxation rates, using (10.77), can be written as,

$$\begin{aligned} {}^{63}W_{\|} &= \frac{2k_BTK^2}{\hbar(1+\alpha)^2\Delta}\left[1.5(1-\alpha)^2\beta(\xi/a)^2 - 1.51(1-\alpha)\beta\ln(\xi/a)\right. \\ &\quad \left. -(0.15-0.90\alpha+0.13\alpha^2)\beta + 4.71 + 18.9a^2\right] \ . \\ {}^{63}W_{\perp} &= \frac{1}{2}{}^{63}W_{\|} + \frac{2k_BTK^2}{\hbar(1+\alpha)^2\Delta}\left[3.02\beta(\xi/a)^2 - 1.51\beta\ln(\xi/a)\right. \\ &\quad \left. -0.60\beta + 11.78\right] \ . \\ {}^{17}W &= \frac{2k_BTK^2C^2}{\hbar\Delta(1+\alpha)^2B^2}\left[0.094\beta\ln(\xi/a)^2 + 0.040\beta + 2.36\right] \\ {}^{89}W &= \frac{2k_BTK^2D^2}{\hbar\Delta(1+\alpha)^2B^2}\left[0.19\beta + 9.43\right] \ . \end{aligned} \tag{10.84}$$

The values of α, β, Δ, and ξ/α can be obtained by fitting the above relations to the experimentally measured relaxation rate and compared with those in a non-interacting Fermi gas. In the free electron gas the dynamic susceptibility at low frequency is proportional to $\omega/v_F q$ and the structure factor,

$$\sum_q S(q,\omega \to 0) = \frac{k_B T}{\hbar\omega} \sum_q \chi''(q,\omega \to 0) = \frac{\pi k_B T}{2\mu_B^2} \chi_o^2 \tag{10.85}$$

where χ_o is the uniform susceptibility at $q = 0$. If Fermi gas is coupled via an isotropic contact interaction E to a nuclear moment, the Knight shift in dimensionless units becomes,

$$K = \frac{E\chi_o}{\mu_B^2} \tag{10.86}$$

and the relaxation rate,

$$W = \frac{3E^2}{4\mu_B^2} \sum_q S(q,\omega \to 0) = \frac{3\pi}{8} K^2 k_B T \ . \tag{10.87}$$

Since $1/T_1 = W$, we write,

$$\Lambda = \frac{8}{3\pi \Gamma_1 K^2 k_B T} = 1 \tag{10.88}$$

which is the Korringa relation at $\omega \to 0, q \to 0$. Using the experimentally measured values, we find,

$$^{63}\Lambda = 2.9 \pm 0.2 \ , \ ^{17}\Lambda = 1.2 \pm 0.2 \quad \text{and} \quad ^{89}\Lambda = 0.15 \tag{10.89}$$

which are not in agreement with (10.88) largely because of neglect of antiferromagnetic correlations. The nuclear relaxation rate of ^{63}Cu from (10.84) is linear in T, $^{63}W_{||} \simeq a + bT$ for large T and $^{63}W_{||} \simeq cT$ for small T. There is a temperature T_x at which the low temperature behaviour changes to high temperature behaviour. This temperature for $YBa_2Cu_3O_7$ is about $T_x \simeq 100$K. The form of temperature dependence of ^{63}W predicted by (10.84) is in resonable agreement with experimental measurements.

We assume that spins are localized at the Cu sites which are antiferromagnetically correlated. The correlation is having two terms, one a normal Fermi liquid type or quasiparticle type and another antiferromagnet type so that the susceptibility can be written as the sum of two terms as,

$$\chi(q,\omega) = \chi_{QP}(q,\omega) + \chi_{AF}(q,\omega) \ . \tag{10.90}$$

The real part of the quasiparticle-like contribution to the susceptibility, χ_{QP} occurs as,

$$\bar{\chi}_o = \chi_{QP}(q = 0, \omega = 0) \ . \tag{10.91}$$

For small ω, the imaginary part of χ_{QP} is related to the real part

$$Im\chi_{QP} \simeq \frac{\pi\omega}{\Gamma}\bar{\chi}_o \ . \tag{10.92}$$

Here Γ is the characteristic spin fluctuation energy. The antiferromagnetic part of the spin-spin correlation function can be modeled around the antiferromagnetic wave vector, $Q = (\pi/a, \pi/a)$ by,

$$\chi_{AF}(q,\omega) = \frac{\chi_Q}{1 + \xi^2(Q-q)^2 - i(\omega/\omega_{SF})} \tag{10.93}$$

where χ_Q is the static susceptibility at the antiferromagnetic wave vector Q, which is related to the static susceptibility at $q = 0$ by the relation,

$$\chi_Q = \bar{\chi}_o(\xi/\xi_o)^2 \tag{10.94}$$

where ξ is the antiferromagnetic correlation length and $1/\xi_o$ is the wave vector at which χ_{AF} becomes larger than χ_{QP}. Writing (10.90) in terms of correlation lengths at zero wave vector and zero frequency,

$$\begin{aligned} \chi_o &= \chi_{QP}(0,0) + \chi_{AF}(0,0) \\ &= \bar{\chi}\left[1 + \frac{1}{2\pi^2}\left(\frac{a}{\xi_o}\right)^2\right] \ . \end{aligned} \tag{10.95}$$

For $(a/\xi_o)^2 \simeq 3$, χ_{QP} is about 86% while χ_{AF} is only 14%. The Knight shifts of all of the nuclei, ^{89}Y, ^{63}Cu and ^{17}O in $YBa_2Cu_3O_{6.63}$ are all having the same temperature dependence.

10.7 Nuclear relaxation in mixed-wave superconductors

The s-wave superconductors have an isotropic gap. In the ground state the electrons are paired with spin singlet and zero wave vector. The isotropic gap may be arranged in a lattice in which case, there is at least one irreducible representation of Γ_1 symmetry which is totally symmetric. The gap may have p wave symmetry in which case there is a change in parity, $(-1)^l$ which for $l = 1$ gives a negative sign for parity. We can take several gaps of p wave symmetry and arrange them in a lattice in which case states of different symmetries according to the irreducible representation of the point group may be obtained. The p wave transforms like the Legendre's polynomial, $P_1(\cos\theta)$. Similarly, the pairing can occur in the d wave which transforms like $P_2^{0,\pm 1,\pm 2}(\cos\theta)$. The gap of the superconductor may accordingly transform like s-wave, p-wave or d wave in most of the systems. The f wave symmetry for the gap may also occur, such as in rare-earth atoms. In the B.C.S. theory the gap transforms like the s-wave symmetry. A trivial modification of the gap in the B.C.S. theory may be made when the gap has a symmetry different from that of the s wave. When there is a phase transition from the normal state to the paired state, there occurs a phase factor in the gap of the superconductor which compensates for the loss of symmetry. In case of a transition from the normal state to the d wave superconductor, there is a phase factor in the gap which reflects the symmetry of the gap. Usually for distinguishing the s-wave from the d wave, it is sufficient to introduce a node which means that at $r = 0$, whereas there is a finite value for the s wave, it is zero for the d wave. If the gap for the s wave is Δ_o, then for $d_{x^2-y^2}$ it can be written as $\Delta_o \cos 2\phi$ and for d_{xy} as $\Delta_o \sin 2\phi$ where ϕ is the usual polar angle. In the case of a mixed $s + id$ character of the gap, there is a phase transition from d wave at low temperatures to s wave at high temperatures. The change from d to s wave may be accompanied by a peak in the specific heat. Instead of the temperature, it is possible to vary some other parameter such as magnetic field or pressure, etc. In the case of $s + d$ symmetry of the gap, both s and d symmetries are degenerate and hence the gaps corresponding to s and that corresponding to d wave are added. The gap may also have $d_{xy} + id_{x^2-y^2}$ type symmetry in which case d_{xy} has a phase factor of $\pi/2$ with respect to $d_{x^2-y^2}$. Similarly, d_{yz} and d_{zx} also have a phase factor of $\pi/2$ and all the fine possible d symmetries, $d_{xy}, d_{yz}, d_{zx}, d_{3z^2-r^2}$ and $d_{x^2-y^2}$ are orthogonal. When magnetic field is varied the symmetry can change from, e.g., d_{xy} to $d_{x^2-y^2}$ at a particualr magnetic field at which a peak also occurs in the specific heat.

We study the nuclear relaxation time of ^{63}Cu for $d+s$ mixed-wave superconductors with b as the degree of admixture with and without scattering of electrons due to impurities. The gap of the superconductors is written in such a way that varying amounts of d-wave contributions can be mixed in the s wave and from that the nuclear relaxation rates are deduced. The calculated value for 80% d wave and 20% s wave without any scattering are found to be in reasonable agreement with those measured for ^{63}Cu in $YBa_2Cu_3O_7$ type superconductors. When experimental value of the electron-scattering time is used to define the gap and the nuclear relaxation rates are calculated, it is found that 40% admixture of d wave into s wave is in agreement with the experimental values of the nuclear relaxation rate at low temperatures. At high temperatures, $T > 0.85T_c$, pure s wave without d wave with experimental values of the scattering time gives agreement with the experimental values of the Gaussian nuclear relaxation time.

The gap parameters Δ of the B.C.S. theory may be replaced by Δf where $f = 1$ for s wave, $\cos 2\phi$ for $d_{x^2-y^2}$ wave and $\sin 2\phi$ for d_{xy} wave. It is sufficient to introduce a cusp by the factor $\cos 2\phi$ for the d wave. the d_{xy} is found largely on the surface of the sample and hence may be ignored or taken up separately so that we are left with $\cos 2\phi$ only for the d wave. In the case of $d + is$, the system changes from s wave at high temperatures to d wave at low temperatures. The high symmetry s wave occurs simultaneously. The gap of the superconductor may be written as,

$$\Delta^* = \Delta_o[(1-b) + \cos 2\phi] \qquad (10.96)$$

where b determines the "degree of admixture" of d wave into the s wave. The wave vector dependence for the d wave may be taken as

$$\Delta^* = (1-b)\Delta_o(T) + (1/2)b\Delta_o(T)(\cos p_x - \cos p_y) \qquad (10.97)$$

where Δ_o has a B.C.S. type temperature dependence and $2\Delta_o = 3.52k_BT_c$.

The effective interaction between nuclear spins, I_o and I_i is,

$$\mathcal{H}' = \sum_{\alpha} J_\alpha(x_i) I_{o\alpha} I_{i\alpha} \tag{10.98}$$

where $\alpha = z$ component of the coupling constant is given by,

$$J_z(x_i) = A_{zz} S_z(x_i) + B \sum_{\delta=1}^{4} S_z(x_{i+\delta})/I_{oz} \ . \tag{10.99}$$

The form of the hyperfine coupling is such that $J_x(x_i)$ and $J_y(x_i)$ are much smaller than $J_z(x_i)$. The decay of the NMR spin-echo envelope for the 1/2 to -1/2 transition of ^{63}Cu(2) can be written as a Gaussian $\exp(-t^2/2T_{2G}^2(x_i)$ with

$$1/T_{2G}^2 = (0.69/8\hbar^2) \sum_i J_z^2(x_i) \tag{10.100}$$

where summation may be performed upto several lattice spacings away from the origin. Since the original B.C.S. theory has only s wave gap, we define a gap parameter b which determines the "degree of admixture" of d wave into s wave. The value of $S_z(x_i)$ is determined by

$$S_z(x_j) = -(1/2) I_{oz} A_{zz} \chi(x_j) + B \sum_{\delta=1}^{4} \chi(x_{j+\delta}) \tag{10.101}$$

with $\chi(x_j) = (1/N)\sum_q \exp(iq.x_j)\chi(q)$. The value of $\chi_o(q)$ depends on the gap of the superconductor and on the coherence factors as given by (1.86). By using the numerical values of the parameters in the single particle energy, $e_p = -2t[\cos p_x + \cos p_y] - \mu$ with $U = 2t$, $t = 10k_BT_c$ and $2\Delta_o = 3.52k_BT_c$ with $f_p = [\exp(\epsilon_p/k_BT) + 1]^{-1}$ we can compute the value fo T_{2G} from (10.100). The effect of scattering of electrons is to reduce the transition temperature and hence the gap is also reduced. The scattering time has been estimated from the microwave surface resistance at 2.95 GHz in a single crystal of $YBa_2Cu_3O_7$. These scattering times are used to find the gap and hence the nuclear rleaxation time. However, it may be noted that the scattering rates have been overstimated by this method.

10.8 Nuclear relaxation measurements

The spin echo amplitudes as a function of time have two terms one decaying as $\exp[-\tau/(T_{2R}/2)]$ and the other decaying as $\exp[-(1/2)\{\tau/(T_{2G}/2)\}^2]$ where T_{2R}^{-1} is the Lorentzian-Redfield term

$$1/T_{2R} = (2 + r)/3T_1 \tag{10.102}$$

and T_{2G} is a Gaussian term. The T_1 is the usual relaxation time and r depends on the crystal structure which has been determined to be 3.7 for 1-2-3 block and 3.3 for 1-2-4 block. The experimental measurements of the nuclear relaxation rate of ^{63}Cu in $YBa_2Cu_3O_{6.982}$, $YBa_2Cu_4O_8$ and $Y_2Ba_4Cu_7O_{15}$ have been performed by Stern et al. Two types of calculations are performed, one with and one without scattering. In Fig. 10.1

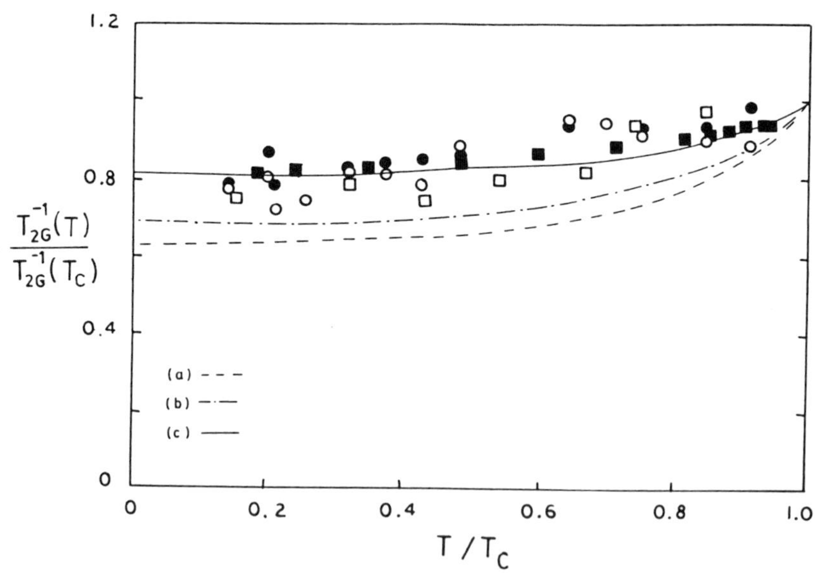

Fig.10.1. The Gaussian nuclear relaxation rate calculated for (a) the degree of admixture, $b = 0.5$, (b) $b = 0.6$ and (c) $b = 0.8$ along with experimental data for 1-2-3 (spin circles), 1-2-4 (filled circles) in $Y_2Ba_4Cu_7O_{15}$, in $YBa_2Cu_3O_{6.98}$ (open squares) and $YBa_2Cu_4O_8$ (filled squares) showing the mixed wave $d + s$ nature of the gap.

we show the calculated values for the degree of admixture $b = 0.5, 0.6$ and 0.8 with no scattering at all along with the experimental data. Here the gap is determined by (10.96) with s wave gap given by $(1 - b)\Delta(T)$ and the d wave component given by the second term of (10.96). It is seen that the experimental values are in reasonable agreement with the calculated values for $b = 0.8$. It is clear that neither pure s nor pure d wave gives nuclear relaxation times which agree with the experimental data. The mixed wave symmetry is clearly indicated. However, there is a large effect of scattering on the Gaussian nuclear relaxation times because of a large effect on the gap. Using experimental values of the scattering times, we calcualte the nuclear relaxation rate shown in Fig. 10.2. For pure s wave, without scatteirng the values are shown by the curve s. The s wave with

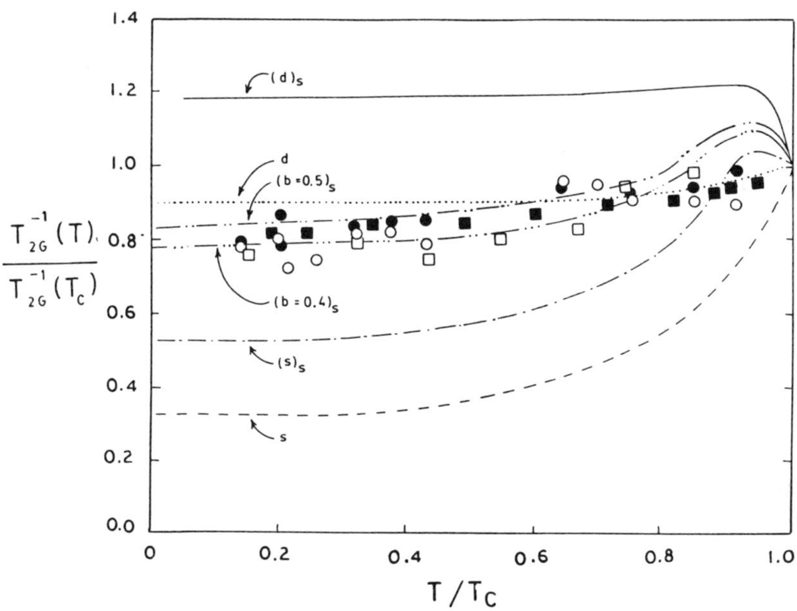

Fig. 10.2. The Gaussian nuclear relaxation rate for the three samples with the same notation as in Fig. 10.1. The calculated curves are pure d-wave (d), pure s wave (s), s wave with scattering $(s)_s$*, mixed* $d + s$ *wave with degree of admixture of 0.4* $(b = 0.4)_s$ *and 0.5* $(b = 0.5)_s$*. At low temperatures, mixed* $d + s$ *with* $b = 0.4$ *agrees with the data. The s wave with scattering agrees with the data only for* $T > 0.85T_c$*.*

scattering time is shown by the curve $(s)_s$. It is found that there is a considerable effect of scattering on the Gaussian nuclear relaxation rate but neither s nor $(s)_s$ agree with the experimental data at low temperatures. The calculated values for pure d wave as well as for the d wave with scattering, $(d)_s$ are also shown. The Gaussian nuclear relaxation rate for the degree of admixture $b = 0.4$ and $b = 0.5$ with scattering which are better than those of the pure s and pure d wave are also shown in Fig. 10.2. Most of the low temperature range of the experimental data are in resonable agreement with $b = 0.4$ with scattering. We can search for the $d+is$ type gap. It is found that for $T > 0.85T_c$ the s wave with scattering agrees with the data. Thus $d + s$ symmetry at low temperatures and s at high temperatures, $T/T_c > 0.85$ is seen.

The Gaussian nuclear relaxation rates of ^{63}Cu in $YBa_2Cu_3O_7$, $YBa_2Cu_4O_8$ and $Y_2Ba_4Cu_7O_{15}$ superconductors have thus been examined from the view point of symmetry of the gap. It is found that the low temperature region agrees with $s + d$ wave symmetry. At high temperatures, s wave with electron scattering taken into account agrees with the data. Therefore $s+d$ symmetry at low temperatures and s wave at high temperatures indicated by $b\Delta_d + (1 - b)\Delta_s + i\Delta_s$ is found. Here b is the "degree of admixture" of d wave into the s wave and Δ_s and Δ_d are the gaps for s and d wave symmetries, respectively. As a function of temperature some element of s wave is always present whereas $s+d$ symmetry occurs at low temperatures.

Suggested References

10.1 W.D. Knight, in Solid State Physics, edited by F. Seitz and D. Turnbull, Academic Press, New York **2**, 93 (1956).

10.2 K.N. Shrivastava, Phys. Stat. Solidi B**115**, K37 (1983); Phys. Rev. B**19**, 1598 (1979).

10.3 D. Beeman and P. Pincus, Phys. Rev. **166**, 359 (1968).

10.4 K.N. Shrivastava, Phys. Rev. B**20**, 2634 (1979).

10.5 K.N. Shrivastava, Phys. Rev. B**20**, 5375 (1979).

10.6 K.N. Shrivastava, Phys. Rev. B**21**, 2702 (1980).This paper deals with the latest theory of $Cu^{2+} - F^{-} - Cu^{2+}$ which has the same electronic configuration as $Cu^{2+} - O^{2-} - Cu^{2+}$.

10.7 K.N. Shrivastava, Phys. Rev. B**34**, 1489 (1986).

10.8 K.N. Shrivastava, J. Phys. (Paris) **49**, C8-2239 (1988).

10.9 K.N. Shrivastava, Bull. Mat. Sci. (India) **14**, 625 (1991).

10.10 K.N. Shrivastava, Solid State Commun. **68**, 1019 (1988).

10.11 K.N. Shrivastava and V. Jaccarino, Phys. Rev. B**13**, 299 (1976).

10.12 F. Mehran and P.W. Anderson, Solid State Commun. **71**, 29 (1989); [This paper has the same result as earlier dated ref.10.8 above].

10.13 S. Chakravarty and R. Orbach, Phys. Rev. Lett. **64**, 224 (1990).

10.14 A.J. Millis, H. Monien and D. Pines, Phys. Rev. B**42**, 167 (1990), [This paper has the same idea as earlier dated paper ref.10.8].

10.15 H. Monien, D. Pines and M. Takigawa, Phys. Rev. B**43**, 258 (1991).

10.16 H. Monien, P. Morthoux and D. Pines, Phys. Rev. B**43**, 275 (1991).

10.17 A.J.Millis and H. Monien, Phys. Rev. B**45**, 3059 (1992).

10.18 D. Thelen, D. Pines and J.P. Liu, Phys. Rev. B**47**, 9151 (1993).

10.19 P.W. Anderson, Phys. Rev. Lett. **3**, 325 (1959); The scattering considered in this paper will actually reduce the T_c so that the results claimed for the superconducting state belong to the normal state only.

10.20 I. Grosu and M. Crisan, Phys. Rev. B**49**, 1269 (1994).

10.21 N. Bulut and D.J. Scalapino, Phys. Rev. Lett. **67**, 2898 (1991).

10.22 R. Stern, M. Mali, J. Roos and D. Brinkmann, Phys. Rev. B**51**, 15478 (1995).

10.23 K.A. Müller, Nature **377**, 133 (1995).

10.24 K.A. Müller and H. Keller, High T_c superconductivity 1996 Ten Years After the Discovery, edited by E. Kaldis, New York: Kluwer Academic Pub. pp. 7-29, (1997).

Appendix A

A. Superconducting compounds.

Many compounds are superconducting. We describe six types of compounds.

(a) Organic superconductors, (b) A-15 compounds, (c) magnetic superconductors, (d) carbon superconductors, (e) heavy-fermion superconductors and (f) the high-temperature superconductors.

(a) Organic compounds.

The organo-metallic compounds which have double bonds may provide electrons to form Cooper pairs by hopping interaction. There are vibrational modes so that a phonon-induced mechanism is feasible. The electrons may also become attractive due to strong correlations in which case a phonon mechanism is not necessary. Some of the molecules which form superconductors are given in Fig. A.1. The transition temperatures of several BEDT-TTF compounds are given in Table A.1.

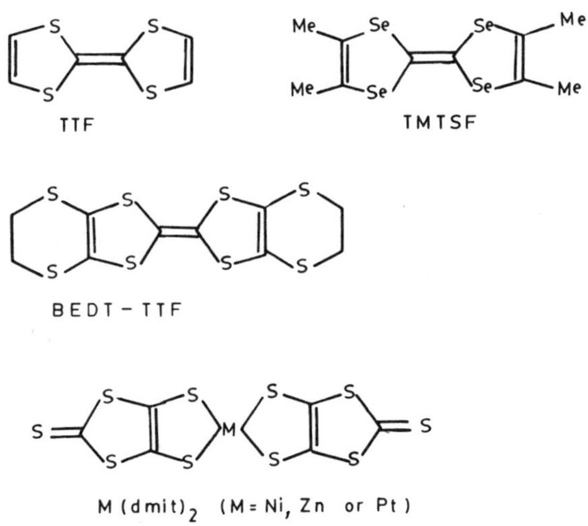

Fig. A.1. The organic compounds, TTF, TMTSF, BEDT-TTF, M (dmit), which form superconducting compounds.

Table A.1. The phase, critical pressure and transition temperatutre of several organo-metallic compounds with BT=BEDT-TTF. BEDT=bis ethylemedithiolo, TTF=tetrathiafulvalene.

S.No.	Formula	Phase	Pressure (k bar)	T_c(K)	Remarks
1.	$(BT)_2ReO_4$	–	4	2.0	
2.	$(BT)_2I_3$	α	0	3.3	I_2 doped
3.	$(BT)_2I_3$	β_L	0	1.5	
4.	$(BT)_2I_3$	β_L	0	2.0	Annealed
5.	$(BT)_2I_3$	β_H	0.5	8.1	
6.	$(BT)_2I_3$	α_t	0	8.1	
7.	$(BT)_2IBr_2$	β	0	2.2 - 3.0	
8.	$(BT)_2AuI_2$	β	0	3.4 - 5.0	
9.	$(BT)_2I_{3-2.5}$	γ	0	2.5	
10.	$(BT)_2I_3$	θ	0	3.6	
11.	$(BT)_2I_3$	κ	0	3.6	
12.	$(BT)_4Hg_{2.89}Br_8$	–	0	4.0	
13.	$(BT)_4Hg_{2.89}Cl_8$	–	12	1.8	
14.	$(BT)_3Cl_2(H_2O)_2$	–	16	2.0	
15.	$(BT)_2Cu(NCS)_2$	κ	0	10.4	

The properties of these superconductors depend on pressure and on annealing as shown in Table A.2. The substance $(BEDT\text{-}TTF)_2$ $Cu(NCS)_2$ crystallizes in monoclinic structure. The $Cu(NCS)_2$ are well separated from BEDT-TTF molecules as shown in Fig. A.2. The trnasition tmeperatures of most of these compounds are about 3K and in no case larger than 10 K. In β-$(BEDT\text{-}TTF)_2I_3$ the upper critical field before annealing at 1 K is about 0.8 Tesla and much higher after annealing along a-axis and much lower along the c axis.

Table A.2. Superconducting parameters of β-$(BT)_2I_3$

S.No.	T_c(K)	P (k bar)	$\xi_{GL\parallel}(0)\mathring{A}$	$\xi_{GL\perp}(0)\mathring{A}$	$\frac{H_{c2\parallel}}{H_{c2\perp}}$	Remarks
1.	1.40	0	508.5	24.3	20.9	Not annealed
2.	2.08	0	354.6	24.5	14.5	Annealed
3.	6.55	1.6	131.0	10.0	13.1	
4.	3.06	3.5	355.0	22.7	15.7	
5.	2.18	5.0	488.0	31.0	15.6	

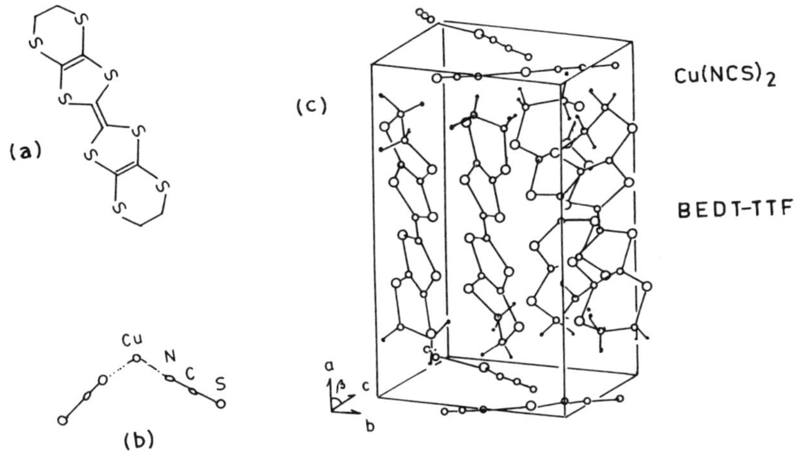

Fig. A.2. Crystal structure of monoclinic (BEDT-TTF)$_2$Cu(NCS)$_2$.

(b) A-15 Compounds.

Compounds of binary composition $A_{3-x}B_{1+x}$ crystalize in many different structures depending on the value of x, temperature and pressure, One of the structures occuring near $x = 0$, A_3B occurs in the form of $A - 15$ type structure shown in Fig. A.3 has a superconducting phase. Nb_3Ge has a superconducting transition temperature of about 23K, Nb_3Ga has 20.7K and Nb_3Sn has 18K. Prior to 1974, these were the largest transition temperatures. Therefore, these materials are used to make superconducting magnets to provide large magnetic fields. The transition temperature of several $A - 15$ type structures are given in Table A.3.

Table A.3. The transition temperatures $T_c(K)$ of $A-15$ compounds.The left column does not have transition metals, the right column has transition metals, three alloys of gold are also included. Notes: q =quenching, p=pressure, f=film.

Ti_3Sb	6.5	Ti_3Ir	4.2
$Zr_{80}Sn^q_{20}$	0.92	Ti_3Pt	0.5
Zr-Pb	0.76	$V_{29}Re_{71}$	8.4
$Zr_{\sim3}Bi^p$	3.4	$V_{50}Os_{50}$	5.7
$V\text{-}Al^t$	11.8	$V_{65}Rh_{35}$	$\simeq$1
V_3Ga	15.9	$V_{63}Ir_{37}$	1.7
V_3Si	17.0	$V_{\sim3}Pd$	0.08
$V_{\sim3}Ge$	6.0- 7.5	V_3Pt	3.7
$V_{\sim3}Ge^t$	6.0-11.0	$Nb_{75}Os_{25}$	1.0
$V_{\sim79}Sn_{\sim21}$	4.3	$Nb_{75}Rh_{25}$	2.6
$V\text{-}Sn^q$	7.0-17.0	$Nb_{72}Ir_{28}$	3.2
$V_{77}As_{23}$	0.2	Nb_3Pt	11.0
$V_{76}Sb_{24}$	0.8	$Ta_{85}Pt_{15}$	0.4
Nb_3Al	19.1	$Cr_{72}Ru_{28}$	3.4
Nb_3Ga	20.7	$Cr_{73}Os_{27}$	4.7
$Nb_{\sim3}In^p$	8.0- 9.2	$Cr_{78}Rh_{22}$	0.07
$Nb_{82}Si^q_8$	4.4	$Cr_{82}Ir_{18}$	0.75
$Nb\text{-}Si^t$	9.3	$Mo_{40}Tc_{60}$	13.4
$Nb\text{-}Si^t$	4.0- 8.0	$Mo_{\sim65}Re^t_{\sim35}$	$\simeq$ 15 ($A15$)
$Nb\text{-}Si^t$	11.0-17.0	$Mo_{75}Os_{25}$	12.7
$Nb\text{-}Ge^q$	6.0-17.0	$Mo_{78}Ir_{22}$	8.5
$Nb\text{-}Ge^t$	23.0	$Mo_{82}Pt_{18}$	4.6
Nb_3Sn	18.0	$W_{\sim60}Re^t_{\sim40}$	11.0
Nb-Sb	2.0	$Ta_{\sim80}Au_{20}$	0.55
$Nb_{\sim3}Bi^p$	3.0	Zr_3Au	0.9
$Ta_{\sim3}Ge^t$	8.0	$V_{76}Au_{24}$	3.0
$Ta_{\sim3}Sn$	8.3	$Nb_{\sim3}Au$	11.5
$Ta_{\sim3}Sb$	0.7		
Mo_3Al	0.58		
Mo_3Ga	0.76		
$Mo_{77}Si_{23}$	1.7		
$Mo_{77}Ge_{23}$	1.8		

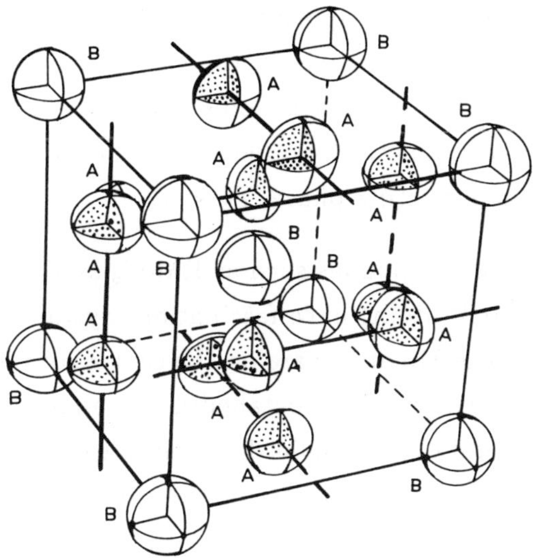

Fig. A.3. The A-15 structure.

(c) Magnetic superconductors.

Magnetic superconductors exhibit some novel features not found in ordinary type-I superconductors. One of these is the reentrant behaviour. Upon cooling from the superconducting phase, $T < T_c$, the material becomes normal again at a low temperature and the superconductivity is destroyed. Very often this normal phase is magnetically ordered. Upon cooling from the normal state the system becomes superconducting below T_c and upon further cooling it becomes magnetically ordered below T_N with $T_N < T_c$ so that the superconducting phase occurs only in a limited range of temperatues, $T_N < T < T_c$. The interaction of conduction electrons with magnetic atoms leads to the formation of a bound state of the electron with the magnetic atom below a certain characteristic temperature called the Kondo temperature. It was found that the resistivity of a magnetic alloy, such as Fe impurity atoms in Cu exhibits a minimum in the resistivity as a function of temperature due to the interaction of conduction electrons of the metal atom with the magnetic moment of the magnetic atom. Such a Kondo effect is found to occur in superconductors containing magnetic impurity atoms at low temperatures so that the superconductivity is destroyed. In the case of magnetic superconductors there is a sublattice of magnetic atoms in addition to the lattice of metallic atoms so that magnetically ordered phase occurs at low temperatures below the superconducting phase, $T_N < T_c$.

The rare-earth compounds such as $RMoS_8$ and RRh_4B_4 (where R represents a rare-earth atom) have fascinating crystal structure. For example, in RMo_6S_8, the Mo atoms are on the centres of the six faces of a cube forming an octahedron. The octahedra are aligned along the [111] direction. This metal-cluster chain determines the conductivity. The eight sulphur atoms are on the corners of the same cube, the faces of which are occupied by metal atoms. The sulphur atoms separate the rare-earth atoms from the metal atoms. The Mo_6S_8 cubes are separated by sheets of rare-earth atoms. The crystal structure of the Chevral-phase compounds $M_xMo_{60}S_8$ (M_x=metal, X =S, Se or Te), is shown in Fig. A.4. In the rare-earth molybdenum chalcogenides there are six metal atoms forming an octahedron corresponding to one rare-earth atom. Therefore, we expect to find similar compounds which have four metal atoms forming a tetrahedron. The rare-earth rhodium borides are of this type. The four Rh atoms form a tetrahedron. The tetrahedron are connected along the [111] direction and a distorted sheet of rare-earth atoms covers the line of tetrahedra. The boron atoms are found in between the rare-earth atoms and the metal Rh atoms. The number of magnetic atoms per unit cell is small. While ferromagnetic superconductors (e.g., $ErRh_4B_4$ and $HoMo_6S_8$) display interesting reentrant behaviour, the antiferromagnetic superconductors (RMo_6S_8 with R=Gd, Tb, Dy and Er and RRh_4S_4 with R=Nd, Sm and Tm) present examples of the coexistence of the two phases along with anomalous behaviour of the second critical field. In $ErRh_4B_4$, having tetragonal structure with $a = 5.299\mathring{A}$ and $c = 7.588\mathring{A}$, a plot of the a.c. magneitc susceptibility, χ_{ac}, and electrical resistance shows a normal to superconducting transition at $T_{c1} = 8.7$K followed by a loss of susceptibility at $T_{c2} = 0.9$K along with the appearance of the ferromagnetic long range order. This is the result at zero magnetic field. At non-zero finite field also the resistance disappears in some temperature interval within the above domain. Neutron scattering data confirm ferromagnetic order and superconductivity and spiral magnetism seems to coexist in a small temperature interval near T_{c2}.

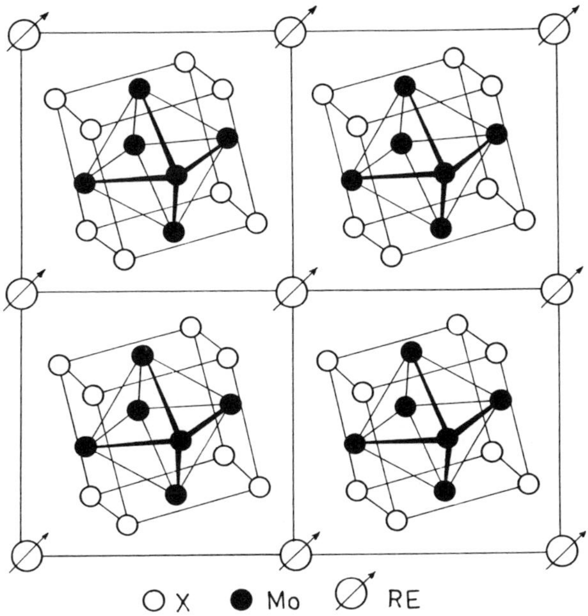

Fig. A.4. Structure of rare-earth MoS_8 which have Mo cube inside sulphur cube which are inside a rare-earth cube.

The low observed magnetic moment, 5.6 μ_B, of Er as compared to the free ion value 9.7 μ_B suggests the importance of crystal field effects or screening by conduction electrons. $HoMo_6S_8$ becomes superconducting at $T_{c1} = 1.3$K in zero field. On further cooling it becomes normal at $T_{c2} = 0.6$K along with the appearance of ferromagnetic long range order. The second transition at T_{c2} is found to show hysteresis. Neutron diffraction results on this system again confirm ferromagnetic order with long-range magnetic correlations. Again the magnetic moment $\mu_{eff} \simeq 6\mu_B$ is considerably smaller than the free ion value $\sim 10.85\mu_B$. In both $ErRh_4B_4$

and $HoMo_6S_8$ one finds sharp specific heat peaks at T_{c2} which is a typical feature of these materials. The antiferromagnetic superconductors provide the most striking case of the coexistence of two kinds of order. These systems (e.g., RMo_6S_8, R=Gd, Tb, Dy and Er; RRh_4B_4 (R=Nd, Sm, Tm) show λ-anomaly in the least capacity and a cusp in the magnetic susceptibility near T_N indicative of the antiferromagnetic ordering of rare-earth magnetic moments in the superconducting state of the system confirmed by the neutron diffraction measurements. The most important result is the anomalous behaviour of the upper critical field H_{c2} as a function of temperature in the vicinity of T_N. For example, RMo_6S_8 (R=Dy, Tb, Gd) show anomalous depression of H_{c2} near but below T_N. In Tb and Gd compounds the maximum occurs at T_N but in D_y compound above T_N but drop in H_{c2} near T_N is pronounced. $NdRh_4B_4$ displays two antiferromagnetic transitions one at T_{N_1}=1.31K and the other at T_{N_2}=0.89K with H_{c2} decreasing at T_{N_1} and then increasing suddenly at T_{N_2}. $TmRh_4B_4$ does not show any appreciable change in H_{c2} at $T_N \simeq$0.4K. For $SmRh_4B_4$, the upper critical field increases below T_N after showing a sharp peak in the slope at T_N=0.87K. If the Kondo temperature, T_K, is less than the superconducting transition temperature and their is exchange interaction between conduction electrons and the magnetic atoms, there is spin non-flip scattering leading to three solutions for T_c which are called T_{c1}, T_{c2} and T_m as shown in Fig. A.5. The superconducting transition temperature without the impurity scattering is called T_{co}. If T_K is considerably lower than T_{co}, the system upon cooling becomes superconducting at an upper transition temperature T_{c1} where pair breaking due to scattering is small. When the temperature is further lowered, there is a transition at T_{c2} at which the material may be supercooled and continue to remain superconducting because of magnetic atoms, $\bar{c}$. The superconductivity is destroyed at T_m. Below T_m the material is dominated by magnetic atoms. Upon heating from T_m the material passes through the metastable region becoming superconducting at T_{c2} and remaining superconducting till at T_{c1} above which the material becomes normal again. Thus the material is normal above T_{c1}, superconducting between T_{c1} and T_{c2} but the nature of the state between T_{c2} and T_m is oscillatory and below T_m the pairs are broken and the material magnetic and normal.

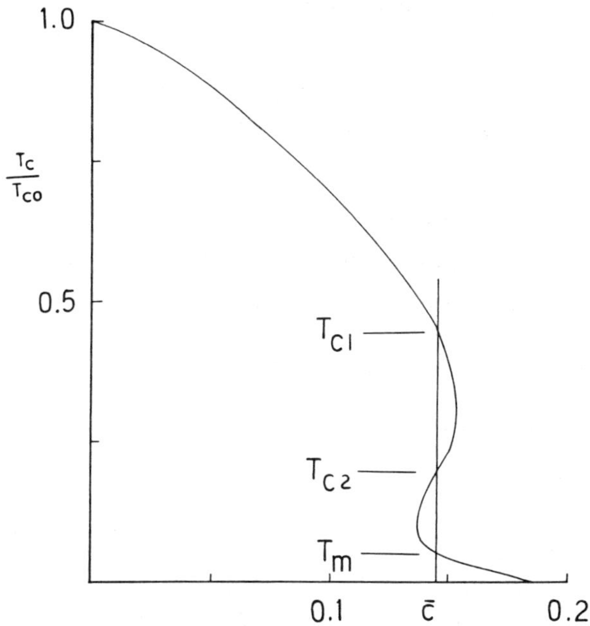

Fig. A.5. The variation of superconducting transition temperature as a function of concentration of magnetic atoms, showing the normal to the superconducting transition temperature T_{c1}. There is an oscillatory state when $T_m < T < T_{c2}$ and magnetically ordered state below T_m.

(d) Carbon superconductors.

One of the forms of carbon has sixty atoms per molecule, C_{60}. A model of 20 hexagons and 12 pentagons forms a spherical molecule with 60 apex points which contain carbon atoms. The C_{60} molecule can be isolated by desolving shoot -produced by grpahite electrodes in benzene. Some of the properties of C_{60} are given in Table A.4. Since the diameter of C_{60} is 7.1 Åit is possible to put atoms inside the case. The atoms located in the central cavity inside the fullerence case are grouped to the left of the symbol ©, whereas atoms which are part of the cage are listed to the right. The C_{60} molecule is then represented by ©C_{60}. When one potassium atom is inside the cage, it is represented by K©C_{60}. When one boron atom substitutes for a carbon in the cage, we write K©$C_{69}B$. The

D atoms outside of C_{60} cage are written as $D(K©C_{59}B)$ which means, outside (inside $©C_{60-x}$ x on surface) so that it is possible to represent molecules obtained from C_{60} chemistry. A large number of molecules can be made using C_{60}. The compounds A_3C_{60} are superconducting and their properties are given in Table A.5, the pressure dependence of the T_c in K_3C_{60} is given by $dT_c/dP = -0.63 \pm 0.08$ K/kbar. In the case of Rb_3C_{60} and K_3C_{60} some of the properties are given in Table A.6. If there are defects in the lattice, it is found that $\log T_c$ varies as the inverse square of the unit cell constant. A plot of the same is shown in Fig. A.6. Thus the carbon superconductors form a class by themselves.

Table A.4. Various physical properties of C_{60}.

S.No.	Physical property	Value
1.	Density	1.7 g/cm^3
2.	Structure	f.c.c.
3.	Nearest neighbor distance	10.04 Å
4.	Case diameter	7.1 Å
5.	Lattice constant	14.198 Å
6.	Index of refraction	2.2 at 630 nm
7.	Infrared	1429, 1183, 577 and 528 cm^{-1}
8.	Bulk modulus	18 G pascals.
9.	Ionization potential	7.6 eV
10.	Cohesive energy per C_{60} mol	1.5 eV
11.	Cohesive energy per C_{60} atom.	7.4 eV
12.	Electrical conductivity	Insulator
13.	Electronic band gap	1.5 eV
14.	Mass of the electron	1.3 m_e

Table A.5. The supercondcting transition temperatures and the lattice constant of the f.c. unit cell

S.No.	Formula	$a_o(\AA)$	$T_c(K)$
1.	Rb_2CsC_{60}	14.493	31.3
2.	Rb_2C_{60}	14.436	29.4
3.	Rb_2KC_{60}	14.364	26.4
4.	$Rb_{1.5}K_{1.5}C_{60}$	14.341	22.15
5.	K_2RbC_{60}	14.299	21.8
6.	K_3C_{60}	14.253	19.28

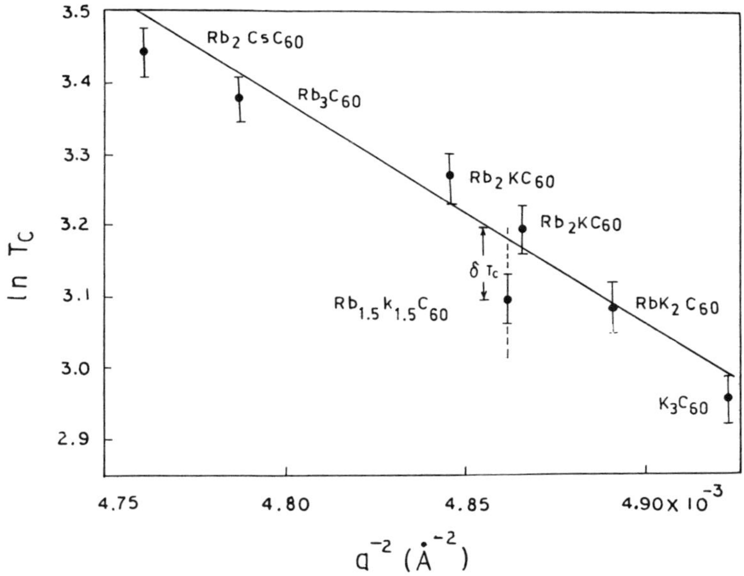

Fig. A.6. The carbon-superconductors $\ln T_c$ *as a function of inverse square of the unit cell constant shows linear behaviour except that there is a deviation near* $Rb_{1.5}K_{1.5}C_{60}$ *indicating that there are lattice defects.*

Table A.6. Superconducting parameters of Rb- and K fullerites.

S.No.	Property	Rb_3C_{60}	K_3C_{60}
1.	T_c	29.6	19.3
2.	$-(dT_c/dP)_{P=0}$K/kbar	0.97	0.78
3.	H_{c1} Oe	260	130
4.	H_{c2} Tesla	78	49
5.	J_c 10^5A/cm^2	15	1.2
6.	ξ_o Å	20±3	26±4
7.	λ_L Å	1680 ± 200	2400±200
8.	$\kappa = \lambda_L/\xi_o$	84	92

(e) Heavy fermion superconductors.

The electronic heat capacity of a solid varies linearly with temperature at low temperatures, $c = \gamma T$, where the constant of proportionality, γ determines the mass of the electron in the conduction band. It is found that in many compounds of Ce or U, this band mass is very large. Not only the effective mass but also the Fermi velocity differes by two orders of magnitude from the free electron values found in ordinary metals. In $CeCu_2Si_2$ the Fermi velocity is found to be $\bar{v}_F = 8.7 \times 10^8$ ms^{-1} and the band mass, $m^* = \hbar \bar{k}_F \bar{v}_F^{-1} = 220$ m$_o$ where m$_o$ the mass of the free electron. The superconducting transition temperature is $T_c \simeq 0.5$K and the London penetration depth approaches a high value of $\lambda_L = 0.2\mu$m, as the temperature approaches zero. However, different samples gave slightly different value as shown in Table A.7.

Table A.7. Properties of different samples of $CeCu_2Si_2$

Sample No.	$T_c(K)$	γ mJ/mol.K^2	$\Delta C/T_{\max}$ mJ/mol. K^2
1.	0.53	830	790
2.	0.47	1130	1000
3.	0.65	1270	1010
4.	0.66	900-1000	660

Here ΔC is defined as $C^{\max}/T - C_n^{\text{extrapol}}/T$ where C_n^{extrapol}/T is extrapolated value of C_n/T at the temperature where the maximum in the specific heat C occurs. The coefficient γ gives the electron mass which is several hundred times the rest mass of the electron. The large specific heat is due to strong correlations amongst electrons. In UBe_{13} a sharp superconducting transition occurs at 0.97 K with $\gamma \simeq 1.1$ J/mol K^2. The variation in T_c from sample to sample is in the range of 0.88 to 0.97 K and that in the specific heat $\Delta C \simeq 1360 - 1680$ in mJ/mol K^2. When Np is added in UBe_{13}, the T_c reduces. For $Np_{0.68}U_{0.32}B_{13}$ the transition temperature is about 0.8 K. UPt_3 has a superconducting transition temperature of 0.54 K with $\Delta C/\gamma T_c \simeq 1.0$. It shows a term in the temperature dependence of specific heat of $T^3 \ln T$ which is caused by the spin fluctuations. The slope of the upper critical field in $CeCu_2Si_2$, $dH_{c2}/dT(T = T_c)$ has been reported between $-10T/K$ to $-23T/K$ with $H_{c2}(o) = 2.4$ T for field perpendicular to the Ce and $H_{c2}(o) \simeq 2.0$ T for field parallel to the Ce planes. In UBe_{13}, $H'_{c2} \simeq -20T/K$ to $-25.7T/K$ has been reported. In UPt_3 for field perpendicular to the c axis, H'_{c2} at T_c is about $-2T/K$. The extreme low temperature indicate that there are strong magnetic correlations.

(f) High temperature superconductors.

The compounds of the formula $La_{2-x}Sr_xCuO_4$ has a superconducting transition temperature in the neighbourhood of 36K. Those near $YBa_2Cu_3O_{7-\delta}$ have a T_c near 92K for certain values of δ. The T_c increases with the increasing number of layers of Cu-O planes and approaches a maximum value of about 120K for $n = 3$ but with further increases of n, the $T_c(n)$ shows peaks which are of smaller heights than reached for $n = 3$. The T_c starts oscillating as a function of number of layers. For $Bi_2Sr_2Ca_2Cu_3O_x$ it is 105K and for $Tl_2Ca_2Ba_2Cu_3O_{10}$ it is 125K. Some of the p type materials are listed in Table A.8. The n type oxides are also included for the sake of comparison. An effort is made to understand the transition temperatures in terms of

Table A.8. Properties of superconducting materials along with carrier density and the specific heat coefficient, γ.

S.No.	Compound	$T_c(K)$	Carrier density	θ Debye (K)	γ mJ/molK^2	Type
1.	$YBa_2Cu_3O_7$	92	3×10^{21}	380	1-10	p
2.	$YBa_2Cu_3O_{6.7}$	60	1.5×10^{21}	380	1-10	p
3.	$La_{2-x}Sr_xCuO_4$	40	1.5×10^{21}	240	1-10	p
4.	$Ba_2Pb_{1-x}Bi_xO_3$	12	4×10^{21}	195	1.6	n
5.	$BaPbO_{3-\delta}$	0.4	2×10^{20}	166	0.6	n
6.	$SrTiO_{3-\delta}$	0.4	1.4×10^{20}	433	1.6	n
7.	$Li_{1+x}Ti_{2-x}O_4$	12	1.3×10^{22}	660	20	n
8.	$CuRh_{2-x}Sn_xSe_4$	3.5	4×10^{21}	218	7	n

charge transfer. The energy of oxygen $2p$ electron excitation is expected to scale with the difference in electronegativity between the anion and cation, so that we expect the charge transfer to be related to the transition temperature. Hence the change in the electronegativity is tabulated along with the superconducting transition temperature in Table A.9. Indeed, large change in electronegativity appears to be linked with large T_c.

Table A.9. Qualitative comparison between change in Pauling electronegativity and the superconducting transition temperature.

S.No.	Compound	Electronegativity Change	T_c
1.	$Li_{0.8}Ti_{2.2}O_4$	1.908	11.5
2.	$Li_{0.8}Ti_{2.2}O_{3.95}$	1.950	11.1
3.	$La_{1.85}Sr_{0.15}CuO_4$	1.959	37
4.	$La_{1.925}Sr_{0.075}CuO_4$	1.950	27
5.	$YBa_2Cu_3O_7$	2.038	91
6.	$YBaCaCu_3O_7$	2.008	77
7.	$Y_{0.75}Sc_{0.25}Ba_2Cu_3O_7$	2.031	87
8.	$Y_{0.5}La_{0.5}Ba_2Cu_3O_7$	2.023	82
9.	$YBa_{1.5}Sr_{0.5}Cu_3O_7$	2.033	85
10.	$YBa_{1.8}Sr_{0.2}Cu_3O_7$	2.036	87
11.	$YBa_{1.6}Sr_{0.4}Cu_3O_7$	2.028	81
12.	$Y_{0.8}Al_{0.2}Ba_2Cu_3O_7$	2.021	80
13.	$Y_{0.6}Al_{0.4}Ba_2Cu_3O_7$	2.004	62
14.	$YBa_2Cu_3O_6F_2$	2.151	155
15.	$YBa_2Cu_3O_{6.67}F_{0.67}$	2.083	118
16.	$LaBa_2Cu_3O_7$	2.008	75
17.	$La_{1.75}Bi_{0.05}Sr_{0.2}CuO_4$	1.961	38
18.	$Tl_2Ca_2Ba_2Cu_3O_y$	2.071	125
19.	$Tl_2CaBa_2Cu_3O_y$	2.093	108
20.	$Tl_5Ba_2Ca_3Cu_6O_y$	2.057	106
21.	$Bi_{2.07}Ca_{0.98}Sr_{1.68}Cu_2O_y$	2.001	70
22.	$Bi_{2.1}Sr_{2.11}Ca_{0.89}Cu_2O_y$	2.026	82
23.	$BaPb_{0.75}Bi_{0.25}O_3$	2.176	11.7
24.	$Ba_{0.6}K_{0.4}BiO_3$	2.194	25
25.	$Rb_{0.33}WO_3$	1.950	2.05

In 1982 it was found by Shrivastava that the valency changes as a function of temperature in the rare-earth intermetallic compounds. In the high temperature superconductors also there is a change in valency due to the change in electron-phonon interaction. The oxygen content is also indicative of a change in the transition temperature. The value of the transition temperature as a function of x and δ $Bi_2Sr_2Ca_{1-x}Y_xCu_2O_{8+\delta}$ is shown in Table A.10.

Table A.10. Composition and oxygen content dependence of transition temperature in Bi containing superconductor as discussed in the text.

S.No.	x	δ	Cu-valency	T_c(K)
1.	0.0	0.23	2.23	70
2.	0.1	0.23	2.18	90
3.	0.3	0.30	2.15	86
4.	0.35	0.30	2.12	74
5.	0.4	0.29	2.09	64
6.	0.5	0.33	2.08	40
7.	0.75	0.41	2.03	< 4
8.	1.0	0.51	2.01	< 4

A series of rare-earth borocarbides, $LuNi_2B_2C$ have superconducting transition temperature of about 16K but LuNiBC is not a superconductor. The superconductor is obtained by adding carbon in the Lu plane. LuNiBC is derived from $LuNi_2B_2C$ by adding another layer of Lu-C. Thus superconducting combination is obtained from an insulator by varying relative ratio of the number of atoms. Similarly in $La_{2-x}Sr_xCuO_4$ superconducting phase is obtained from the insulating phase by varying x. The compounds, RNi_2B_2C with R =Y, Lu, Tm, Er, Ho and Dy are supercondcuting with moderately high values of the transition temperature, $T_c \sim 16$K. There is an antiferromagnetic phase at low temperatures and there is possibly a phase in which superconductivity and magnetic order seem to coexist. There is a soft spin wave mode which is slightly incommensurate with the crystal lattice along the a direction. The fraction of the Fermi surface that can participate in nesting is ~ 4.4 per cent.

The dependence of the transition temperature of a superconductor on the number n of Cu-O layers per unit cell has been examined in ref. A.9. We have examined the electron-phonon interaction from the view point of its dependence on the crystal structure with oxygen defects which destroy the translational symmetry. Introducing the wave vector in the coupling constant and transforming the electron-phonon interaction as in the B.C.S. theory and linearizing the transition temperature leads to a T_c dependent on a spherical Bessel function,

$$T_c = 110 - 70.68 j_o(n) \quad K\,.$$

We predict from Fig. A.7 that there is a maximum at $n = 3$ and the next maximum occurs at $n = 7$ in the family of superconductors, $(Bi/Tl)_2Ca_{n-1}(Sr/Ba)_2Cu_nO_{2n+4}$ where $n =$ integer.

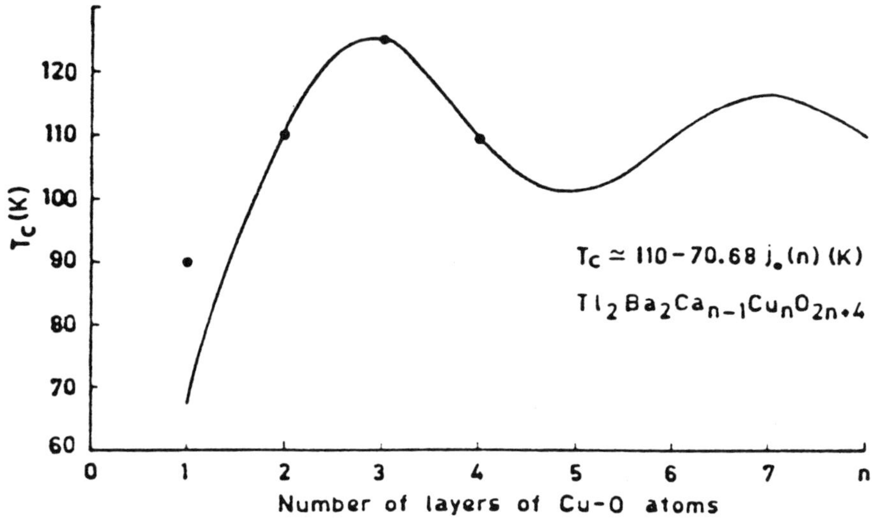

Fig. A.7. The transition temperature as a function of number of layers of Cu-O atoms showing a maximum at $n = 3$ but reduction at $n = 4, 5$ and 6. The next maximum occurs at $n = 7$ but the value is less than at $n = 3$. More detailed theory is given in ref. A.9.

Hg substituted in the high temperature superconductors leads to compounds of large transition temepratures.

The compounds $HgBa_2Ca_{n-1}Cu_nO_{2n+2+\delta}$ ($n \geq 1$) are superconducting with transition temperatures as large as 135K for $HgBa_2Ca_2Cu_3O_{8+\delta}$. The transition temperature increases with increasing number of Cu-O layers upto $n = 3$, e.g.,

Hg(1201) (T_c=94K), Hg(1212) (T_c=128K), Hg(1223) ($T_c \simeq$133-135K), Hg(1234) (T_c=123K) and Hg (1245) (T_c=110K) exactly as predicted.

In recent years many superconducting compounds have been synthesized.

Suggested References

A.1 H. Uryama, H. Yamochi, G. Saito, S. Saito, A. Kawamoto, J. Tanaka, T. Mori, Y. Maruyama and H. Inokuchi, Chem. Lett. **1988**, 463 (1988).

A.2 M. Tokumoto, in The Physics and Chemistry of Organic Superconductors, eds. G. Saito and S. Kagoshima, Springer Verlag, Berlin, 1990.

A.3 K. Murata, M.M. Takumoto, H. Anzai, H. Bando, K. Kajimura, T. Ishiguro and G. Saito, Synth. Metals. **13**, 3 (1986); Synth. Metals **19**, 151 (1987).

A.4 J. Muller, Rept. Prog. Phys. **43**, 641 (1980).

A.5 E.S. Machlin and S.H. Whang, J. Phys. Chem. Solids **37**, 555 (1976); **39**, 607 (1978).

A.6 K. Tanigaki, T.W. Ebbesen, S. Saito, J. Mitzuki, J.S. Tsai, Y. Kubo and S. Kuroshima, Nature **352**, 222 (1991).

A.7 R.M. Fleming, A.P. Ramirez, M.J. Rosseinsky, D.W. Murphy, R.C. Haddan, S.M. Zahurak and A.V. Makhija, Nature **352**, 787 (1991).

A.8 J.E. Schrieber, D.L. Overmyer, H.H. Wang, J.M. Willims, K.D. Carlson, A.M. Kini, U. Wlep and W.-K. Kwok, Physica C**178**, 137 (1991).

A.9 N. M. Krishna and K.N. Shrivastava, Supercond. Sci. Technol. **10**, 278 (1997).[Describes the oscillations in the transition temperature as a function of number of Cu-O layers].

A.10 U. Rauchschwalbe, et. al., Phys. Rev. Lett. **49**, 1448 (1982).

A.11 F. Steglich et. al., Phys. Rev. Lett. **43**, 1892 (1979).

A.12 G.R. Stewart, Rev. Mod. Phys. **56**, 755 (1984).

A.13 W.E. Pickett, Rev. Mod. Phys. **61**, 433 (1989).

A.14 K. N. Shrivastava, Phys. Repts. **200**, 51 (1991).

A.15 T. Siegrist, H.W. Zandbergen, R.J. Cava, J. Krajewski, and W.F. Peck, Jr., Nature **367**, 254 (1994).

A.16 K.N. Shrivastava, Phys. Lett. A**91**, 33 (1982).

A.17 W.A. Groen, D.M. deLeeuw and L.F. Feiner, Physica C**165**, 55 (1990).

A.18 L.J. De Jongh, Physica C**152**, 171 (1988).

A.19 D.A. Nepela and J.M. McKay, Physica C**158**, 65 (1989).

A.20 R. Nagarajan, C. Mazumdar, Z. Hossain, S.K. dhar, K.V. Gopalakrishnan, L.C. Gupta, C. Godart, B.D. Padalia and R. Vijayaraghavan, Phys. Rev. Lett. **72**, 274 (1994).

A.21 R.J. Cava, et. al., Nature **367**, 252 (1994).

A.22 S.B. Dugdale, et al, Phys. Rev. Lett. **83**,4824 (1999).

A.23 S.N. Putilin, et al, Nature **362**, 226 (1993).

A.24 A. Schilling, M. Cantoni, J.D. Guo and H.R. Ott, Nature **363**, 56 (1993).

A.25 J. A. Hodges, I. Gameson, P.P. Edwards, A.P. Kharel and A. Porch, Phys. Rev. B**55**, 12776 (1997).

A.26 H. Zhang, L.L. Cheng, X.C. Qin and Y. Zhao, Phys. Rev. B**61**, 1618 (2000).

A.27 K. N. Shrivastava, Bull. Am. Phys. Soc.**39**, 1127 (1994).

A.28 H. J. Kaufmann, O. V. Dolgov and E. K. H. Salje, Phys. Rev. B**58**,9479 (1998).

Appendix B

B.1. Isotope effect.

Prior to the development of B.C.S. theory of superconductivity, it was found that the transition temperature of a superconductor depends on the inverse square root of the isotopic mass. By making a mixture of different quantities of different isotopic masses, it is possible to define the average mass per atom. The transition temperatures can then be related to the inverse square root of the average mass per atom. The B.C.S. were thus motivated to use the electron-phonon interaction which is linked to the mass per atom. It was found that electrons became attractive by exchange of phonons. It is possible to exchange some other quasiparticles but in a material like mercury there are no candidates other than phonons. Therefore the B.C.S. obtained the transition temperature of a superconductor as, (1.96),

$$k_B T_c \propto 1.13\hbar\omega \exp(-1/N(o)V) \ . \tag{B.1}$$

The potential energy of a harmonic oscillator is written as $\frac{1}{2}kx^2$ where the force constant k is related to the classical frequency $\omega = (k/m)^{1/2}$ with m as the mass per atom. Therefore,

$$T_c \propto \frac{1}{m^{1/2}} \tag{B.2}$$

is readily understood. However, there may be mass dependence in the density of state $N(0)$ or in the attractive potential V. Since the square root of mass did not emerge properly in the experimental measurements of transition temperatures, it was decided to write

$$T_c \propto \frac{1}{m^{\alpha}} \tag{B.3}$$

in which the value of α may be determined by experiments very often $\alpha = 0.5$ was not found and smaller values of α were noted. Some of the reduction in α may be understood by including the electron-electron repulsive interaction in addition to the attractive part of the electron-phonon interaction. The displacement of the harmonic oscillator is proportional to $(\hbar/2m\omega)^{1/2}$. Therefore the square of the amplitude of the oscillation of the harmonic oscillator varies as $\hbar/2m\omega$. We can eliminate ω as $(k/m)^{1/2}$ so that the square of the amplitude becomes $\hbar/2m(k/m)^{1/2} \propto 1/m^{1/2}$. Therefore mass dependence comes from V in (B.1) so that (B¿2) is not strictly correct. Similarly, the electronic density of states $N(0)$ in (B.1) depend on the single particle frequency and hence on the mass dependent shift. Therefore, (B.2) need not be strictly observed. The attractive potential

$$\frac{D^2\hbar\omega}{(\epsilon_k - \epsilon_{k'})^2 - (\hbar\omega_{k-k'})^2} \simeq -\frac{D^2}{\hbar\omega} \simeq -\frac{D^2 m^{1/2}}{\hbar k^{1/2}} \qquad (B.4)$$

appears to depend on mass but in a simple model $D^2 \simeq D_o^2\hbar/2m\omega \simeq D_o^2\hbar/2m(k/m)^{1/2} \simeq D_o^2\hbar/2k^{1/2}$ so that (B.4) becomes independent of mass and (B.2) becomes more likely. However, in the derivation of (B.1) the potential is kept independent of single particle frequency so that the bound state equation can be integrated. The solution of the bound state equation can not be obtained for any arbitrary form of the frequency dependence of the attractive potential. The Fermi distribution has the energy ratio $\hbar\omega_D/k_BT_c$ when evaluated at $T = T_c$ and when the entire attractive range of single particle energy is taken to be $\hbar\omega_D$. The Debye frequency may be taken in the form $\hbar\omega_D = \hbar(k_D/m)^{1/2}$ except that the force constant is evaluated at the cutoff frequency so that the Fermi term $1 - 2f_k$ may also depend on mass. However, $1-2f_k \simeq 1$ when $f_k << 1$, is independent of the mass per atom. From the bound state equation $\Delta_k \propto 1/2\epsilon_k \propto 1/2\omega_D \sim \frac{m^{1/2}}{2k^{1/2}}$ when $\epsilon_k \sim \hbar\omega_D$. Since $\Delta_k \sim k_BT_c$, the transition temperature varies as the square root of mass, $T_c \propto m^{1/2}$ which is different from (B.2). Since deviations are expected from both these results, we write,

$$T_c \propto m^{-\alpha(1-\zeta)}. \qquad (B.5)$$

One of the samples of Pb has 25.8% Pb^{204}, 33.9% Pb^{206}, 14.2% Pb^{207} and 26.1% Pb^{208}. In this way several samples can be made and $T_c(m)$ can be determined. One of the studies showed $\alpha = 0.461 \pm 0.025$. Molybdenum isotopes of mean mass 92.37, 96.04, 97.94 and 99.24 together with natural Mo mass 95.95 show $\alpha = 1/3$. The values of α for some of the materials are given in Table B.1. It is thought that

Table B.1. The values of the isotope effect exponent α of eq.(B.3) for several metals

S.No.	Material	α
1.	Zn	0.45±0.05
2.	Cd	0.50±0.10
3.	Sn	0.47±0.02
4.	Hg	0.50±0.03
5.	Pb	0.48±0.01
6.	Tl	0.50±0.10
7.	Nb_3Sn	0.08±0.02
8.	Mo_3Ir	0.33±0.03

deviations from the value of $\alpha = 0.5$ are caused by the repulsive Coulomb interaction between electrons. The values of α given in Table B.1 for pure metals suggest that the electron-phonon interaction is important for the mechanism of superconductivity. The samples of $Y_{1-x}Pr_xBa_2Cu_3O_7$, $YBa_{2-x}La_xCu_3O_7$ and $YBa_2(Cu_{1-x}Co_x)_3O_7$ have been prepared with ^{16}O as well as ^{18}O. In this case $\alpha \simeq 0.06$ is found which does not agree with the value of 0.5 expected for the electron-phonon interaction. Apparently, the value of α can be controlled by doping. Large T_c is associated with small α and small T_c with large α. For $T_c \simeq 40$K, the value of α is about 0.5.

B.2. Proximity isotope effect.

We consider a film of a superconductor of which the resistivity is measured to determine the transition temperature. When a thin film of a normal metal is deposited on the superconducting film, there is a change in the transition temperature as well as in the isotope-effect exponent. When the thickness of the normal metal, L_N is smaller than the normal coherence length, ξ_N there is a pronounced effect on the exponent of the isotope effect. Here the normal state coherence length is defined in terms of the Fermi velocity, v_F^N as, $xi_N = \hbar v_F^N / k_B T$. Since the thickness of the normal film is smaller than ξ_N, the normal electrons can reach the superconducting regions and hence reduce the superconducting transition temperature by enhancing the scattering of the superconducting electrons. The electrons can tunnel through the surface barrier so that the scattering rate can be calculated from the tunneling matrix element.

$$\Gamma_{ik} = |T_{ik}|^2 \rho_k V_k \tag{B.6}$$

where Γ_{ik} is the scattering rate, T_{ik} is the matrix element of tunneling from the state $|i>$ to state $<k|$, ρ_k is the density of states per unit volume and V_k is the volume. The states $|i>$ and $<k|$ are in the normal or in the superconducting film, $|i>=|N>$ or $|S>$. Similarly $<k|=<N|$ or $<S|$ for the normal or superconducting part of the film. The matrix element T_{iK} is taken for the tunneling from the normal state to the superconducting state and vice versa but neither from normal to normal nor from superconductor to superconductor. The tunneling rate is determined from $\Gamma = \Gamma_{SN} + \Gamma_{NS}$ where Γ_{SN} is the tunneling rate from normal to the superconductor and Γ_{NS} is the similar rate for tunneling from superconductor to the normal part of the film. The transition temperature of the superconducting film without the normal surface is T_{c0} and it changes to T_c when a normal film is deposited on the superconducting surface,

$$T_c = T_{co}(\pi T_{co}/2\gamma u)^\rho \tag{B.7}$$

with

$$\rho = \rho_N L_N / \rho_s L_s \tag{B.8}$$

where $\gamma = 0.577$ is the Euler's constant. If $\Gamma \gg \omega_D$ the Debye cutoff frequency, the tunneling rate is so fast that phonons play no role, then $u = \omega_D$. If $\Gamma \ll \omega_D$, $u \simeq \Gamma$. Thus the value of u is between $\hbar\Gamma/2k_B$ and $\hbar\omega_D/k_B$. When the mass m is changed to isotopic mass m^* in the superconducting film, there is a change in mass per atom, $\Delta m = m^* - m$. Similarly, there is a change in the transition temperature from T_{co} for m to T^*_{co} for $m^*, \Delta T_{co} = T^*_{co} - T_{co}$ for which the exponent is α_o,

$$\alpha_o = -\frac{m}{\Delta m}\frac{\Delta T_{co}}{T_{co}}. \tag{B.9}$$

The presence of the normal film changes the T_{co} due to scattering which is slightly different from ΔT_c. Therefore, the value of α_o is also changed by the presence of the normal film. In the above expression, change is obtained from T_{co}. Therefore, in the case of normal film quoting, the change should occur from T_c so that the exponent of the isotope effect may be written as,

$$\alpha = -\frac{m}{\Delta m}\frac{\Delta T_c}{T_c} \quad . \tag{B.10}$$

If the superconducting film is covered by the normal film then the value of T_c is modified by the proximity effect and the value of the critical temperature for the sandwich is described by (B.7). Then the value of the shift ΔT_c is determined not only by ΔT_{co} but also by the value of the parameter ρ given by (B.8) which has the properties of the normal density of states and the normal length. In the range $k_B T_c \ll \Gamma \ll \omega_D$, $u \approx \Gamma$, the isotope substitution leads to a shift of T_{co} by the amount ΔT_{co}. The equation (B.7) contains $T_{co}^{\rho+1}$. Therefore its variation has $(\rho+1)T_{co}^{\rho}\Delta T_{co}$. Accordingly finding the variation of (B.7), we find,

$$\Delta T_c = \left(\frac{\pi}{2\gamma\Gamma}\right)^{\rho} \Delta(T_{co}^{\rho+1}) = (\rho+1)\Delta T_{co}\left(\frac{\pi T_{co}}{2\gamma\Gamma}\right)^{\rho} . \tag{B.11}$$

Therefore,

$$\frac{\Delta T_c}{T_c} = \frac{\Delta T_{co}}{T_{co}}(1+\rho) \tag{B.12}$$

and hence the exponent of the isotope effect becomes,

$$\alpha = \alpha_o[1 + (\rho_N L_N)/(\rho_s L_s)]. \tag{B.13}$$

Therefore, a decrease in T_c due to tunneling from the normal film causes an icrease in the exponent of the isotope effect. The value of α can be increased by increasing the density of states in the normal state and by increasing the thickness of the normal film. If $\alpha_o = 0.5, \rho_N/\rho_s = 0.8$, $L_N/L_s = 0.5$ then $\alpha = 0.7$. The increase in α is thus based on the density of states and hence on the dimensionality, d. For n dimensional layered compounds the normal density of states depends on the single-particle energy. Therefore, the value of α depends on the dimensionality of the system.

In the opposite limit, $\Gamma \gg \omega_D$, the exponent of the isotope effect is small because both T_{co} and ω_D are affected by the isotope substitution. Since the proximity effect leads to a change in the value of the isotope coefficient it is clear than the exponent of the isotope effect changes significantly when normal quanting is done to the superconductor.

B.3. Magnetic impurities isotope effect.

We consider a superconductor with magnetic impurities. For field larger than the lower critical field, H_{c1}, it shows vortices and becomes normal when the applied field is larger than the upper critical field. There is considerable scattering of electrons by the magnetic atoms so that the transition temperature is much less than that of the pure superconductor, $T_c = T_{co} - \hbar/2\tau_{AG}k_B$. Hence T_{co} is the transition temperature of the superconductor without magnetic atoms and τ_{AG} is the Abrikosov-Gorkov scattering time due to magnetic atoms. Another way of saying that there is a reduction in the transition temperature is by use of digamma function, $\psi(x)$. The reduction in T_{co} is then described by

$$ln(T_{co}/T_c = \psi[0.5 + \gamma_s] - \psi(0.5) \qquad (B.14)$$

where $\gamma_s = \Gamma_s/2\pi T_c$, $\Gamma_s = \tau_{AG}^{-1}$. It is assumed that the scattering rate is a linear function of the concentration of magnetic atoms so that $\Gamma_s = \Gamma_{so}n_m$ when n_m is the concentration of magnetic impurities and Γ_{so} is the value of Γ_s when $n_m = 1$ which means that there is only one mangetic atom in the superconductor and we calculate the scattering rate with only one magnetic atom and multiply the same by the concentration of magnetic atoms. Using (B.9), (B.10) and (B.14) we obtain,

$$\alpha = \alpha_o[1 - \psi'(0.5 + \gamma_s)\gamma_s]^{-1} \ , \qquad (B.15)$$

where α is the exponent of the isotope effect in the presence of magnetic impurities. This equation is valid in the broad range except for a very small region near the critical concentration, n_{cr} where the condition $\Delta T_c/T_c \ll 1$ is not satisfied and T_c is close to zero. When the concentration of magnetic atoms exceeds n_{cr}. There is not superconducting phase. The presence of magetic impurities leads to an increase of the exponent of the isotope effect, $\alpha > \alpha_o$, since $\psi' > 0$. For small concentration of magnetic impurites, the change in α is proprotional to the concentration of magnetic atoms, n_m. In the region $\gamma_s \gg 1$, we can use the asymptotic expansion for the digamma function so that the change in α becomes proportional to n_m^2. For the critical concentration $n_m \simeq n_{cr}$, the transition temperature approaches zero, $T_c \to 0$ and hence the dependence of α on n_m also reduces. In this way the T_c reduces whether we introduce a non-magnetic proximity film or magnetic impurities and α seems to increase. Thus we have no way to increase the T_c or to reduce α.

B.4. Adiabatic concentration isotope effect.

We consider the product of electronic wave function and the vibrational wave function and two such wave functions which cross as a function of distance. The wave function determines the concentration of carriers which depends on mass because of the dependence of the matrix element of the distance-dependent hamiltonian and the single particle energy. The transition temperature depends on the carrier concentration and hence implicit mass dependence appears in the T_c. In a three-dimensional system all of the modes contribute equally but in the case of a two-dimensional system the amplitude of one of the modes is weak and may be much less than these in the plane. Therefore oscillations cross each other at a particular position and hence become degenerate so that the degree of freedom describing the motion corresponds to electronic term crossing. An atom has two equilibrium positions so that there are degenerate states. There may be an oscillation of a certain symmetry with respect to which the degeneracy is removed. Then the system distorts itself to acquire an equilibrium position in which that symmetry is missing. This is called the Jahn-Teller effect which introduces the mass effect in the electronic properties. The Jahn-Teller effect may be static in which the symmetry is entirely removed or it may be dynamic in which the symmetry is only dynamically removed in a time dependent manner. The oscillations occur without any change in temperature so that the change in symmetry is not associated with heating or cooling. The coordinate dependent hamiltonian is of the form $\mathcal{H}_r = T_r + V(r, R)$ where T_r is the kinetic energy and $V(r, R)$ is the potential energy which depends on the coordinates of the electrons as well as those of the atoms. The total wave function can be written as

$$\psi(r, R, t) = a(t)\psi_1(r, R) + b(t)\psi_2(r, R) \qquad (B.16)$$

where ψ_1 and ψ_2 are the products of electronic and vibrational wave functions that correspond to two different electronic states,

$$\begin{aligned} \psi_1(r,R) &= \psi_{1e}(r,R)\phi_1(R) \\ \psi_2(r,R) &= \psi_{2e}(r,R)\phi_2(R) \end{aligned} \tag{B.17}$$

where $\psi_{1e}(r,R)$ corresponds to one electronic state which is a function of electron coordinates r as well as a function of atomic coordinate, R. The vibrational wave function associated with the first electronic state is $\phi_1(R)$ with similar interpretation of $\psi_2(r,R)$ for the second electron. We introduce the anisotropy in the vibrational wave function so that the amplitude of vibrations along the z direction is very different from that in the xy plane. The vibrational wave function can then be factorized as,

$$\phi_1(R) = \varphi_1(x,y)\varphi_1(z) \tag{$B.18$}$$

with similar factors for $\phi_2(R)$. The electronic parts ψ_{1e} and ψ_{2e} are non-degenerate so that the vibrational parts are split by the energy ϵ. In the harmonic approximation.

$$\varphi_1(z) = \left(\frac{2\pi^2\hbar}{m\omega_1}\right)^{-1/2} \exp[-(z-z_{o1})^2/2(2\pi\hbar/m\omega_1)]$$

and

$$\varphi_2(z) = \left(\frac{2\pi^2\hbar}{m\omega_2}\right)^{-1/2} \exp[-(z-z_{o2})^2/2(2\pi\hbar/m\omega_2)] \tag{$B.19$}$$

where m is the atomic mass and z_{o1} and z_{o2} are the z coordinates of the two atoms in the equilibrium. The matrix element of a transition may be written as,

$$V_{12} \simeq L_o F_{12} \tag{$B.20$}$$

where

$$L_o = <\psi_2|\mathcal{H}_r|\psi_1> \tag{B.21}$$

$$F_{12} = <\varphi_2(z)|\varphi_1(z)> \quad . \tag{B.22}$$

The Franck-Condon factor depends on the mass through the mass dependence of the wave function in (B.19) but the value of z in $\varphi_1(z)$ is the same as in $\varphi_2(z)$ otherwise the momentum can not be conserved. The value of $b(t)$ of (B.15) is given by,

$$|b(t)|^2 = b_o^2[1 - \cos 2\{(\epsilon^2/4) + H_{12}^2\}^{1/2} t] \qquad (B.23)$$

where

$$H_{12} = < 1|\mathcal{H}_r|2 > \qquad (B.24)$$

$$b_o^2 = \frac{L_o^2 F_{12}^2}{2[(\epsilon^2/4) + L_o^2 F_{12}^2]} \; . \qquad (B.25)$$

When the splitting ϵ becomes zero, the above probability becomes 0.5. The probability (B.23) oscillates with frequency $\{(\epsilon^2/4) + H_{12}^2\}^{1/2}$ with the average value given by (B.25). The potential is assymmetric along z direction as in (B.18). In case of large asymmetry, $L_o F_{12} \ll \epsilon$ so that $L_o F_{12}$ in the denominator of (B.25) can be neglected for the purpose of determining the carier concentration,

$$n_{p1} \propto \tilde{b}^2 = 2L_o^2 |F_{12}|^2/\epsilon^2 \qquad (B.26)$$

which depends on the inverse square of the splitting energy, ϵ, and hence on the atomic mass. The exponent of the isotope effect thus has two terms which depend on mass. The explicit dependence on mass comes from the mass dependence of the phonon frequency and there is considerable implicit mass dependence due to other quantities. Therefore, we can write $\alpha = \alpha_1 + \alpha_2$ where $\alpha_1 = -(m/T_c)(\partial \tau_c/\partial \omega)(\partial \omega/\partial m)$ describes the isotope effect because of the mass dependence of the transition temperature. The contribution α_2 is the implicit effect of the mass dependence of the carrier concentration and the dependence of the transition temperature on the carrier concentration, i.e.,

$$\alpha_2 = -\frac{m}{T_c}\frac{\partial T_c}{\partial n}\frac{\partial n}{\partial m} \; . \qquad (B.27a)$$

According to eqs.(B.21) and (B.26) the dependence $n(m)$ is determined by the Franck-Condon factor (B.22). From (B.20), (B.22) and (B.26) and the dependence $\epsilon^2 \propto 1/m$, we expect

$$\alpha_2 \propto \gamma_A (n/T_c)(\partial T_c/\partial n) \qquad (B.27b)$$

where the constant of proportionality γ_A has a weak logarithmic dependence on m. In this way, the isotope effect can be written in terms of implicit as well as explicit contributions of mass to the transition temperature.

B.5. Spin Fluctuations in isotope effect.

We consider the spin fluctuations by means of imaginary part of the wave vector and frequency dependent susceptibility. We have already considered the g shift in Chapter 5 and the Knight shift in Chapter 10 which depend on the response function. We add the spin dependent interaction in the B.C.S. hamiltonian with the coupling constant,

$$V_a^{sf}(\omega) = g_{sf}^2 \chi(q, \omega)$$

where the potential depends on the wave vector q and the spin-fluctuation, (sf), coupling constant is g_{sf}. The effect of addition of this interaction is to change the value of the gap energy, Δ which varies as a function of phonon frequency. In the phonon theory the gap increases with increasing the phonon frequency. We expect that at small frequencies, the effect of the spin fluctuation is to reduce the gap with increasing phonon frequency. Therefore, the transition temperature is expected to reduce with increasing phonon frequency at small frequencies and increase with increasing phonon frequency at large frequencies. The phonon frequency depends on the mass of an atom as in the classical oscillator. Therefore, the exponent of the isotope effect also changes due to spin fluctuations.

B.6. Penetration depth isotope effect.

We arbitrarily assume that the London penetration depth depends on the mass per atom upon substitution of isotopic mass,

$$\lambda_L \sim m^{-\beta} \qquad (B.28)$$

to be compared with (B.3) which describes the change in the transition temperature. The argument is that the electron-phonon interaction changes the electron single-particle energy in a mass-per-atom dependent way so that the electron concentration acquires acquires a mass dependence. Then the derivative of the electron concentration with respect to the mass becomes nonzero and (B.27) gives a small value. In Chapter 6, we have discussed the origin of the penetration depth which involves the electron concentration. We now imply that $n_s = n_s(m)$. According to the electromagnetic theory

$$\lambda^2 = \frac{m_e c^2}{4\pi n_s e^2} \tag{B.29}$$

where m_e is the effective mass of the electron, n_s is the superconeducting density of charge carriers. New the phase transition

$$n_s = n_s^o[1 - (T/T_c)^4] \tag{B.30}$$

which goes to zero at $T = T_c$. At $T = 0, n_s = n_s^o$ is the concentration of superconducting quasiparticles. Therefore, we can write the exponent of (B.28) as,

$$\beta = -\frac{m}{\lambda_L}\frac{\partial \lambda_L}{\partial n_s}\frac{\partial n_s}{\partial m} = \frac{m}{2n_s}\frac{\partial n_s}{\partial m}. \tag{B.31}$$

In (B.30) the concentration of quasiparticles depends on the transition temperature and hence on the mass through the exponent α. Therefore, there is a B.C.S. type contribution to β which may arise from the variation of phonon frequency with the atomic mass as in classical oscillator. The isotopic mass leads to a shift in T_c and hence in λ_L which is valid near T_c. In addition to this effect, there is another effect arising from the isotope dependence of the charge-carrier concentration. From eq.(B.30) and (B.31) we can separate the two types of contributions to the mass dependence of the exponent of the penetration depth as,

$$\beta = \beta_{na} + \beta_{ph} \ , \tag{B.33}$$

where the zero-temperature concentration of superconducting quasiparticles depends on the electron-phonon interaction and hence on the isotopic mass so that,

$$\beta_{na} = \frac{m}{2n_s^o}\frac{\partial n_s^o}{\partial m} \tag{B.33}$$

and

$$\beta_{ph} = \frac{m}{2[1-(T/T_c)^4]}\frac{\partial}{\partial m}[1-(T/T_c)^4] \tag{B.34}$$

because of the B.C.S. requirement of the dependence of T_c on the mass. Since the electrons are paired in the superconducting state, the number of pairs is half the number of electrons. From (B.27b) and (B.33), $\beta_{na} \simeq -\gamma_A/2$ so that

$$\alpha_{na} = -2\beta_{na}\frac{n}{T_c}\frac{\partial T_c}{\partial n}. \tag{B.35}$$

Apparently, there is a critical doping. For underdoped region in the high temperature superconductors, $\partial T_c/\partial n > 0$ so that α_{na} and β_{na} have opposite signs. From (B.29) and (B.30), there is an effect of the isotopic substitution on T_c and hence on the penetration depth. The magnetic impurities also change the T_c and hence have effect on λ_L. The density of states depend on the gap of the superconductor. Therefore, the penetration depth depends on the gap of the superconductor so that the exponent β is affected by the impurity scattering.

We thus find that the exponent of the isotope effect which is 0.5 in the B.C.S. theory for metals like Hg may be changed in going from simple metals to more complicated materials which have phonons of more involved dispersion relations or magnetic excitations.

Suggested References

B.1 J.W. Garland Jr., Phys. Rev. Lett. **11**, 114 (1963).

B.2 W.I. McMillan, Phys. Rev. **167**, 331 (1968).

B.3 R.W. Shaw, D.E. Mapother and D.C. Hopkins, Phys. Rev. **121**, 86 (1961).

B.4 B.T. Matthias, T.H. Geballe, E. Corenzwit and G.W. Hull, Jr., Phys. Rev. **129**, 1025 (1963).

B.5 V.Z. Kresin, A. Bill, S.A. Wolf and Yu. N. Ovchinnikov, Phys. Rev. B**56**, 107 (1997).

B.6 W. McMillan, Phys. Rev. **175**, 537 (1968).

B.7 V. Kresin, Phys. Rev. B**25**, 157 (1982).

B.8 J. Carbotte et al, Phys. Rev. Lett. **66**, 1789 (1991).

B.9 S.P. Singh, R.K. Pandey and P. Singh, J. Supercond. **9**, 269 (1996).

B.10 V.Z. Kresin and S.A. Wolf, Phys. Rev. B**49**, 3652 (1994).

B.11 T.S. Nunner, J. Schmalian and K.H. Bennemann, Phys. Rev. B**29**, 8859 (1999).

B.12 A. Bill, V.Z. Kresin and S.A. Wolf, Phys. Rev. B**57**, 10814 (1998).

B.13 Y. Yokoya, Phys. Rev. B**56**, 6107 (1997).

B.14 H.B. Schüttler and C.H. Pao, Phys. Rev. Lett. **75**, 4504 (1995).

B.15 C.H. Pao and H.B. Schüttler, Phys. Rev. B**60**, 1283 (1999).

B.16 A Bussmann-Holder, A.R. Bishop, L. Genzel and A. Simon, Phys. Rev. B**55**, 11751 (1997).

B.17 Y.N. Chiu, Phys. Rev. B**55**, 6022 (1997).

B.18 M.S. Fuhrer, K. Cherrey, A. Zettl, M.L. Cohen and V.H. Crespi, Phys. Rev. Lett. **83**, 404 (1999).

B.19 G.V.M. Williams, J.L. Tallon, J.W. Quilty, H.J. Trodahl and N.E. Flower, Phys. Rev. Lett. **80**, 377 (1998).

B.20 A. Shengelaya, G. Zhao, C.M. Aegerter, K. Conder, I.M. Savic and H. Keller, Phys. Rev. Lett. **83**, 5142 (1999).

B.21 G. Zhao, K. Ghosh, H. Keller and R.L. Greene, Phys. Rev. B**59**, 81 (1999).

B.22 D.J. Pringle, J.L. Tallon, B.G. Walker and H.J. Trodahl, Phys. Rev. B**59**, 11679 (1999).

Appendix C: Symmetries

C.1. Gap.

In the original B.C.S. theory the potential due to the electron-phonon interaction has been assumed to be isotopic so that the bound state equation comes out to be in a particularly integrable form. The transition temperature can then be obtained in a simple form. The relation between the gap and the transition temperature assumes that the potential is independent of wave vector, and the solutions are obtained for isotopic three dimensional systems. It has been observed that magnetism plays an important role so that the superconducting state may be a triplet instead of a B.C.S. singlet. In that case, the superconducting state will have the same symmetry as that of the crystallographic point group. The critical exponents are sensitive functions of the dimensionality according to (9.17). The London-penetration depth varies with temperature with an exponent characteristic of dimensionality of the system indicating that the superconducting state may depend on the dimensionality and hence on the crystal structure. The gap of the superconductor involves the symmetry of the potential as well as that of the fermion wave-vectors which are summed according to the symmetry of the Fermi surface. The superconducting order parameter may be written as,

$$\begin{aligned} \Delta(k) &= \psi(k) i\sigma_y \qquad (S=0) \\ \Delta(k) &= i\sigma d(k)\sigma_y \qquad (S=1) \end{aligned} \tag{C.1}$$

for the singlet and the triplet, respectively. The coordinate part $\psi(k)$ or $d(k)$ is either even or odd with respect to change of k to $-k$. For the S state $\psi(k) = f(k) = f(-k)$. In the case of cubic point group, for $S = 1, L = 1$, cubic harmonics can combine to make $J = 0, 1$ or 2. For a tetragonal lattice, five even parity irreducible representations are discussed below. The extended S-wave,

$$\Delta_s(k) = \Delta_o(\cos k_x + \cos k_y) \tag{C.2}$$

is a basis state of symmetry $\Gamma_1^+(A_{1g})$ and the d-waves are

$$\Delta_{d_{x^2-y^2}}(k) = \Delta_o(\cos k_x - \cos k_y)$$

$$\text{and} \quad \Delta_{d_{xy}}(k) = \Delta_o \sin k_x \sin k_y \tag{C.3}$$

which belong to $\Gamma_3^+(B_{1g})$ and $\Gamma_4^+(B_{2g})$ irreducible representations of D_{4h}. $\Delta_{d_{yz}}$ and $\Delta_{d_{zx}}$ transform like Γ_5^+ with parity shown by superscript +. Instead of carrying the details of the wave vector dependence, we some times introduce only the characteristics of a node in the gap. The gap Δ is replaced by Δf where,

$$\begin{aligned} f &= 1 && (s \text{ wave}) \\ &= \cos 2\phi && (d_{x^2-y^2} \text{ wave}) \\ &= \sin 2\phi && (d_{xy} \text{ wave}) . \end{aligned} \tag{C.4}$$

There is a fine structure in the superconducting state due to the orbital states which transform like the irreducible representations of the point group symmetry. Such a group of states does not occur for the singlet B.C.S. type superconducting state in which case, the gap transforms like the totally symmetric irreducible representation of the point group symmetry of the lattice. There are two types of superconductors, the B.C.S. type which have $S = 0, L = 0$ and the triplet type which have $S = 1$ and finite L but there are pairs of electrons in both the cases. That means that in the B.C.S. theory the electrons form pairs with both the electrons within the conduction band which is usually spherically symmetric whereas in the present theory, the electrons can be in crystal-field orbital states of any symmetry. We consider the reduction of the orbital part of the gap function independent of the spin components. There is a totally symmetric state which transforms like the Γ_2 irreducible representation of the point group. When this is a spin as well as an orbital singlet we recover the B.C.S. state and when it is a triplet we find a different type of superconductivity which will not show Meissner effect in all of its components and hence the components may phase separate. The component $L = 1, M_L = 0$ may exhibit Meissner effect but the components

$L = 1, M_L = \pm 1$ are unlikely to show the Meissner effect. We consider the electron-phonon interaction to give rise to attraction between electrons for three types of superconducting states. We write the unperturbed hamiltonian for the system of electrons and phonons in a metal as,

$$\begin{aligned} \mathcal{H}_o &= \sum_k (\epsilon_{k\uparrow} c^\dagger_{k\uparrow} c_{k\uparrow} + \epsilon_{k\downarrow} c^\dagger_{k\downarrow} c_{k\downarrow} + \epsilon_{-k\uparrow} c^\dagger_{-k\uparrow} c_{-k\uparrow} + \epsilon_{-k\downarrow} c^\dagger_{-k\downarrow} c_{-k\downarrow}) \\ &+ \sum_q (\hbar\omega_q a^\dagger_q a_q + \hbar\omega_{-q} a^\dagger_{-q} a_{-q}) \end{aligned} \tag{C.5}$$

where $\epsilon_{\pm k\sigma}$ are the single-electron energies with wave vector $\pm k$ and spin $\sigma =\uparrow$ or $\downarrow$, $c^\dagger_{k\sigma}(c_{k\sigma})$ are the creation (annihilation) operators for the electrons. Similarly, $a^\dagger_q(a_q)$ are the operators for the phonons of single-particle energy $\hbar\omega_q$. The electron-phonon interaction is given by,

$$\begin{aligned} \mathcal{H}' &= \sum_{k,q} [D_q c^\dagger_{k+q\uparrow} c_{k\uparrow}(a_q - a^\dagger_{-q}) + D^*_q c^\dagger_{k\uparrow} c_{k+q\uparrow}(a^\dagger_q - a_{-q}) \\ &+ D_q c^\dagger_{k+q\downarrow} c_{k\downarrow}(a_q - a^\dagger_{-q}) + D^*_q c^\dagger_{k\downarrow} c_{k+q\downarrow})(a^\dagger_q - a_{-q})] \end{aligned} \tag{C.6}$$

so that the hamiltonian of our system becomes,

$$\mathcal{H} = \mathcal{H}_l + \mathcal{H}' \ . \tag{C.7}$$

we perform a canonical transformation,

$$\tilde{\mathcal{H}} = e^{-s}\mathcal{H}e^{s} = \mathcal{H}_o + \mathcal{H}' + [\mathcal{H}_o, s] + [\mathcal{H}', s] + \frac{1}{2}[\mathcal{H}_o, [\mathcal{H}_o, s]] + \cdots \tag{C.8}$$

which is a generalization of the Bogoliubov transformation to eliminate the phonon variables in which the generating function s is found by setting,

$$\mathcal{H}' + [\mathcal{H}_o, s] = 0 \ . \tag{C.9}$$

The transformed hamiltonian is then,

$$\tilde{\mathcal{H}} = H_o + \frac{1}{2}[\mathcal{H}', s] \ . \tag{C.10}$$

From (C.7) and (C.9) we find that,

$$s = s_\uparrow + s_\downarrow \tag{C.11}$$

where

$$\begin{aligned} s_\uparrow &= \sum_{k,q} D_q(\epsilon_{k\uparrow} - \epsilon_{k+q\uparrow} + \hbar\omega_q)^{-1} c^\dagger_{k+q\uparrow} c_{k\uparrow} a_q \\ &-\sum_{k,q} D_q(\epsilon_{k\uparrow} - \epsilon_{k+q\uparrow} - \hbar\omega_{-q})^{-1} c^\dagger_{k+q\uparrow} c_{k\uparrow} a^\dagger_{-q} \\ &+\sum_{k,q} D^*_q(\epsilon_{k+q\uparrow} - \epsilon_{k\uparrow} - \hbar\omega_q)^{-1} c^\dagger_{k\uparrow} c_{k+q\uparrow} a^\dagger_q \\ &-\sum_{k,q} D^*_q(\epsilon_{k+q\uparrow} - \epsilon_{k\uparrow} + \hbar\omega_{-q})^{-1} c^\dagger_{k\uparrow} c_{k+q\uparrow} a_{-q} \end{aligned} \tag{C.12}$$

and $s_\downarrow$ may be obtained from $s_\uparrow$ by reversing the spin. The transformed hamiltonian then appears with the single particle energies in (C.5) slightly shifted and the effective interaction appears as,

$$\mathcal{H} = \mathcal{H}_o + \mathcal{H}_s + \mathcal{H}_t \tag{C.13}$$

where the singlet pairing interaction becomes,

$$\begin{aligned} \mathcal{H}_{s_\uparrow} &= \sum_{k,k',q} 2\hbar\omega_q D_q^2 \left[(\epsilon_{k+q\uparrow} - \epsilon_{k\uparrow})^2 - (\hbar\omega_q)^2\right]^{-1} c^\dagger_{k'+q\downarrow} c^\dagger_{k\uparrow} c_{k'\downarrow} c_{k+q\uparrow} \\ &+\sum_{k,k',q} 2\hbar\omega_q D_q^2 \left[(\epsilon_{k\uparrow} - \epsilon_{k+q\uparrow})^2 - (\hbar\omega_q)^2\right]^{-1} c^\dagger_{k'\downarrow} c^\dagger_{k+q\uparrow} c_{k'+q\downarrow} c_{k\uparrow} \\ &-\sum_{k,k',q} 2\hbar\omega_q D_q D_{-q} \left[(\epsilon_{k+q\uparrow} - \epsilon_{k\uparrow})^2 - (\hbar\omega_q)^2\right]^{-1} c^\dagger_{k'\downarrow} c^\dagger_{k\uparrow} c_{k'-q\downarrow} c_{k+q\uparrow} \\ &-\sum_{k,k',q} 2\hbar\omega_q D_q D_{-q} \left[(\epsilon_{k\uparrow} - \epsilon_{k+q\uparrow})^2 - (\hbar\omega_q)^2\right]^{-1} c^\dagger_{k'-q\downarrow} c^\dagger_{k+q\uparrow} c_{k'\downarrow} c_{k\uparrow} \end{aligned} \tag{C.14}$$

with $\mathcal{H}_s = \mathcal{H}_{s\uparrow} + \mathcal{H}_{s\downarrow}$. As before, $\mathcal{H}_{s\downarrow}$ may be written from the above by reversing the spin indices. In writing the above, we have assumed that the phonon frequencies are independent of the direction of the wave vector, $\hbar\omega_q = \hbar\omega_{-q}$. Similarly, the triplet pairing interaction is found to be,

$$\begin{aligned} \mathcal{H}_{t_\uparrow} &= \sum_{k,k',q} 2\hbar\omega_q D_q^2 \left[(\epsilon_{k+q\uparrow} - \epsilon_{k\uparrow})^2 - (\hbar\omega_q)^2\right]^{-1} (c^\dagger_{k'+q\uparrow} c^\dagger_{k\uparrow} c_{k'\uparrow} c_{k+q\uparrow} \\ &+ c^\dagger_{k'\uparrow} c^\dagger_{k+q\uparrow} c_{k'+q\uparrow} c_{k\uparrow}) \end{aligned}$$

$$-\sum_{k,k',q} 2\hbar\omega_q D_q D_{-q}\left[(\epsilon_{k+q\uparrow}-\epsilon_{k\uparrow})^2-(\hbar\omega_q)^2\right]^{-1}$$
$$\times(c^\dagger_{k'\uparrow}c^\dagger_{k\uparrow}c_{k'-q\uparrow}c_{k+q\uparrow}+c^\dagger_{k'-q\uparrow}c^\dagger_{k+q\uparrow}c_{k'\uparrow}c_{k\uparrow} \qquad (C.15)$$

with $\mathcal{H}_{t\uparrow}+\mathcal{H}_{t\downarrow}$. We note that the coupling constant of the singlet pairing terms of (C.14) is exactly equal to that of the triplet pairing terms. The two are attractive for one and the same condition, i.e., $\epsilon_{h+q\uparrow}-\epsilon_{k\uparrow}<\hbar\omega_q$ which means that $\epsilon_F-\hbar\omega_D<\epsilon_k$ and $\epsilon_{k\pm q}<\epsilon_F+\hbar\omega_D$. In general both (C.14) and (C.15) occur with one and the same amplitude so that the triplets as well as the singlets occur. If we leave out the triplet terms and consider only the singlets, the B.C.S. theory may be recovered with onset transition temperature T_c. However, depending on which of the term are kept (C.14) or (C.15), either a normal to the singlet Cooper state or a normal to the triplet state transition is possible. The spin triplet has orbital components transforming like the irreducible representation of the point group symmetry of the lattice. The spin singlet can be in the $L=0$ state but the spin triplet must have finite L and hence the symmetry of the state is important. We consider the attractive interaction between electrons so that the hamiltonian of our system is defined as,

$$\begin{aligned}\mathcal{H} &= \sum_{k\sigma}(\epsilon_{k\sigma}-\epsilon_{F\sigma})c^\dagger{}_{k\sigma}c_{k\sigma}\\ &+\sum_{k,k'}\sum_{\sigma_1,\sigma_2,\sigma_1',\sigma_2'} V_{\sigma_1,\sigma_2,\sigma_1',\sigma_2'}(k,k')c^\dagger{}_{-k,\sigma_1}c^\dagger{}_{k,\sigma_2}c_{k'\sigma_1'}c_{-k',\sigma_2'} \qquad (C.16)\end{aligned}$$

where $c^\dagger{}_{k\sigma}(c_{k\sigma})$ are the creation (annihilation) operators of the electrons in the conduction band. For attractive interaction, the gap of the superconductor is defined as,

$$\Delta_{\sigma_1\sigma_2}(R,k)=\begin{pmatrix}\Delta^*_{\uparrow\uparrow}(R,k) & \Delta^*_{\uparrow\downarrow}(R,k)\\ \Delta^*_{\downarrow\uparrow}(R,k) & \Delta^*_{\downarrow\downarrow}(R,k)\end{pmatrix} \qquad (C.17)$$

which can be written as

$$\Delta_{\sigma_1\sigma_2}(k) = -\sum_{k'}[V_{\sigma_2\sigma_1\uparrow\uparrow}(k,k')<c_{k'\uparrow}c_{-k'\uparrow}>$$

$$+V_{\sigma_2\sigma_1\uparrow\downarrow}(k,k') < c_{k'\uparrow}c_{-k'\downarrow} > +V_{\sigma_2\sigma_1\downarrow\uparrow}(k,k') < c_{k'\downarrow}c_{-k'\uparrow} >$$
$$+V_{\sigma_2\sigma_1\downarrow\downarrow}(k,k') < c_{k'\downarrow}c_{-k'\downarrow} >] \ . \qquad (C.18)$$

The parity P and the time reversal operator, T, are defined as,

$$P\Delta_{\sigma_1\sigma_2}(k) = \Delta_{\sigma_1\sigma_2}(-k) \qquad (C.19)$$
$$T\Delta_{\sigma_1\sigma_2}(k) = \begin{pmatrix} \Delta_{\downarrow\downarrow}(-k) & -\Delta_{\downarrow\uparrow}(-k) \\ -\Delta^*_{\uparrow\downarrow}(-k) & \Delta^*_{\uparrow\uparrow}(-k) \end{pmatrix} . \qquad (C.20)$$

The fermion antisymmetry requires that,

$$\Delta_{\sigma_1\sigma_2}(k) = -\Delta_{\sigma_2\sigma_1}(-k) \ . \qquad (C.21)$$

The rotation operator is given by

$$R\Delta_{\sigma_1\sigma_2}(k) = \sum_{\sigma_1',\sigma_2'} \left[D^{(1/2)}(R)\right]_{\sigma_1'\sigma_1} \left[D^{(1/2)}(R)\right]_{\sigma_2'\sigma_2} \Delta_{\sigma_1',\sigma_2'}(R,k) \qquad (C.22)$$

where $D^{(1/2)}(R)$ is the rotation matrix for spin 1/2, which is needed to describe the symmetries of the point group. For two electrons in the p-wave, $l = 1$, the parity, $(-1)^l$ is odd and the gap changes sign upon parity operation,

$$P(\Delta_{\sigma_1\sigma_2}(k) = -\Delta_{\sigma_1\sigma_2}(k) \ . \qquad (C.23)$$

For $l = 1$ the gap is written as the scalar product of two vectors as,

$$(\Delta_{\sigma_1\sigma_2}(k) = -i\vec{d}(k).(\vec{\tau}\tau_y)_{\sigma_1\sigma_2}, \ , \qquad (C.24)$$

where $\tau_j (j = x, y, z)$ are the Pauli spin matrices. The rotation operator for spin one is now given by

$$Rd_i(k) = \sum_j [D^{(1)}(R)]_{ji} d_j(k) \ . \qquad (C.25)$$

The gap is determined from the solution of,

$$\Delta_{\sigma_1\sigma_2}(k) = -\sum_{\sigma_1'\sigma_2'} N(o) \int_o^{\omega_D'} d\omega_{k'}$$

$$\times \quad \frac{V_{\sigma_2\sigma_1\sigma_1'\sigma_2'}(k,k')\Delta_{\sigma_1'\sigma_2'}(k')\tanh\{(\hbar/2k_BT)[\omega_{k'}^2 - \Delta^2_{\sigma_1'\sigma_2'}(k')]^{1/2}\}}{[\omega_{k'}^2 - \Delta^2_{\sigma_1'\sigma_2'}(k')]^{1/2}} \tag{C.26}$$

where $\omega_D' = [\omega_D^2 + \Delta^2_{\sigma_1\sigma_2}/\hbar^2]^{1/2}$ with ω_D as the Debye frequency and $N(o)$ is the density of states at the Fermi surface. The transition temperature may be obtained from (C.26) by replacing T by T_c and setting the gap equal to zero. In this way the singlet as well as the triplet transition temperatures are obtained. Thus there are three values of the T_c for the triplet and one for the singlet. For $L = 1$, the gap belongs to an orbital triplet which can be represented by the functions $p_x(k), p_y(k)$ and $p_z(k)$. The vector $\vec{d}$ in (C.24) then may be described by

$$d(k) = \sum_{ij} d_{ij}\hat{\imath}p_j(k) \tag{C.27}$$

where $\hat{\imath}$ is a unit vector $(\hat{\imath}, \hat{\jmath} = \hat{x}, \hat{y}, \hat{z})$ which may be substituted in (C.25). Eq.(C.22) gives the prescription for the rotation of spin 1/2 while (C.25) gives the same for spin 1. For definiteness, the form of $d(k)$ required for (C.25) is taken to be linear in $\hat{x}, \hat{y}$ and $\hat{z}$ as in (C.27). We first consider the transformation properties in the tetrahedral group. The functions p_x, p_y and p_z transform like the T irreducible representation of the tetrahedral group. The $\hat{x}, \hat{y}$ and $\hat{z}$ are also basis functions of the T representation. Therefore, the gap belongs to TXT which in the tetrahedral group is reduced to

$$TXT = A + E + 2T \ . \tag{C.28}$$

i.e., the gap belongs to a set of functions, a totally symmetric orbital singlet transforming like the A irreducible representation of the tetrahedral group, an orbital doublet transforming like E and two orbital triplets each transforming like the T irreducible representation of the tetrahedral group. Each irreducible representation belongs to a certain eigen value of Δ_A, Δ_E. Δ_T and $\Delta_{T'}$, which determines the transition temperature of the order of

$$k_BT_c \approx 1.14\hbar\omega_D \exp[-1/N(o)\Delta_i] \ \ (i = A, E, T, T') \tag{C.29}$$

This within a given bound state there are several phase transitions all of which are superconducting. In a tetrahedral symmetry, one of the superconducting phases is nondegenerate and totally symmetric, one phase is doubly degenerate and two are triply degenerate. The eigen functions are

$$\begin{aligned} d(A) &= 3^{-1/2}(\hat{x}p_x + \hat{y}p_y + \hat{z}p_z) \\ d(E_z) &= 6^{-1/2}(2\hat{z}p_z - \hat{x}p_x - \hat{y}p_y) \\ d(E_{xy}) &= 2^{-1/2}(\hat{x}p_x - \hat{y}p_y) \\ d(T,x) &= \hat{y}p_z \\ d(T,y) &= \hat{z}p_x \\ d(T,z) &= \hat{x}p_y \end{aligned} \tag{C.30}$$

There are two functions of symmetry T in (C.28). There occur several phase transitions from one symmetry to another. This type of phase transitions have been observed in the paired state of ^3He. The $^3He - A$ is doubly degenerate $|\pm 1>$ and there is a S phase $^3He - S$ corresponding to $|0>$. In the S phase of ^3He, $S = 1, M_s = 0$ so that it is similar to B.C.S. type superconductivity in as much as specific heat is concerned. The $^3He - A$ and $^3He - B$ are well known but the S phase of ^3He, which we call as $^3He - S$ is relatively not so well known. It is of interest to learn the orbital states of electron pairs in various symmetries which represent various phases within the bound state. We write the product representation for the tetrahedral T, the octahedral O, and the axially symmetric trigonal D_3 and the tetragonal D_4 symmetry point groups as,

$$\begin{aligned} &\text{Tetrahedral T:} && A + E + 2T \\ &\text{Octahedral O:} && A_1 + E + E + T_1 + T_2 \\ &\text{Trigonal } D_3\text{:} && 2A_1 + A_2 + 3E \\ &\text{Tetragonal } D_4\text{:} && 2A_1 + A_2 + B_1 + B_2 + 2E \ . \end{aligned} \tag{C.31}$$

We have thus found p-wave superconducting states in four different point group symmetries. In general, all stable states have real order parameter.

However, complex order parameters are possible in which case the time reversal symmetry is broaken particularly for degenerate states. Each group representation has its associated T_c and that normally only one representation will be relevant. For $S = 0, L = 0$ there is only one totally symmetric B.C.S. state. Because of the Pauli principle, when $S = 1$ the orbital angular momentum $L = 0$ is not allowed in atoms. Therefore, $S = 1, L = 1$ arises for which $L = 1$ must have the states of the same symmetry as one of the irreducible representations of the point group symmetry of the crystal. It is not necessary that the totally symmetric Γ_1 be the ground state so that superconductivity may arise in any one or more of the states of any orbital symmetry which depends on the atomic structure or on the chemical bonding. Then there is only one T_c which corresponds to the ground state of the "Cooper pairs" in a crystal. Thus there should be one of the levels of (C.31) which broadens into a band where pairing occurs so that the superconducting state obeys the crystal symmetries.

C.2. London penetration depth.

The London penetration depth depends on the superfluid density derived from the concentration of electrons as discussed in Chapter 6. Since the superconductivity arises from levels which are classified according to the irreducible representations of the point group symmetry, the superfluid density reflects the symmetry of the paired state.

Accordingly, the London penetration depth depends on the symmetry. Therefore, we can calculate the London penetration depth for different symmetries and compared the calculated values with those measured as a function of temperature. It is then possible to measure the symmetry of the superconducting state. The diamagentic current of the B.C.S. theory is used to define the London-penetration depth as,

$$\left(\frac{\lambda(o)}{\lambda(T)}\right)^2 = 1 - \frac{2}{\hbar k_B T}\int_o^\infty dE \frac{N(E)}{N(o)} f_E[1 - f_E] \qquad (C.32)$$

with $N(E)$ as the quasi-particle density of states which is normalized to the normal state single-particle density of states $N(o)$. For s-wave symmetry as in (C.4) the gap is constant so that the density of states is written as,

$$\frac{N(E)}{N(o)} = Re\frac{\omega}{(\omega^2 - \Delta_o^2)^{1/2}} \tag{C.33}$$

which gives,

$$\frac{\lambda(T) - \lambda(o)}{\lambda(o)} \simeq \left(\frac{2\pi\Delta}{k_B T}\right)^{1/2} \exp(-\Delta/k_B T) \qquad (s \text{ wave}) \tag{C.34}$$

because the number of quasiparticles is exponentially small at low temperatures. For a gap with a node at an energy ω which is small compared to the maximum anti-node value of the gap, $\omega \ll \Delta_o$, which we call as the d-wave gap,

$$\frac{N(E)}{N(o)} \simeq \frac{\omega}{\Delta_o} \ . \tag{C.35}$$

Accordingly for the d-wave symmetry, the penetration depth varies with temperature as,

$$\frac{\lambda(T) - \lambda(o)}{\lambda(o)} \simeq \ln(2)\frac{k_B T}{\Delta_o} \qquad (d \text{ wave}). \tag{C.36}$$

In real systems, there is considerable scattering of electrons by the impurities. The transition temperatures are considerably reduced by the impurity scattering, $T_c = T_{co} - \dfrac{\hbar}{2\tau_{AG} k_B}$. The scattering rate is $\Gamma = \hbar/2\tau_{AG}$. Therefore we can define a characteristic energy, $k_B T^* = \left(\dfrac{\hbar\Delta_o}{2\tau_{AG}}\right)^{1/2}$. For small impurity concentration, n_i, the scattering rate is linearly proportional to the impurity concentration, $\Gamma = n_i/\pi N(o)$, so that we can assume that the contribution is proportional to ω^2 or,

$$N(\omega) \simeq N(o) + a\omega^2 \tag{C.37}$$

with $N(o) \simeq (\Gamma/\Delta_o)^{1/2} N_o$. Above this energy $N(\omega) \simeq (\hbar\omega/\Delta_o)N_o$ so that,

$$\lambda(T) - \lambda(o) = \frac{1}{2}\left(\frac{\Gamma}{\Delta_o}\right)^{1/2} \lambda(o) + \lambda(o)\ln(2)\frac{T^2}{T + T^*}. \tag{C.38}$$

Thus a d-wave gap with strong scattering gives a T^2 term in the change in the penetration depth as long as $T < T^*$. When $T > T^*$, the change in λ is linear in T just as if there is no scattering. Thus the penetration depth varies as T^2 at low temperatures and as T at high temperatures when scattering is taken into account.

We ignore the impurity scattering and treat the density of states by using the complete elliptic integral so that the London penetration depth is found to become,

$$\begin{aligned} \left|\frac{\lambda_{x^2-y^2}(o)}{\lambda_{x^2-y^2}(T)}\right|^2 &= 1 - \frac{4}{\pi k_B T}\left[\int_o^{\Delta_{x^2-y^2}} dE f_E(1-f_E) \right. \\ &\times \left(\frac{E}{\Delta_{x^2-y^2}}\right) K\left(\frac{E}{\Delta_{x^2-y^2}}\right) \\ &\left. + \int_{\Delta_{x^2-y^2}}^{\infty} dE f_E(1-f_E) K\left(\frac{\Delta_{x^2-y^2}}{E}\right)\right], \quad \text{(C.39)} \end{aligned}$$

The temperature dependence of the gap is determined by

$$\gamma^{-1} = 2\pi k_B T < |f|^2 >^{-1} \sum_n \frac{|f|^2}{\omega_n^2 + \Delta^2 |f|^2} \qquad (C.40)$$

where f is given by (C.4) and γ is the dimensionless coupling constant, ω_n is the Matsubara frequency and the sum is cut off at $\omega_n = \epsilon_c$. Due to the crystal field in metals the Fermi surface becomes anisotropic so that at low temperatures the ground state is of $d_{x^2-y^2}$ symmetry. However, in such a case upon warming from low temperatures, the ground state may change from $d_{x^2-y^2}$ to s. The London penetration depth for d_{xy} may be written from (C.39) by replacing $\Delta_{x^2-y^2}$ by Δ_{xy}. However, the symmetry occurs only near the surface of the sample. In the Landau model, the higher temperature phase is more symmetric than the low temperature phase. Therefore, we predict that at higher temperatures, the superconductor will become the s-wave type. Therefore, the $d_{x^2-y^2}$ state should change to s state. This is possible only when the order parameter is complex, $d_{x^2-y^2} + is$. The result is in confirmity with the calculations of the minimization of free energy. For complex order parameter, the penetration depth is given by

$$\left|\frac{\lambda_{x^2-y^2,s}(o)}{\lambda_{x^2-y^2,s}(T)}\right|^2 = 1 - \frac{4}{\pi k_B T}\left[\int_{x^2-y^2}^{\Delta_{x^2-y^2,s}} dE f_E(1-f_E)\frac{E}{\Delta_s}\right.$$

$$\times K\left(\frac{(E-\Delta^2_{x^2-y^2}f^2)^{1/2}}{\Delta_s}\right)$$
$$+\int_{\Delta_{x^2-y^2,s}}^{\infty} dE f_E(1-f_E)\frac{E}{(E^2-\Delta^2_{x^2-y^2}f^2)^{1/2}}$$
$$\times K\left(\frac{\Delta_s}{(E^2-\Delta^2_{x^2-y^2}f^2)^{1/2}}\right)\Bigg] , \qquad \text{(C.41)}$$

where

$$\Delta_{x^2-y^2,s} = (\Delta^2_{x^2-y^2}+\Delta^2_s)^{1/2} \qquad \text{(C.42)}$$
$$f_E = (e^{E/k_BT}+1)^{-1} . \qquad \text{(C.43)}$$

The surface microwave resistivity is proportional to $\mu_o\omega\lambda_{ab}(o)$ where the permeability μ_o and the microwave frequency ω are known so that the penetration depth can be determined.

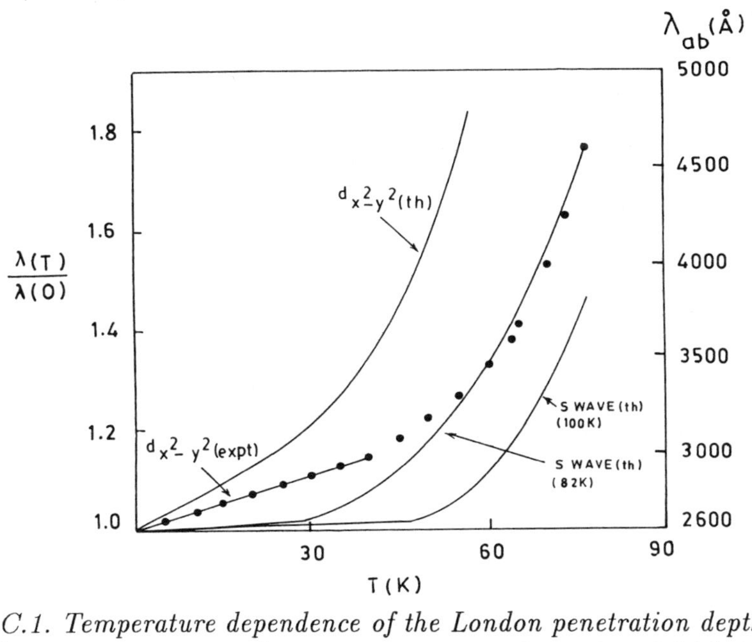

Fig. C.1. Temperature dependence of the London penetration depth showing d wave symmetry at low temperatures and s wave at high temperatures. The curves marked as $d_{x^2-y^2}$ *(th), s wave(th 100K), and s wave (th 82K) are all theoretically calculated where* T_c *is given in the parenthesis. The dots are deduced from the experimental measurements.*

The experimental measurements of the a.c. resistivity have been performed by Jacobs et al using a single crystal of $Bi_2Sr_2CaCu_2O_{8+\delta}$ from which the penetration depth has been deduced as a function of temperature. A few experimental points of the London penetration depth are shown in Fig. C.1. The experimental values are linear upto a temperature of $T_{x^2-y^2,s} \simeq 40$K. It is found that the theoretical expression (C.39) for $d_{x^2-y^2}$ state is also linear,

$$\frac{\lambda_{x^2-y^2}(T)}{\lambda_{x^2-y^2}(o)} \simeq 1 + 0.69 k_B T/\Delta_{x^2-y^2}(o) \quad . \tag{C.44}$$

The expression (C.36) obtained for a node is also linear. Therefore, it is identified that the low-temperature behaviour is caused by a $d_{x^2-y^2}$ type state. The experimentally measured slope for $d_{x^2-y^2}$ is $d\delta\lambda_{x^2-y^2}/dT \simeq 10$ Å/K. The theoretical value of the slope $d[\lambda(T)/\lambda(o)]/dT \simeq 15.3$ K^{-1} is larger than the experimental value. The theoretically calculated value of $\lambda_{x^2-y^2}(T)/\lambda_{x^2-y^2}(o)$ for $T_c = 82$ K and $\lambda(o) = 2600$ Åis also shown in Fig. C.1. This value depends on temperature a bit more strongly than the experimentally deduced values. Using $\Delta_s = 1.77 k_B T_c (1 - T/T_c)^{1/2}$ we find $\lambda_s(T)/\lambda_s(o)$ for two different values of $T_c = 82$ and 100 K which are also shown in Fig. C.1. These calculated values of $\lambda_s(T)/\lambda_s(o)$ are less than the experimentally measured but for $T > 60$ K only s wave agrees with the experimental data. Thus we see that for $T < 40$ K, the order parameter has $d_{x^2-y^2}$ symmetry becoming s wave for $60 < T < 82$ K. The occurrence of multiple symmetries is permitted by (C.31). According to (C.41) both the s-wave and the $d_{x^2-y^2}$-wave type gaps occur at all temperatures with the average value given by (C.42) for a given value of the ratio $\Delta_{x^2-y^2}/\Delta_s$ so that the calculated $\lambda(T)$ for the mixed wave occurs in between the two curves. $\lambda_{x^2-y^2}(T)$ and $\lambda_s(T)$. The theory may be constructed in such a way that at low temperatures only the $d_{x^2-y^2}$ phase occurs and at elevated temperatures only the s phase occurs with a phase boundary separating the two phases.

C.3. Time reversal.

If we change the time, t to $-t$, we reverse the time. Changing the sign of the current in a coil, reverses the direction of the magnetic field.

The energy levels of $\mu.H = g\mu_B H.S$ are $\pm\frac{1}{2}g/\mu_B H$. For $+H$ the levels are $\pm\frac{1}{2}g/\mu_B H$ and for $-H$, these are $\mp\frac{1}{2}g/\mu_B H$. If H and S both change sign, the sign of the product $H.S$ remains unchanged. Therefore, the eigen values remain unchanged with respect to change in sign of the current and hence with respect to change in sign of time, t. This is known as the invariance of the hamiltonian with respect to time reversal. If there is an odd number of electrons, the atomic spin is $\frac{1}{2}, \frac{3}{2}, \frac{5}{2}$, etc., which corresponds to the eigenvalues of $\pm\frac{1}{2}$; $\pm\frac{1}{2}$, $\pm\frac{3}{2}$; $\pm\frac{1}{2}$, $\pm\frac{3}{2}$, $\pm\frac{5}{2}$ etc. Therefore, if the applied magnetic field is zero, there is always at least a doubly degenerate level corresponding to $\pm\frac{1}{2}$. This degeneracy can not be removed by an electric field.

Kramers theorem states that there always will be atleast a twofold degenerate state in the presence of electric fields having any kind of symmetry provided that the number of electrons, n, is odd. Wigner has shown that this degeneracy is related to the invariance of the system under time reversal. If we interchange t and $-t$, the linear momentum p and electron spin s change their sign. In a free ion, the hamiltonian is invariant under time reversal, for the terms $p^2/2m$ and $\vec{L}.\vec{S} = (\vec{r} \times \vec{p}).\vec{S}$ are invariant if $\vec{p} \to -\vec{p}$ and $\vec{S} \to -\vec{S}$. In a magnetic field the state J splits into $2J+1$ components leaving no degeneracy because the magnetic moment operator as well as the field in the Hamiltonian is reversed on time reversal. We denote the time reversal operator by K such that the hamiltonian is invariant,

$$K[\mathcal{H}(r,p,S)]K^{-1} = \mathcal{H}(r,-p,-S) = \mathcal{H}' \ . \qquad (C.45)$$

Since time and space are independent variables, $\vec{r}$ does not change by reversing time. The operator which reverses time will turn out to be $K = is_y C$ where C is the operator which replaces the function by its complex conjugate and s_y is the y-component of the Pauli spin matrices,

$$s_x = \begin{pmatrix} 0 & 1 \\ 1 & 0 \end{pmatrix}, \quad s_y = \begin{pmatrix} 0 & -i \\ i & 0 \end{pmatrix}, \quad s_z = \begin{pmatrix} 1 & 0 \\ 0 & -1 \end{pmatrix}. \qquad (C.46)$$

The hamiltonian of a one-electron system is given by,

$$\mathcal{H}(r,p,s) = \mathcal{H}(r,p).1 + \mathcal{H}_x(r,p).s_x + \mathcal{H}_y(r,p).s_y + \mathcal{H}_z(r,p).s_z \ . \qquad (C.47)$$

Since C commutes with is_y and $C = C^{-1}$, we have

$$(C\mathcal{H}C^{-1})\varphi = C\mathcal{H}\varphi^* = \mathcal{H}^*\varphi \tag{C.48}$$

and

$$C\mathcal{H}C^{-1} = \mathcal{H}^* \ . \tag{C.49}$$

When we take the complex conjugate, s_y changes sign but s_x and s_z remain unchanged so that

$$C\mathcal{H}C^{-1} = \mathcal{H}^* = \mathcal{H}_o(r,-p).1+\mathcal{H}_x(r,-p).s_x-\mathcal{H}_y(r,-p).s_y+\mathcal{H}_z(r,-p).s_z. \tag{C.50}$$

Therefore,

$$\begin{aligned} K\mathcal{H}K^{-1} &= \{is_yC\}\mathcal{H}\{C^{-1}(is_y)^{-1}\} \\ &= \mathcal{H}_o(r,-p).1 - \mathcal{H}_x(r,-p).s_x - \mathcal{H}_y(r,-p).s_y - \mathcal{H}_z(r,-p).s_z \ . \end{aligned} \tag{C.51}$$

For n electrons, the corresponding operator of time reversal is

$$K = i^n s_{y_1} \cdots s_{y_n} C \ . \tag{C.52}$$

If two wave functions ψ and $\psi_1 = K\psi$ corresponding to $\mathcal{H}(r,p,s)$ and $\mathcal{H}(r,-p,-s)$ are linearly independent, then the energy levels are degenerate. We assume that

$$\psi_1 = K\psi = a\psi \tag{C.53}$$

where a is a complex number, i.e., ψ and ψ_1 are linearly dependent, then

$$K^2\psi = K\psi_1 = Ka\psi = a^*K\psi = |a|^2\psi \tag{C.54}$$

but

$$K^2 = (i^n s_{y_1} \cdots s_{y_n})^2 = i^{2n}.1 = (-1)^n.1 \tag{C.55}$$

and therefore

$$|a|^2 = (-1)^n \tag{C.56}$$

which can be true only for even n. Hence for odd n, $K\psi$ and ψ are necessarily independent and the energy state must be degenerate.

We consider that η_1 and η_2 are two complex order parameters with the same symmetry properties as that of the corresponding pair wavefunctions. That means that both are superconducting and hence two superconducting phases exist. The time reversal transforms the order parameter η_1 to its complex conjugate η_1^* and similarly η_2 to η_2^*. The order parameters are of the form of

$$\begin{aligned} \eta_j &= |\eta_j| \exp(i\phi_j) \ , \\ \text{and} \qquad \eta_j^* &= |\eta_j^*| \exp(-i\phi_j). \end{aligned} \tag{C.57}$$

Therefore, the effect of the time reversal operator is equivalent to changing ϕ_j to $-\phi_j$. The product $I_c.\phi_o$ is invariant with respect to change in sign of the critical current and unit flux. If I_c changes sign, then ϕ_o also changes sign and the sign of the product $I_c.\phi_o$ remains unchanged. When we change the sign of ϕ_j the sign of $\cos\phi_j$ also changes. Therefore $\cos\phi$ is not invariant with respect to change of sign of ϕ and hence not invariant with respect to time reversal.

The superconducting order parameter consists of two components, s and $d_{x^2-y^2}$ which we write as $s + id_{x^2-y^2}$. The wave function is then

$$\psi = \eta_1\psi_1(k) + \eta_2\psi_2(k). \tag{C.58}$$

If time-reversal symmetry is conserved, then η_1, η_2 have a common phase factor which is equal to that in η_1^*, η_2^* or

$$\frac{\eta_1}{\eta_2} = \frac{\eta_1^*}{\eta_2^*} \qquad \text{(conserved } \mathcal{T}) \ , \tag{C.59}$$

and the state is at least two fold degenerate since η_1, η_2 and η_1^*, η_2^* have the same free energy, otherwise, $\eta_1/\eta_2 \neq \eta_1^*/\eta_2^*$ and the order parameter breaks the time reversal symmetry. For an interface between two superconductors, A and B, both with the order parameters $\eta_1^A, \eta_1^B, \eta_2^A$ and η_2^B and the wave functions,

$$\begin{aligned} \psi^A &= \eta_1^A\psi_1(k) + \eta_2^A\psi_2(k)) \\ \text{and} \qquad \psi^B &= \eta_1^B\psi_1(k) + \eta_2^B\psi_2(k) \ , \end{aligned} \tag{C.60}$$

the Josephson current, J and phase relation, at a given point on the interface has the form,

$$\begin{aligned} J &= J_{c11}\sin(\phi_1^B-\phi_1^A)+J_{c12}\sin(\phi_1^B-\phi_2^A) \\ &\quad +J_{c21}\sin(\phi_2^B-\phi_1^A)+J_{c22}\sin(\phi_2^B-\phi_2^A). \ , \end{aligned} \tag{C.61}$$

The energy for a uniform interface of area s is given by,

$$\begin{aligned} E_J &= -\frac{\phi_o S}{2\pi c}\left[J_{c11}\cos(\phi_1^B-\phi_1^A)+J_{c12}\cos(\phi_1^B-\phi_2^A)\right. \\ &\quad \left. +J_{c21}\cos(\phi_2^B-\phi_1^A)+J_{c22}\cos(\phi_2^B-\phi_2^A)\right] \ . \end{aligned} \tag{C.62}$$

Suppose that $|1>$ is a s wave, $|s>$, and $|2>$ is a d wave, $|d>$, then in A superconductor s and d are orthogonal so that $\phi_1^A-\phi_2^A=\frac{\pi}{2}$. Similarly in the B superconductor, $\phi_1^B-\phi_2^B=\frac{\pi}{2}$. The current density J and the interface energy density ϵ_J depend only on one phase difference through the interface, $\varphi=\phi_1^B-\phi_1^A$,

$$J(\varphi) = \tilde{J}_c\sin(\varphi-\tilde{\alpha}) \tag{C.63}$$

$$\epsilon_J(\varphi) = -\frac{\phi_o S\tilde{J}_o}{2\pi c}\cos(\varphi-\tilde{\alpha}) \tag{C.64}$$

with

$$\tilde{J}_c = (J_{c11}+J_{c22})^2+(J_{12}-J_{c21})^2]^{1/2} \tag{C.65}$$

$$\tan\tilde{\alpha} = \frac{J_{c12}-J_{c21}}{J_{c11}+J_{c22}}. \tag{C.66}$$

The phase shift $\tilde{\alpha}$ corresponds to φ which minimizes the interface energy. It can assume any value and depends only on the relative magnitude of the different coupling components. In the case of $\mathcal{T}$-invariant combination of η_1 and η_2 the value of $\tilde{\alpha}$ is strictly either 0 or π.

We show in Fig. C.2 a triangular superconducting film as an inclusion in another superconducting film. The line 1-2 works as a boundary. If the tunneling current crosses this line there is a change in the phase of the current. We assume that the change in phase, α, of the current in crossing the linear 1-2 and 1-3 is π and while crossing the line 2-3 is zero. The lines 1-2 and 1-3 represent π junctions whereas 2-3 represents a zero junction, the junction energy is

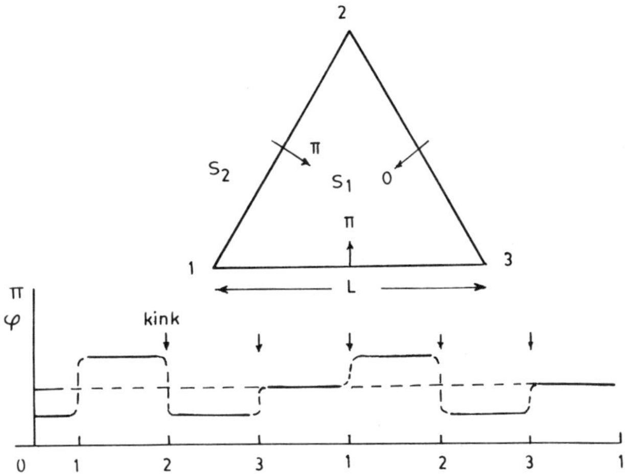

Fig. C.2. A superconducting triangular sheet, s_1, inside another superconducting film, s_2, forming junctions at the boundaries 1-2, 2-3 and 3-1. There is a phase shift in going from s_1 to s_2. The phase shift is indicated to be π or 0. The triangle is linearized and the plot of $\varphi(\tilde{x})$ is shown to have kinks at the corners of the triangle. In the T-violating case, there is freedom of choosing values different from 0 or π.

$$E_J(\varphi) = -\frac{I_c \phi_o}{2\pi c} \cos(\varphi - \alpha) \tag{C.67}$$

in which $\alpha = 0$ for zero junction and $\alpha = \pi$ for a π-junction. We can map the triangle into a one-dimensional axis with periodic boundary conditions for the coordinate $\tilde{x}$. The length of a side of the triangle is L so that in the linear projection, the periodicity occurs for $\tilde{x} + 3L = \tilde{x}$. The spatial variation of φ is determined by the sine-Gordon equation,

$$(\partial^2/\partial x^2)\varphi = \lambda_J^{-2}\sin[\varphi - \alpha(\tilde{x})] \qquad (C.68)$$

where λ_J is the Josephson penetration depth

$$\lambda_J = \phi_o c/8\pi^2 d I_c. \qquad (C.69)$$

Here d is the width of the interface. The value of α is a constant which is different from 0 or π, for the distance from 1 to 2, from 2 to 3 and from 3 to 1 but it changes at the points 1, 2 and 3. These are the corners of the triangle. The variation $\varphi(\tilde{x})$ at the points 1, 2 and 3, $(d/d\tilde{x})\varphi$ tends to infinity and hence these points are called "kinks". Each segment is characterized by the phase shift $\tilde{\alpha}$ which is neither 0 nor π but has intermediate values, $0 < \tilde{\alpha} < \pi$. There are kinks in φ at all of the corners connecting different values of $\tilde{\alpha}$ in each segment. Therefore the vortices at the corners have fluxes,

$$\phi = \phi_o \frac{\tilde{\alpha}_r - \tilde{\alpha}_l}{2\pi} \qquad (C.70)$$

where $\tilde{\alpha}_r$ is the value of $\tilde{\alpha}$ on the segment on right and α_l on the left. Thus the kinks have fractional fluxes and these are $\mathcal{T}$ violating. Let us set $\alpha = 0$ in (C.67) so that we can choose the value of φ in such a way that it exceeds $\pi/2$. The energy $E_J(\varphi)$ then changes sign and hence the magnetization becomes positive as in the case of paramagnetic Meissner effect (PME), discussed in Chapter 4 after eq.(4.53) onwards. In the Fig. C.2, the regions 1-2 and 3-1 thus have PME whereas 2-3 does not.

Suggested References

C.1. G.E. Volovik and L.P. Gor'kov, Pis'ma Zh. Eksp. Teor. Fiz. **39**, 550 (1984); JETP Letters **39**, 674 (1984).

C.2. K.N. Shrivastava, Phys. Lett. A**113**, 437 (1986).

C.3. N.M. Krishna and K.N. Shrivastava, Physica B**230**, 939 (1997).

C.4. P.W. Anderson, Phys. Rev. B**30**, 4000 (1984).

C.5. D.J. Scalapino, Phys. Rep. **250**, 329 (1995).

C.6. T. Jacobs, S. Sridhar, Q.Li, G.D. Gu and N. Koshizuka, Phys. Rev. Lett. **75**, 4516 (1995).

C.7. D.H. Wu, J.S.A. Bridgewater, A. Gokirmak and S.M. Anlage, Phys. Rev. Lett. **81**, 2890 (1998).

C.8. V.N. Prigodin, Phys. Rev. Lett. **74**, 1566 (1995).

C.9. D.B. Bailey, M. Sigrist and R.B. Laughlin, Phys. Rev. B**55**, 15239 (1997).

C.10. C.M. Varma, Phys. Rev. Lett. **83**, 3538 (1999); Phys. Rev. B**55**, 14554 (1997).

C.11. D.F. Agterberg, Phys. Rev. Lett. **80**, 5184 (1998).

C.12. A.V. Balatsky, Phys. Rev. Lett. **80**, 1972 (1998).

C.13. J.R. Kirtley, P. Chaudhari, M.B. Ketchen, N. Khare, S.Y. Lin and T. Shaw, Phys. Rev. B**51**, 12057 (1995).

Appendix D: Pseudogap

D. Pseudogap.

Any gap in the dispersion relation of single quasiparticles, other than that of the B.C.S. theory of pairing of electrons in the conduction band is called a pseudogap. Therefore, the pseudogap may arise from the anisotropy in ferromagnets, the anisotropy in antiferromagnets, spin fluctuations, spin-density waves, charge-density waves or Bose condensation. The phonon disperson relation is usually isotropic without any gap for acoustic phonons but in one-or two-dimensional metals, it is possible to introduce charge-density waves, (CDW), which are caused by the lattice distortions which introduce a gap in the dispersion relation. Therefore in the phonon dispersion relation there is an anomaly when the phonon wave vector, q, is equal to twice the Fermi wave vector, $q = 2k_F$. The average value of the phonon operator is zero $< a_q >=< a^{\dagger}{}_q >= 0$ as long as the symmetry is conserved in which case the mean equilibrium position of the oscillator is maintained. However, in the case of a distortion which breaks the symmetry, we introduce a gap in the dispersion relation, so that there is a nonzero value of the phonon operators,

$$< a_q >= \left(\frac{\Delta}{V_{ph}}\right) \delta_{q,2k_F} \qquad (D.1)$$

where V_{ph} is the energy of the electron-phonon interaction and Δ is the gap energy. Similarly, when the coupling constant depends on the spin, introduction of a distortion produces spin-density waves (SDW). The delta function demands that the phonon wave vector be equal to twice the Fermi wavevector so that there is a problem of nesting of the Fermi surface and hence commensurate and incommensurate phase transitions arise.

There is a characteristic temperature, T^* at which antiferromagnetic correlations occur. This temperature is higher than the superconducting transition temperature and is associated with a gap so that experimental measurements of the gap of the superconductor continue over from one gap to the other. The superconducting gap is smaller than the pseudogap, $\Delta < \Delta^*$ and $T_c < T^*$. This problem is quite distinct from that of the onset temperature at which the superconductivity first appears, $T_c < T_{\text{onset}} < T^*$. The pseudogap is related to the scattering rate such that,

$$\frac{1}{\tau} = \frac{2c}{\hbar} k_B T^* \qquad (D.2)$$

where $c \simeq 1$. Due to dynamical character of the antiferromagnetic short-range order, τ^{-1} generates a resonant coupling of states $\vec{k}$ at the Fermi surface and states $\vec{k} + Q\,[Q = (\frac{\pi}{a}, \frac{\pi}{b})]$ and determines the pseudogap, $\Delta^* \simeq k_B T^*$, in the density of states. This pseudogap exhibits itself in the susceptibility and hence in the nuclear relaxation rates in several high temperature superconductors. The system is thermodynamically paramagnetic above T^* and breaks spin-ordering symmetry at this temperature.

The electrons move in a potential with wave vector $2k_F$ at which there is anomaly in the phonon wave vector so that the distortion in the lattice causes a gap in the dispersion relation. This gap is proportional to the amplitude of the spin-density wave, SDW. One of the components of the spinor represents the electrons which are moving towards left with spin up and the other component represents the electrons which are moving towards right with spin down with Fermi velocity v_F. The Hamiltonian of these electrons can be written by using the Pauli spin matrices, $\sigma_3 = \sigma_z$, $\sigma_\pm = \sigma_1 \pm i\sigma_2$. It is assumed that the SDW is incommensurate with the lattice and so $\Delta(x)$ is complex. In the case of charge density waves, the electron-phonon interaction does not depend on spin and hence the coefficient of the gap in the case of CDW is independent of the spin. However, in the case of spin-density waves, the coefficient of the gap has terms which depend on $\sigma_\pm$ so that the Hamiltonian can be written as,

$$H = \int dx \psi^+ \left[-iv_F \sigma_3 \partial/\partial x + \frac{1}{2} \{\Delta(x)\sigma_+ + h.c.\} \right] \psi \ . \qquad (D.3)$$

The gap $\Delta(x)$ fluctuates due to thermal and quantum effects. In the Gaussian approximation, the correlation function is of the form,

$$\langle \Delta(x) \Delta^*(x') \rangle = \psi^2 \exp\left[-|x - x'|/\xi_z(T)\right] \ , \qquad (D.4)$$

where $\xi(T)$ is the correlation length along the chains and ψ is the root-mean-square fluctuation in the order parameter. For a one dimensional system, the coherence length diverges at the transition temperature, T^*, in the present case,

$$\psi^2 \propto T\xi_z(T) \simeq T\xi_z\left(1-\frac{T}{T^*}\right)^{-\nu} \quad . \tag{D.5}$$

Taking into account the coupling between chains with a cutoff wave vector q_o,

$$\psi^2 \propto q_o T \frac{\xi_z^2(T)}{\xi_x(T)\xi_y(T)} \quad . \tag{D.6}$$

so that the SDW amplitude is finite when the correlation length diverges. As the transverse correlation length ξ_x or ξ_y decreases, the ψ increases. When a large magnetic field is applied along the c-axis along which the conductivity is minimum, the coherence length along the b direction, ξ_b reduces. If ω_c is the cyclotron frequency, the reduction is determined by the factor,

$$\frac{\hbar\omega_c}{k_B T} = \frac{ebv_F B}{T} \equiv \frac{B}{B_o}\frac{T_{MF}}{T} \tag{D.7}$$

where b is the unit cell size along the b direction, v_F the Fermi velocity and T_{MF} is the mean field transition temperature. In the limit of large correlation lengths, $\xi_z \gg v_F/\psi$, near T_{SDW}, the density of states is zero at the Fermi energy and suppressed on an energy scale of the order of ψ, i.e., there is a pseudogap. As $\xi_z\psi/v_F$ decreases the pseudogap gradually fills in. In CDW systems, at zero field, there is a pseudogap due to fluctuations. The ratio of the SDW temperature to that of the mean field theory, T_{SDW}/T_{MF} is a function of ψ/T_{MF}. As the fluctuation increases T_{SDW} decreases. For $\psi > T_{MF}$ the SDW is not formed. The pseudogap in the high-temperature superconductors is most likely caused by the charge-density waves or by the spin-density waves.

Suggested References

D.1 G. Preosti, Y.M. Vilk and M.R. Norman, Phys. Rev. **B59**, 1474 (1999).

D.2 A.S. Alexandrov, V.N. Zavaritsky, W.Y. Liang and P.L. Nevsky, Phys. Rev. Lett. **76**, 983 (1996).

D.3 C. Panagopoulos, J.R. Cooper, T. Xiang, Y.S. Wang and C.W. Chu, Phys. Rev. B**61**, 3808 (2000).

D.4 J. Ranninger and A. Romano, Phys. Rev. Lett. **80**, 5643 (1998).

D.5 Q. Chen, I. Kosztin, B. Janko and K. Levin, Phys. Rev. B**59**, 7083 (1999).

D.6 M.J. Rice and S. Strässler, Solid State Commun. **13** 1389 (1973).

D.7 P.A. Lee, T.M. Rice and P.W. Anderson, Phys. Rev. Lett. **31**, 462 (1973).

D.8 J. Demsar, B. Podobnik, V.V. Kabanov, Th. Wolf and D. Mihailovic, Phys. Rev. Lett. **82**, 4918 (1999).

D.9 Ch. Renner, B. Revaz, K. Kadowaki, I. Maggio-Aprile and Φ. Fischer, Phys. Rev. Lett. **80**, 3606 (1998).

D.10 T. Sato, T. Yokoya, Y. Naitoh, T. Takahashi, K. Yamada and Y. Endoh, Phys. Rev. Lett. **83**, 2254 (1999).

D.11 C.M. Varma, Phys. Rev. Lett. **83**, 3538 (1999).

D.12 B. Janko, I. Kosztin, K. Levin, M.R. Norman and D.J. Scalapino, Phys. Rev. Lett. **82**, 4304 (1999).

D.13 D.R. Temprano, J. Mesot, S. Janssen, K. Conder, A. Furrer, H. Mutka, K.A. Müller, Phys. Rev. Lett. **84**, 1990 (2000).

D.14 J. Schmalian, S. Grabowski and K.H. Bennemann, Phys. Rev. B**56**, 509 (1997).

D.15 H. Kontani and K. Veda, Phys. Rev. Lett. **80**, 5619 (1998).

D.16 B. Janko, J. Maly and K. Levin, Phys. Rev. B**56**, 11407 (1997).

Appendix E: Relativistic effects

E.1. Gap.

In the appendix C we have discussed the symmetries of the gap of the superconductor in terms of s wave or d wave, which are the solutions of the non-relativistic Schrödinger equation. In this appendix, we shall discuss the gap in terms of Dirac matrices which are the solutions of the relativistic Dirac equation which has the kinetic energy linear in p so that the first derivative $\partial/\partial x$ occurs symmetrically with the linear time derivative $\partial/\partial t$ in E. In the case of Cooper pairs with angular momentum, $J = 0$ and parity $(-1)^l = +$ for $l = 0$, the gap consistent with Fermi statistics is,

$$\Delta(n) = \Delta_1\gamma_5 + \Delta_2 n.\gamma\gamma_o\gamma_5 + \Delta_3\gamma_o\gamma_5 \tag{E.1}$$

where γ_μ are the Dirac matrices, n is the unit vector along the direction of the wave vector and Δ_i are the gap parameters. The Dirac matrices anticommute,

$$\gamma_\mu\gamma_\nu + \gamma_\nu\gamma_\mu = 2\delta_{\mu\nu} \tag{E.2}$$

for $\mu = 1, 2, 3$ or 4 with $\gamma_o = \gamma_4$ and $\gamma_5 = \gamma_1\gamma_2\gamma_3\gamma_4$. There are five quantities which are Lorentz invariant,

$$\begin{array}{rl} 1 & \text{4scalar,} \\ \gamma_\mu & \text{4vector,} \\ \frac{1}{2i}(\gamma_\mu\gamma_\nu - \gamma_\nu\gamma_\mu) & \text{4tensor,} \\ i\gamma_5\gamma_\mu & \text{4axial vector,} \\ \gamma_5 & \text{4pseudoscalar ,} \end{array} \tag{E.3}$$

The 4-vector γ_μ mix in the same way as the four components $(\vec{x}, ict)$. The axial vector and pseudoscalar have the same properties as the vector and scalar under a velocity transformation. They differ under a reflection, the axial vector not changing the sign and the pseudoscalar changing sign. An axial vector has opposite parity to a vector, and a pseudoscalar opposite parity to a scalar. An example of a pseudoscalar is the socalled scalar triple product $A.B \times C$, the dot product of the vecotr A and the axial vector from the cross product of the vectors B and C. A relativistic gap may be formed from the following,

$$\begin{aligned}
&1 \ , \\
&\gamma_\mu \ , \\
&\frac{1}{2i}(\gamma_\mu\gamma_\nu - \gamma_\nu\gamma_\mu) \ , \\
&(i\gamma_5\gamma_\mu) \ , \\
&\gamma_5 \ .
\end{aligned} \tag{E.4}$$

In the non-relativistic limit

$$\begin{aligned}
1 &\rightarrow 1 \ , \\
\gamma_\mu &\rightarrow 1 \ , \\
\frac{1}{2i}(\gamma_\mu\gamma_\nu - \gamma_\nu\gamma_\mu) &\rightarrow \sigma \ , \\
(i\gamma_5\gamma_\mu) &\rightarrow \sigma \ , \\
r_5 &\rightarrow 0 \ .
\end{aligned} \tag{E.5}$$

The interaction $(1-\gamma_5)\gamma_\mu$ violates parity and has been written as $(g_v - g_A\gamma_5)\gamma_\mu$ where g_v and g_A are the adjustable parameters which describe the amount of vector or the axial vector part of the interaction. The gap of the superconductor may also be written by means of a suitable combination of the matrices given by (E.3).

E.2. Cherenkov-Josephson radiation.

When a relativistic particle passes through a medium, the medium gets polarized. After some time the polarization relaxes to the unpolarized state by coherent emission of radiation called the Cherenkov radiation. For this phenomenon to occur, the velocity of the particle, v should be larger than the phase velocity of light in the medium,

$$c > v\frac{c}{\sqrt{\epsilon(\omega)}}. \tag{E.6}$$

The Cherenekov radiation per unit distance along the path of the particle is,

$$\frac{dE}{dx} = \frac{(ze)^2}{c^2} \int_{\epsilon(\omega)>1/\beta^2} \omega \left(1 - \frac{1}{\beta^2 \epsilon(\omega)}\right) d\omega. \qquad (E.7)$$

The integrand gives the differential spectrum in the frequency. Here ze is the atomic charge and $\beta = v/c$. At large distances from the path, the field becomes transverse. The direction of propagation is given by $\vec{E} \times \vec{B}$. The angle of emission of Cherenkov radiation relative to the velocity of the particle is given by

$$\cos\theta_c \frac{1}{\beta\sqrt{\epsilon(\omega)}}. \qquad (E.8)$$

Therefore, the Cherenkov radiation is completely linearly polarized in the plane containing the direction of observation and the path of the particle.

In a long one-dimensional Josephson junction (LJJ), there are solitons which describe the vortex motion but their velocity does not exceed the phase velocity of light so that the Cherenkov radiation does not occur. The solitons are the solutions of non-linear Schrödinger equation which means that the usual equation $\mathcal{H}\psi = E\psi$ is not valid and there are terms with $\sin\psi$, $|\psi|^2$ or $|\psi|^3$ to generate nonlinearity. In the present case, the quasiparticles in the long Josephson junction are described by the sine-Gordon equation which is nonlinear because of the sine term. The highest velocity of the soliton, $\bar{C}_o$ is found to be less than that of the linear electromagnetic waves and hence Cherenkov effect is not found in a single Josephson junction. When two Josephson junctions are coupled by the coupling S, there are two solutions for the soliton velocity

$$\bar{C}_\pm = \bar{C}_o/(1 \pm S)^{1/2} \qquad (E.9)$$

where + sign shows the in-phase mode and − sign indicates the out-of-phase mode. As long as $-1 < S < 0$, the soliton velocity does not exceed the velocity of light in the medium. We consider a combination of N solitons in one LJJ and M solitons in the other LJJ. Such a combination of two LJJs is described by the notation $[N/M]$. The positive numbers are used for solitons and negative numbers are used for antisolitons, N or $M < 0$. For the symmetric and antisymmetric modes the system of two coupled sine-Gordon equations split into two independent sine-Gordon equations so that the maximum velocity of the soliton is almost equal to the velocity of the linear waves and hence Cherenkov radiation does not occur.

In symmetric [1|0] junctions, the soliton modes of the vortex motion can be faster than in the nonlinear Josephson junction. The asymmetric-junction-vortex soliton can also be faster than in the case of coupled Josephson junction. Therefore, there is a possibility that the vortex velocity will become equal to $c/[\epsilon(\omega)]^{1/2}$ although may not exceed as required in (E.6). Therefore, even if Cherenkov radiation does not occur, there is a possibility of observing a "Cherenkov step" in the current (voltage) characteristics of annular Josephson junction rings for which $v \simeq c/[\epsilon(\omega)]^{1/2}$. We consider a ring-shaped annular junction. The oxide is deposited on the surface of the annular ring and then another layer of the superconductor is deposited on the oxide so as to make annular Josephson junction. The experimental material is (Nb-Al-AlO$_x$)$_2$-Nb which forms a long Josephson junction. The radius of the ring is R so that the length of the Josephson junction is $2\pi R = \pi D$ where D is the diameter of the ring. The distances are normalized in units of Josephson penetration depth, λ_J so that the length of the junction becomes $\pi D/\lambda_J = L/\lambda_j \simeq 10$. A small bias current, I, is applied to the junction so that some vortices are trapped in the ring. With increasing I, the velocity of the vortex-increases and can reach the relatisvistic region. The solitonic velocity of the vortex is determined from the relations,

$$\bar{C}_o = V_1 \pi D/\phi_o \simeq 0.03C \qquad (E.10)$$

where $V_1 \simeq 52\mu$V is the asymptotic value of the voltage, ϕ_o is the unit flux and c is the velocity of light. A large critical current suggests that no vortices are trapped so that the configuration is [0|0]. Under certain circumstance, there is a step in the current (voltage) characteristics. The voltage at the step is proportional to the total number of vortices trapped in the system minus the number of pinned vortices. Cherenkov step can appear if the soliton velocity v is equal to the phase velocity of the emitted linear e.m. waves,

$$v = \frac{\omega(k)}{k} = \bar{C}_o \left[\frac{1}{1-S} + \left(\frac{1}{\lambda_J k} \right)^2 \right]^{1/2} \qquad (E.11)$$

where ω is the angular frequency of the soliton waves and k is their wave vector. The annular LJJ of length $L = \pi D$ with periodic boundary conditions gives the wave vectors as,

$$k_n = \frac{2\pi}{L} n \qquad (n = 1, 2, \cdots \infty). \qquad (E.12)$$

If the vortex velocity is equal to that of the velocity of light in the medium, the vortices may resonate with the absorption or emission of light. The resonant wave vector is determined by

$$v_n = \frac{\omega(k_n)}{k_n} = \left[\bar{C}_-^2 + \left(\frac{L}{2\pi \lambda_J n} \right)^2 \right]^{1/2} . \qquad (E.13)$$

It is possible that such a soliton condition arises in the case of two coupled sine-Gordon equations which are given below:

$$\begin{aligned} \frac{1}{1-S^2}\phi^A_{xx} - \phi^A_{tt} - \sin\phi^A - \frac{S}{1-S^2}\phi^B_{xx} &= \alpha\phi^A_t - \gamma \ , \\ \frac{1}{1-S^2}\phi^B_{xx} - \phi^B_{tt} - \frac{1}{J}\sin\phi^B - \frac{S}{1-S^2}\phi^A_{xx} &= \alpha\phi^B_t - \gamma \ , \end{aligned} \qquad (E.14)$$

which describe the dynamics of the superconducting phase differences, in the configuration [1|0] for $J = j_c^A / j_c^B < 1$, the soliton velocity is larger than $\bar{C}_p$. At this condition Cherenkov radiation occurs which can be detected by the usual current (Voltage) characteristics. The onset of the Cherenkov radiation is called the Cherenkov step because the current is expected to become constant and independent of the soliton velocity. At resonance $\phi^{A,B}$ oscillates causing plasma waves.

Therefore, under certain circumstances in coupled Josephson junctions with certain amount of flux trapping, the vortex velocity can exceed that in the single Josephson junction. Therefore, it is possible to accelerate the vortices so that the condition of the Cherenkov radiation is satisfied. Such effects are observable in the current (voltage) characteristics. The gap of the superconductor is then described by a relativistic theory or a suitable combination of Lorentz invariant products of Dirac matrices.

Selected References

E.1 D. Bailin and A. Love, Nucl. Phys. B**190**, 175 (1981); B190, 751 (1981); B**205**, 119 (1982).

E.2 D. Bailin and A. Love, Phys. Repts. B**107**, 325 (1984).

E.3 M. Alford, J. Berges, K. Rajagopal, Phys. Rev. Lett. **84**, 598 (2000).

E.4 A. Wallraff, A.V. Ustinov, V.V. Kurin, I.A. Shereshevsky and N.K. Vdovicheva, Phys. Rev. Lett. **84**, 151 (2000).

E.5 K. Capelle, E.K.V. Gross and B.L. Gyorffy, Phys. Rev. Lett. **78**, 3753 (1997).

E.6 G. Hechtfischer, R. Kleiner, A.V. Ustinov and P. Muller, Phys. Rev. Lett. **79**, 1365 (1997).

E.7 E. Goldobin, A. Wallraff, N. Thyssen and A.V. Ustinov, Phys. Rev. B**57**, 130 (1998).

E.8 U.M. Krasnov and D. Winkler, Phys. Rev. B**60**, 13179 (1999).

E.9 V.V. Kurin and A.V. Yulin, Phys. Rev. B**55**, 11659 (1997).

Appendix F: Soft vortices

F.1. Soft vortices.

The lattice distortions occuring at characteristic temperatures are accompanied by charge-density waves. The frequency associated with the lattice distortion is finite at zero temperature but varies with temperature till at the characteristic temperature at which it becomes zero. The materials containing such lattice distortions are soft and hence the mode of which the frequency vanishes is called a soft mode. The distortions are associated with the coordinates of atoms which have finite inertial mass.

The vortices have a large magnetic field at the centre which reduces as we go away from the centre. The vortex lattice is said to melt at a characteristic temperature. In this appendix we discuss the distortions in the vortex lattice which are the cause of vortex-lattice waves which means that the vortex density oscillates so that energy flows in the form of waves. The vortices have certain frequencies which vanish at characteristic temperatures, associated with critical exponents which are associated with the dimensionality of the system. The distance between vortices as a function of temperature and as a function of magnetic field shows critical exponents. The vortex frequencies show similar behaviour as a function of temperature and field. The apex angle also varies with temperature and field and the critical behaviour is clearly visible in the temperature dependence of the penetration depth. In some of the systems, the lattice distortions occur in the proximity of magnetic ordering. Therefore, in such systems both the critical temperature as well as the magnetic ordering temperature shows cirtical behaviour. We determine the critical temperatures and critical exponents using the vortex lattice in superconductors.

In the case of soft modes the higher temperature phase has higher symmetry and lower temperature has lower symmetry. In the case of vortex lattices, we predict that the displacement of vortices should vary as,

$$\delta x = (\delta x)_o[1 - (T/T_c)]^{1/2} \qquad (F.1)$$

where mean field value for the exponent of 1/2 has been chosen. Instead of choosing the distance between vortices, we may describe the soft vortex by softening of an angle of the vortex lattice. In the Abrikosov flux lattice, the square lattice has the lowest energy but it changes to a triangular lattice upon the application of a magnetic field. Therefore, in the case of softening of the Abrikosov lattice, one of the angles is predicted to vary as

$$\beta = \beta_o \left[1 - \frac{B}{B_o}\right]^{1/2} \tag{F.2}$$

where β is the apex angle, β_o is a constant which is the value of the angle at zero external field, and B_o is the critical induction. This means that the square as well as the triangular flux lattice can get completely destroyed and the vortices become completely randomized. Similarly, the distances between vortices vary with applied magnetic field as

$$\delta x = [\delta x]_o[1 - (B/B_o)]^{1/2}. \tag{F.3}$$

Thus it is clear that the distances between vortices as well as the angles between vortices become soft and go to zero at a certain characteristic temperature and at a certain field for soft vortex lattices. Accordingly, the frequency of a soft vortex mode varies as,

$$\omega = \omega(o)[1 - (T/T_c)]^2 \tag{F.4}$$

as a function of temperature going to zero at $T = T_c$. It follows that the frequency also varies as a function of magnetic field as,

$$\omega = \omega(o)[1 - (B/B_o)]^{1/2} \tag{F.5}$$

which is zero at $B = B_o$. So far we have considered only one T_c or only one B_o. We now consider a situation in which two transition temperatures occur. There is a Neel temperature very near T_c such that $T_N < T_c$. In such a case, the displacement of vortices given by (F.1) is zero at T_c as well as at T_N so that the vortices are required to become soft at both the T_c as well as T_N. The soft vortex mode is thus confined within the superconducting region. Therefore, the apex angle is predicted to vary as,

$$\beta = \beta_1 + \beta_2 \tag{F.6}$$

where β_1 becomes soft at $T = T_N$ and β_2 becomes soft at T_c so that,

$$\beta_1 = \beta_{10}[(T/T_N) - 1]^{1/2} \tag{F.7}$$

and

$$\beta_2 = \beta_{20}[1 - (T/T_N)]^{1/2} \tag{F.8}$$

where β_{10} and β_{20} are constants. Therefore, the soft vortices are associated with superconductivity. It is also possible that the temperature and magnetic field at which the vortices become soft are not the same as those associated with superconductivity, i.e.,

$$\begin{aligned} T_s &\neq T_c \\ T_s &\neq T_N \\ B_o &\neq B_{c2} \end{aligned} \tag{F.9}$$

in which case a phase transition occurs in the vortex lattice.

F.2. Penetration depth and soft vortices.

We have discussed the London penetration depth in Chapter 6. We now consider the modification in the London penetration depth caused by the soft vortices. We have shown that the vortices are subject to a viscous force which contributes to the current. In general, the penetration depth can be complex and can change when we apply magnetic field to the vortices. In the normal state, there is the skin depth upto which the magnetic field can penetrate in the solid. The London penetration depth in the superconducting phase depends on the normal state skin depth. The equation of motion of vortex has been solved in Chapter 6 to define a penetration depth. In addition to the visocity term in the force, we use the harmonic oscillations of the vortex. We assume that the vortex has an effective mass so that the equation of motion which descirbes the current can be written as,

$$M\frac{dv}{dt} + \eta v + kx = \frac{1}{c}J\phi_o \tag{F.10}$$

where M is the mass of the vortex, dv/dt is the acceleration, η is the coefficient of viscosity of the vortex of velocity v, force constant k and the critical current J. It was found in Chapter 6, eq. 6.26, that at large magnetic fields, the effective London penetration depth varies as,

$$\lambda_{eff} \approx B^{1/2} \left[\phi_o/(4\pi\omega)\right]^{1/2} \left[\left\{ \left(\frac{k}{\omega}\right) - \omega M \right\}^2 + \eta^2 \right]^{-1/4} . \qquad (F.11)$$

The original London penetration depth as, e.g., in eq.(1.6) is independent of magnetic field as well as temperature. However, we have found that the penetration depth changes when magnetic field is applied to a vortex. The eq.(F.11) gives the field dependence of the penetration depth in one of the approximations. For large viscosity, $\eta \gg (k/\omega) - \omega M$, we retain only one term in the large bracket, and substitute (F.5) in (F.11) and leave out the remaining factors in a constant, so that the effective London penetration depth becomes,

$$\lambda_{eff} = k_1 \frac{B^{1/2}}{\omega_o^{1/2}[1 - (B/B_o)]^{1/4}} \qquad (F.12)$$

where the constant k_1 can be determined by comparing (F.12) with (F.11). If peaks are absent in the resistivity, the mass term in (F.10) is small and hence the force which depends on the acceleration is neglected. The London penetration depth varies with field with a divergence at $B = B_o$ and the critical exponent in this case is 1/4.

F.3. Experimental measurements.

The neutron diffraction measurements on the flux-line lattice in $TmNi_2B_2C$ as a function of temperature at a fixed magnetic field have been peformed by Paul et al. One of the measurements is given in Fig. F.1 at $H = 3$ kOe. For $10 > T > 4$K, the apex angle becomes soft. Therfore according to (F.8), β is plotted as a function of $[1 - (T/T_c)]^{1/2}$ where the exponent of 1/2 is assumed on the basis of mean-field theory.

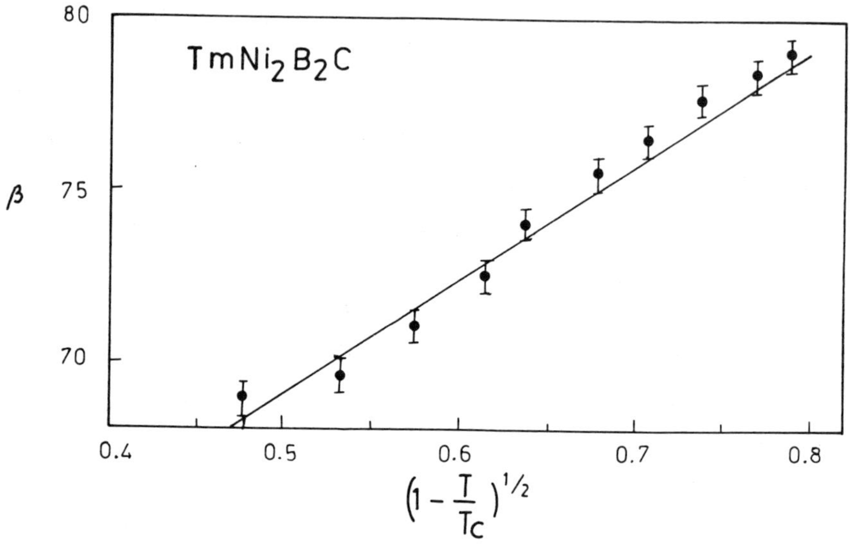

Fig. F.1: The apex angle for $10 > T > 4K$, as a function of $[1-(T/T_c)]^{1/2}$ showing soft vortex behaviour in TmNi$_2$B$_2$C. The value of the angle is zero at T_c.

The temperature dependence of the apex angle fits with the soft mode interpretation very well. In order to see, if the angle becomes soft also at T_N we have plotted in Fig. F.2 the angle β in the range $1.5 < T < 4$K as a function of $[(T/T_N) - 1]^{1/2}$. The measured values fit very well and the angle does become soft at T_N. Thus $TmNi_2B_2C$ shows soft vortices with the apex angle having zero value at T_c as well as at T_N.

The soft vortices which are due to two minima in the free energy when one parameter is varied have been identified in YNi_2B_2C. In the present case, the parameter is the magnetic field. When field is varied, the system changes from one minimum value to another. The higher field minimum has higher symmetry and lower field minimum has a lower symmetry. Therefore, the two states have a critical field, B_o. Since both the phases are superconducting $B_o < B_{c2}$. The experimental values of the apex angle,

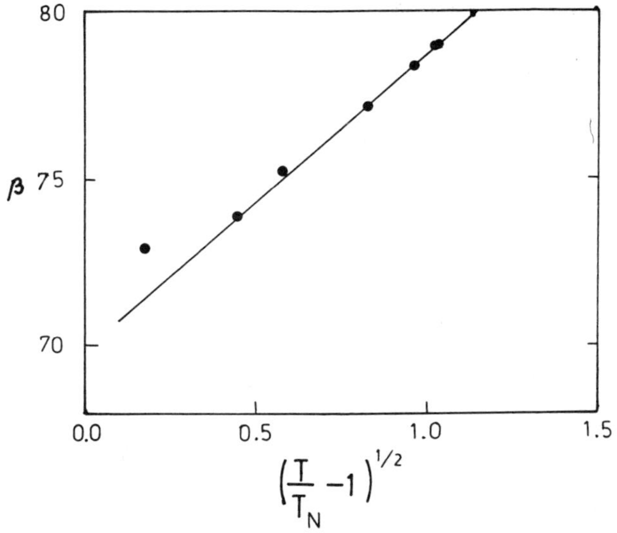

Fig. F.2: The apex angle for $1 < T < 4K$ *in* $TmNi_2B_2C$ *showing soft vortex behaviour as a function of* $[(T/T_N) - 1]^{1/2}$ *which approaches zero at* T_N.

β, of the rhombic unit cell of YNi_2B_2C are compared with (F.2). It is found that the data fit very well with,

$$\beta = 8[1 - (B/B_o)]^{1/2} + 52 \qquad \text{(deg)} \qquad (F.13)$$

with $B_o = 1130 \pm 70$ Oe as shows in Fig. F.3. Therefore, we conclude that there is a soft vortex with critical field of B_o. The angle β has a large value at $B = 0$. There is a component in the angle which reduces with increasing field and becomes zero at $B = B_o$.

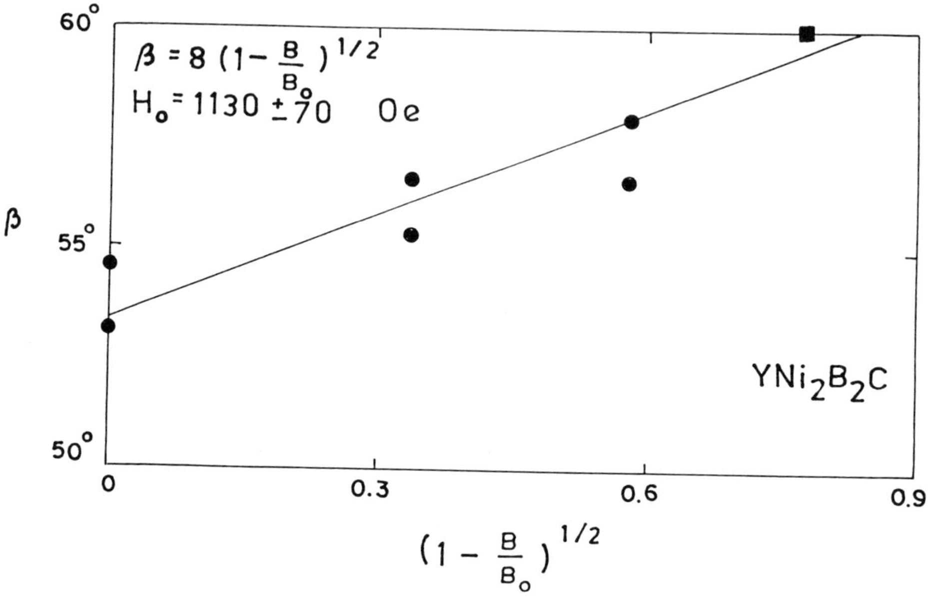

Fig. F.3: The apex angle β of the rhombic unit cell as a function of $[1-(B/B_o)]^{1/2}$ showing linear dependence. The experimental points are taken from the small angle neutron scattering measurements performed on YNi_2B_2C.

In Chapter 6, the penetration depth as a function of mangetic field in $ErNi_2B_2C$ was explained only for magnetic fields larger than ~ 250 G. For small fields, the penetration depth increases slowly with increasing magnetic field but the measured values do not agree with the predicted $B^{1/2}$ or B^2. The changes in the penetration depth at small magnetic fields are assigned to soft modes. According to eq. (F.12) for magnetic fields smaller than $\sim$ 250G, the change in London penetration depth as a function of magnetic field as shown in Fig. F.4 agrees with the expression,

$$\delta\lambda_L = c_1 B^{1/2}[1-(B/B_o)]^{-1/4}$$

for $c_1 = 4.73$ and $B_o = 25 \pm 10$ G. Thus the measured values are in agreement with the scaling theory of soft modes.

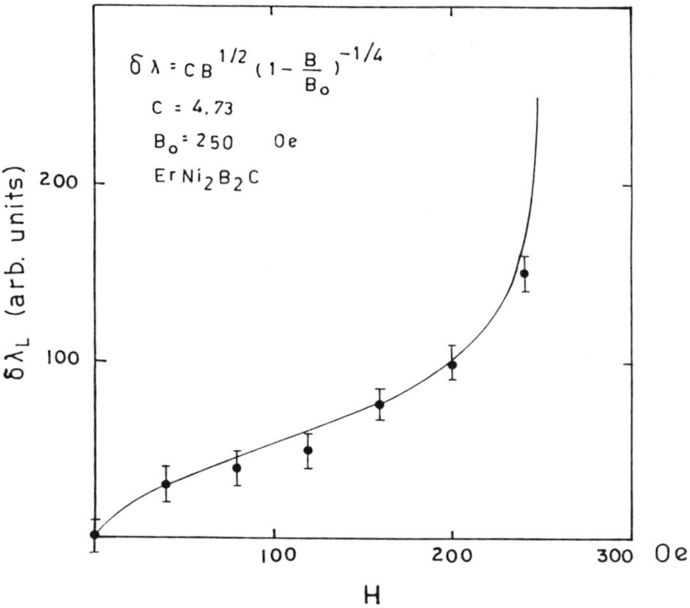

Fig. F.4: The change in London penetration depth as a function of magnetic field calculated for a soft vortex with critical field $B_o = 250G$. *There is a divergence in the calculated value at 250 G. The experimental points are deduced from the measurements carried out by Eskildsen et al using* $ErNi_2B_2C$.

The free energy as a function of field has two minima separated by a barrier. When the applied magnetic field energy, $B^2/8\pi$ is equal to that of the barrier, there is a transition from one minimum value to another accompanied by a soft vortex. In the apex angle such a soft vortex gives rise to a critical behaviour which we have described by the mean field theory. A critical field exists such that the London penetration depth diverges at this field, $\lambda_L \propto B^{1/2}[1 - (B/B_o)]^{-1/4}$. For $B > B_o$, the vortex lattice has the symmetry of a square lattice while for $B < B_o$ the structure is hexagonal.

Suggested References

F.1 P.K. Ghosh and K.N. Shrivastava, Physica C**306**, 233 (1998).

F.2 K.N. Shrivastava, Solid State Commun. (2000).

F.3 D. McK. Paul, C.V. Tomy, C.M. Aegerter, R. Cubitt, S.H. Lloyd, E.M. Forgan, S.L. Lee, M. Yethiraj, Phys. Rev. Lett. **80**, 1517 (1998).

F.4 V.G. Kogan, M. Bullock, B. Harmon, P. Miranovic, P.L. Gammel, and D.J. Bishop, Phys. Rev. B**55**, 8693 (1997).

F.5 V.G. Kogan, P. Miranovic, Lj. Dobrosavljevic - Grujic, W.E. Pickett and D.K. Christen, Phys. Rev. Lett. **79**, 741 (1997).

F.6 M.R. Eskildsen, P.L Gammel, B.P. Barber, V. Yaron, A.P. Ramirez, D.A. Huse, D.j. Bishop, C.Boole, C.M. Lieber, S. Oxx, S. Sridhar, N.H. Andersen, K. Mortensen, P.C. Canfield, Phys. Rev. Lett. **78**, 1968 (1997).

F.7 P.K. Ghosh, N. M. Krishna and K. N. Shrivastava, J. Phys.: Condens. Matter **9**, L 663 (1997).

Appendix G: Specific heat

G.1. Normal state.

The specific heat of a normal solid was calculated by Albert Einstein in 1910 by assuming that all of the atoms have a single frequency. However, in a real-three-dimensional solid, there is a frequency spectrum from zero to any characteristic value with a continuum and a reasonable cutoff. The summation over all of the frequencies requires the knowledge of wave vector and hence a dispersion relation. The calculation of the lattice energy, U, is then straight forward and the specific heat at constant volume is given by,

$$C_v = \frac{\partial U}{\partial T}. \quad , \qquad (G.1)$$

The energy of one phonon is $\hbar\omega_k$ and the number of phonons of wave vector k is n_k. The hamiltonian of a harmonic oscillator has conjugate variables p and x. Hence due to the uncertainty relation both the p as well as the x can not be determined so that there is some energy. The energy of the harmonic oscillator is therefore minimized at $\frac{1}{2}\hbar\omega$. Therefore, the number of excitations of wave vector k becomes $(n_k + \frac{1}{2})$. The energy of the lattice can therefore be written as,

$$U = \sum_k (n_k + \frac{1}{2})\hbar\omega_k. \qquad (G.2)$$

The summation over k requires the knowledge of dimensionality. For a one-dimensional solid,

$$\sum_k = \frac{L}{2\pi}\int dk \quad , (d = 1) \qquad (G.3)$$

because the zero is restricted by $\pm\pi/a$ where a is the distance between atoms in the lattice of length L. In the case of a two-dimensional solid,

$$\sum_k = 2\left(\frac{L}{2\pi}\right)^2 \int kdk \qquad (d = 2) \qquad (G.4)$$

where a factor of 2 arises from $\int \sin\theta d\theta$, and for a three-dimensional solid,

$$\sum_k = 3\left(\frac{L}{2\pi}\right)^3 .4\pi \int k^2 dk = \frac{12\pi V}{(2\pi)^3}\int_o^{k_D} k^2 dk, \quad , \qquad (d = 3). \quad (G.5)$$

There are three integrals one each along x, y and z axes for which we have written $\int\int\int dk_x dk_y dk_z = \int k^2 dk$. We should add the similar results for two transverse and one longitudinal branch of the phonon so that we put a factor of 3 in the above expression. For the integral $\int d(\cos\theta) = 2$ and $\int d\phi = 2\pi$, we have multiplied the above by 4π and replace L^3 by V, the volume of the lattice. The lowest value of the wave vector of a phonon is assumed to be zero for an isotropic solid and the maximum value is fixed by fixing the number of atoms, N in the solid by,

$$N = \int_o^{k_D} n_k = \frac{3V}{2\pi^2} k^2 dk. \tag{G.6}$$

To evaluate the lattice energy in three dimensions, we substitute (G.5) in (G.2) to find,

$$U = \frac{3V}{2\pi^2} \int \left(n_k + \frac{1}{2} \right) \hbar\omega_k k^2 dk \tag{G.7}$$

which requires the knowledge of dispersion relation. For this purpose we assume a simple non-interacting system of phonons for which the dispersion is,

$$\hbar\omega_k = \hbar v k. \tag{G.8}$$

Substituting (G.8) in (G.7) we find,

$$U = U_o + U(T) \tag{G.9}$$

where the zero-point excitation gives the first term

$$U_o = \frac{3V\hbar v}{16\pi^2} k_D^4 \tag{G.10}$$

and the temperature dependent part of the lattice energy in three dimensions is,

$$U(T) = \frac{3V\hbar v}{2\pi^2} \int \frac{k^3 dk}{e^{\hbar\omega/k_B T} - 1}. \tag{G.11}$$

We make use of the dispersion relation (G.8) to eliminate the wave vector so that the above integral can be written as,

$$U(T) = \frac{3V\hbar}{2\pi^2 v^3} \int_o^{\omega_D} \frac{\omega^3 d\omega}{e^{\hbar\omega/k_B T} - 1}. \tag{G.12}$$

where we have replaced the maximum wave vector by the maximum cutoff frequency. Making the change of variables, $\hbar\omega/k_BT = x$, it is clearly seen that at low temperatures,

$$U(T) = \frac{3V\hbar}{2\pi^2 v^3}\left(\frac{k_BT}{\hbar}\right)^4 \int_o^X \frac{x^3dx}{e^x - 1} \qquad (d = 3). \tag{G.13}$$

For very large value of $X = \hbar\omega_D/k_BT$, at sufficiently low temperatures, the limit can be replaced by infinity so that U varies as T^4 in three dimensions and hence the specific heat $C_v = \partial U/\partial T$ varies as T^3. The temperature dependence is essentially a consequence of wave vector dependence. Therefore, when we use (G.3) or (G.4) we find that the specific heat varies with temperature T in one dimension and as T^2 in two dimensions. This calculation requires wave propagation and hence the wave vector dependence appropriate to that of a pure crystal lattice.

For a one-dimensional chain of atoms, we substitute (G.3) in (G.2) so that,

$$U = \frac{L}{2\pi}\int (n_k + \frac{1}{2})\hbar\omega_k dk \tag{G.14}$$

which also has a zero-point vibrational contribution as well as a temperature dependent part so that

$$U = U(T = 0, d = 1) + U(T, 1) \tag{G.15}$$

with

$$U(0, 1) = \frac{L\hbar}{4\pi}\int \omega dk. \tag{G.16}$$

We use the simple dispersion relation, $\omega = vk$ so that the zero-point vibrational contribution for the lattice energy of a one-dimensional solid becomes

$$U(T, 1) = \frac{L\hbar}{4\pi}\int_o^{k_D} n_k\omega_k dk = \frac{L\hbar v}{2\pi}\int_o^{k_D} \frac{kdk}{e^{\hbar\omega/k_BT} - 1}. \tag{G.18}$$

We make the change of variables, $\hbar vk/k_BT = x$ so that

$$U(T, 1) = \frac{L\hbar v}{2\pi}\left(\frac{k_BT}{\hbar v}\right)^2 \int_o^X \frac{xdx}{e^x - 1} \tag{G.19}$$

shows that in a one-dimensional chain of atoms, the lattice energy varies as T^2 and hence the specific heat is linear in temperature, $C_v \sim T$ for $d = 1$. In the case of a two-dimensional sheet of atoms which are found in layered structures, we substitute (G.4) in (G.2) to find,

$$U = \frac{2L^2\hbar}{(2\pi)^2}\int (n_k + \frac{1}{2})\omega_k k dk \qquad (G.20)$$

which also has a temperature dependent part as well as a zero-point contribution. For two-dimensional materials the zero-point contribution is given by,

$$U(T=0, d=2) = \frac{L^2\hbar v}{(2\pi)^2}\int k^2 dk = \frac{L^2\hbar v}{3(2\pi)^2}k_D^2 \ , \qquad (G.21)$$

and the temperature dependent part is found to be

$$U(T,2) = \frac{2L^2\hbar}{(2\pi)^2}\int n_k\omega_k k dk. \qquad (G.22)$$

We use the dispersion relation to eliminate the frequency so that,

$$U(T,2) = \frac{2L^2\hbar v}{(2\pi)^2}\int \frac{k^2 dk}{e^{\hbar v k/k_B T} - 1}. \qquad (G.23)$$

Making the change of variables $\hbar v k/k_B T = x$ we find,

$$U(T=2) = \frac{2L^2\hbar v}{(2\pi)^2}\left(\frac{k_B T}{\hbar v}\right)^3 \int_o^X \frac{x^2 dx}{e^x - 1} \ , \qquad (G.24)$$

which shows that for a layered structure the lattice energy varies as T^3 and hence the specific heat for a two-dimensional system varies as T^2. Thus the specific heat of a solid varies as T, T^2 or T^3 for $d = 1, 2$, or 3, respectively, as long as there is a perfect lattice. However, in the case of a two-level impurity, there is a potential barrier so that at this energy level separation, a peak is found in the specific heat.

Consider two energy levels caused by impurities in the lattice. In this case there is no wave vector space. Therefore, we define a partition function,

$$z = \sum_i \exp(-\epsilon_i/k_B T). \qquad (G.25)$$

The summation is over the Boltzmann factor for all the states, i. The probability of occurrence of the ith state is given by,

$$P(\epsilon_i) = \frac{\exp(-\epsilon_i/k_B T)}{z} \qquad (G.26)$$

which summed over all the states is unity, $\sum_i P(\epsilon_i) = 1$. The average energy of the system is

$$U = <\epsilon> = \sum_i \epsilon_i P(\epsilon_i) \qquad (G.27)$$

$$U = \frac{\sum_i \epsilon_i \exp(-\epsilon_i/k_B T)}{z}. \qquad (G.28)$$

In the case of two levels, the energy of the lower level is zero while that of the upper level is ϵ so that the partition becomes,

$$z = \exp(-O/k_B T) + \exp(-\epsilon/k_B T) = 1 + \exp(-\epsilon/k_B T). \qquad (G.29)$$

The average energy is

$$U = \frac{\epsilon \exp(-\epsilon/k_B T)}{1 + \epsilon(-\epsilon/k_B T)} \qquad (G.30)$$

A plot of $\partial U/\partial T$ from this expression displayed in Fig. G.1 as a function of $k_B T$ shows a peak at about $k_B T_s \simeq 0.42\epsilon$, i.e., at a temperature of $T_s \simeq 0.42\epsilon/k_B$.

We shift the energy scale from 0 at lower energy level and ϵ at higher level to $-(\frac{1}{2})\epsilon$ at lower level and $+(\frac{1}{2})\epsilon$ at higher level so that,

$$z = \exp(-\epsilon/2k_B T) + \exp(-\epsilon/2k_B T) = 2\cosh(\epsilon/2k_B T). \qquad (G.31)$$

and the average energy becomes,

$$\begin{aligned} <\epsilon> &= \left[\left(-\frac{1}{2}\right)\epsilon \exp(\epsilon/2k_B T) + \left(\frac{1}{2}\epsilon\right)\exp(-\epsilon/2k_B T)\right]/z \\ &= -\frac{1}{2}\epsilon \tanh(\epsilon/2k_B T) \ . \end{aligned} \qquad (G.32)$$

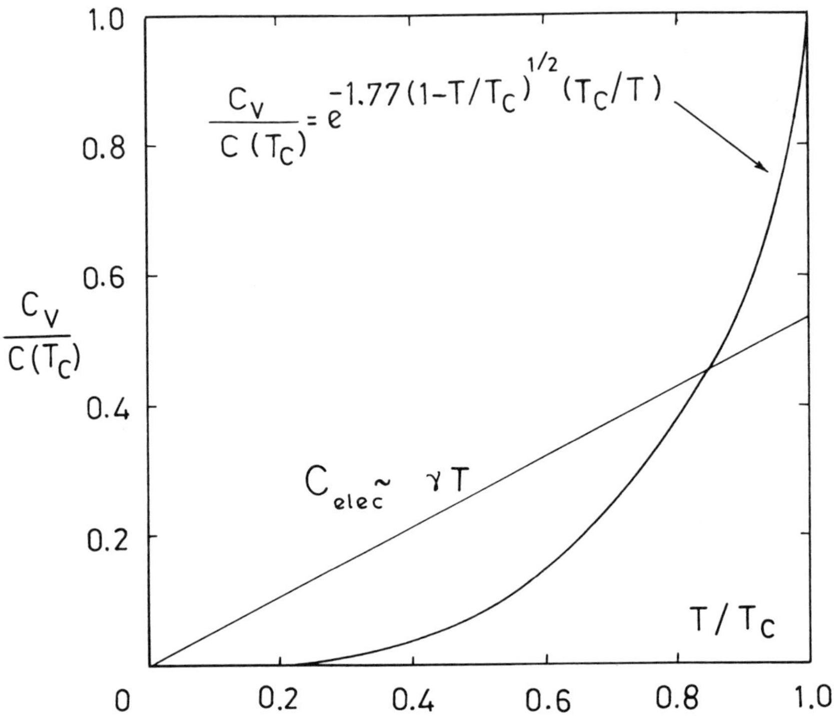

Fig. G.1: The specific heat as a function of temperature for two levels shows a peak at T_s.

The specific heat is defined by $C_v = \partial < \epsilon > / \partial T$ at a constant volume so that,

$$C_v = \left(\frac{\epsilon}{k_B T}\right)^2 \frac{\exp(\epsilon/k_B T)}{[\exp(\epsilon/k_B T) + 1]^2}. \qquad (G.33)$$

This specific heat peak due to impurity levels is called the Schottky anomaly. In solids, there is a gap in the dispersion relation due to the anisotropy. The specific heat then shows a Schottky type anomaly due to the gap in the dispersion. In superconductors there is a gap due to pairing. Hence anomaly in the specific heat occurs near the transition temperature.

G.2. Superconducting state.

The transition from normal to the superconducting state is accompanied by an anomaly in the temperature dependence of the specific heat due to the gap in the dispersion relation. Far below the transition temperature, the specific heat varies as,

$$C_v \propto \exp(-\Delta/k_B T) \tag{G.34}$$

where 2Δ is the gap.

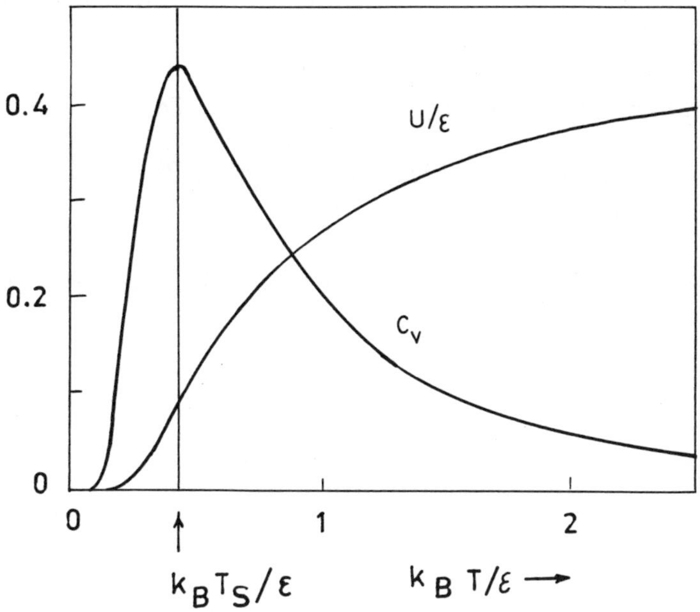

Fig.G.2: The specific heat of a superconductor for $T \ll T_c$ varies exponentially as $\exp(-\Delta/k_B T)$.

In Chapter 3, we have shown that the flux lattice melts. This melting is reflected by the intensity of the microwave absorption where liquid like minimum in the absorption intensity as a function of temperature is found. In the critical current, the flux lattice melting is associated with the minimum at the flux-lattice melting temperature. If there is a Josephson energy, J, in the xy plane and another value along the z direction, and a magnetic field is applied the two-dimensional vortex liquid orders at T_{FM} into a lattice of flux lines along the c axis and of hexagonal symmetry in the ab-plane via a first-order thermodynamic phase transition. Within a small range of temperature near the flux-lattice melting, the specific heat varies as, $C_v \propto (< H^2 > - < H >^2)/k_B T^2$.

Suggested References

G.1. P. Kumar, D. Hall and R.G. Goodrich, Phys. Rev. Lett. **82**, 4532 (1999).

G.2. X. Hu, S. Miyashita and M. Tachiki, Phys. Rev. Lett. **79**, 3498 (1997).

G.3. S.H. Simon and P.A. Lee, Phys. Rev. Lett. **78**, 1548 (1997).

G.4. D.F. Agterberg, T.M. Rice and M. Sigrist, Phys. Rev. Lett. **78**, 3374 (1999).

G.5. V.B. Geshkenbein, L.B. Ioffe and A.J. Millis, Phys. Rev. Lett. **80**, 5778 (1998).

Appendix H: Quantum Hall effect

H.1. Hall effect.

In a semiconductor, if an electric field is applied along the x direction and a magnetic field is applied along the z direction, a current appears along the y direction. This current is called the Hall current. While performing the experiment, a Hall probe is made from a thin sheet in the $x-z$ plane of the semiconductor and the voltage is measured along the y direction so that the resistivity along the y direction is small and a large voltage is developed along this direction. This voltage is called the Hall voltage. The resistivity of this system is obviously very anisotropic. The force on the electrons, is, $m\frac{d\vec{v}}{dt} = \frac{d\vec{p}}{dt} = \hbar\frac{d\vec{k}}{dt}$ where m is the mass of the electron, v is the velocity and hence dv/dt is the acceleration. The linear momentum is, $\vec{p} = \hbar\vec{k}$, and $\vec{k}$ is the wave vector. The charge of the electron is $-e$ and hence the Lorentz force is $-e[\frac{1}{c}\vec{v}\times\vec{B}]$. The electric voltage applied is E so that the total force on the electron becomes,

$$F = \hbar\frac{d\vec{k}}{dt} = -e[\vec{E} + \frac{1}{c}\vec{v}\times\vec{B}]\,. \tag{H.1}$$

When $B = 0$,

$$\hbar d\vec{k} = -eEdt\,, \tag{H.2}$$

the electric voltage is applied for a certain time t during which the momentum changes so that the wave vector changes from $j(o)$ to $k(t)$. Integrating the above equation,

$$\hbar\int_{k(o)}^{k}(t)dk = -eE\int_{o}^{\tau}dt \tag{H.3}$$

leads to

$$k(t) - k(o) = -eE\tau/\hbar\,. \tag{H.4}$$

Thus the electric voltage changes the wave vector by the amount $\delta k = k(t) - k(o)$ or the linear momentum,

$$m\vec{v} = \hbar\delta\vec{k} = -eE\tau/\hbar\,. \tag{H.5}$$

Therefore, the velocity of the electron upon the application of an electric voltage becomes,

$$v = -eE\tau/m \,. \tag{H.6}$$

The current due to the electric field is then

$$j = nev = ne^2E\tau/m \,. \tag{H.7}$$

According to Ohm's law,

$$j = \sigma E \qquad \text{or} \qquad E = j(1/\sigma) \,. \tag{H.8}$$

Here n is the number of electrons per unit volume and σ is the conductivity. The inverse of the conductivity is the resistivity.

$$\sigma = \frac{j}{E} = \frac{ne^2\tau}{m} \,. \tag{H.9}$$

We assume that the magnetic field is along the z direction so that $B = B_z$ and $B_x = B_y = 0$. Then many terms in the cross product in (H.1) become zero,

$$-\frac{e}{c}\vec{v} \times \vec{B} = -\frac{e}{c}\begin{vmatrix} i & j & k \\ v_x & v_y & v_z \\ 0 & 0 & B_z \end{vmatrix} \,, \tag{H.10}$$

so that we can write the three components of the left hand side of (H.1) as,

$$\begin{aligned} m\left(\frac{d}{dt} + \frac{1}{\tau}\right) v_x &= -e\left[E_x + \frac{1}{c}v_y B_z\right] \\ m\left(\frac{d}{dt} + \frac{1}{\tau}\right) v_y &= -e\left[E_y - \frac{1}{c}v_x B_z\right] \\ m\left(\frac{d}{dt} + \frac{1}{\tau}\right) v_z &= -eE_x \,. \end{aligned} \tag{H.11}$$

In the steady state, in a static electric field, the time derivatives are zero so that,

$$\begin{aligned} v_x &= -\frac{e\tau}{m}E_x - \omega_c \tau v_y \\ v_y &= -\frac{e\tau}{m}E_y + \omega_c \tau v_x \end{aligned}$$

$$v_z \;=\; -\frac{e\tau}{m}E_z \tag{H.12}$$

where $\omega_c = eB_z/mc$ is the cyclotron frequency. We have already noted that the applied electric field is along x direction. Therefore $\vec{j}$ is also along the x direction. The applied field is along the z direction. Therefore, the Hall field is along $\vec{j} \times \vec{B}$. We set $v_y = 0$ so that from the second relation in (H.12) we obtain E_y in which we substitute v_x of the first relation so that, upon using the expression for the cyclotron frequency, we obtain,

$$E_y = -\frac{eB_z\tau}{mc}E_x \; . \tag{H.13}$$

From (H.7) we can write $j_x = ne^2\tau E_x/m$. Therefore, we define the Hall coefficient as,

$$R_H = \frac{E_y}{j_x B_z} = -\frac{1}{nec} \; . \tag{H.14}$$

Therefore, by measuring B, E_y and j_x, we can determine n, the concentration of electrons. From a graph of y voltage as a function of magnetic field along z direction, we can measure the carrier concentration depending on the type of semiconductor used. We can write the right-hand side of (H.14) as,

$$\frac{E_y}{j_x} = -\frac{1}{nec}B_z \; . \tag{H.15}$$

Since E_y is along y direction and the current j_x is along the x direction, the usual Ohm's law is not applicable. Therefore, we define a quantity, called the Hall resistance as,

$$\rho_H(x,y) = -\frac{B_z}{nec}. \tag{H.16}$$

When the electron concentration is n_1 and the hole concentration is n_2, we write the equation (H.15) as

$$\sigma_{yx} = \frac{(n_2 - n_1)ec}{B_z} = \frac{n_s ec}{B_z} \tag{H.17}$$

from which it is clear that a plot of transverse conductivity as a function of field determines the concentration of carriers.

H.2. Quantization.

Usually the linear momentum of the electron, p is written as $\vec{p} - \frac{e}{c}\vec{A}$ when electromagnetic field is considered. Here $\vec{A}$ is the vector potential of the electromagnetic field. We assume that $B = B_z$, $A = A_x = yB_z$ so that the x component of the kinetic energy is changed to

$$\frac{1}{2m}p_x^2 = \frac{\hbar^2}{2m}\left[\frac{1}{i}\frac{\partial}{\partial\, x} - \frac{eB_z y}{\hbar c}\right]^2 . \qquad (H.18)$$

In the y component of p, we need not consider the magnetic field so that the y component is

$$\frac{1}{2m}p_y^2 = -\frac{\hbar^2}{2m}\frac{\partial^2}{\partial y^2} \qquad (H.19)$$

for the electrons in the xy plane, the Schrödinger equation may be written as,

$$\frac{\hbar^2}{2m}\left\{\left[\frac{1}{i}\frac{\partial}{\partial x} - \frac{eB_z y}{\hbar c}\right]^2 - \frac{\partial^2}{\partial y^2}\right\}\psi = E\psi \qquad (H.20)$$

where there is no potential term in $(calH_o + V)\psi = E\psi$. We take a plane wave solution of the Schrödinger equation as,

$$\psi = e^{ikx}\varphi(y) . \qquad (H.21)$$

Substituting (H.21) into (H.20) we find

$$\frac{\hbar\omega_c}{2}\left\{-l^2\frac{\partial^2}{\partial y^2} + \left(\frac{y}{l} - lk\right)^2\right\}\varphi = E\varphi \qquad (H.22)$$

where we have introduced the magnetic field measured as a distance,

$$l = \left(\frac{\hbar c}{eB}\right)^{1/2} . \qquad (H.23)$$

The equation (H.22) is that of a harmonic oscillator except that the potential has been shifted by shifting the value of y. Therefore, the solution of this equation has Hermite's polynomials in the eigenfunctions and the eigen values are equally spaced. These are called the Landau levels defined by

$$E_{nk} = \hbar\omega_c\left(n + \frac{1}{2}\right) \qquad (H.24)$$

independent of k. The wave functions are

$$\varphi_{nk}(y) \propto H_n(y/l - lk)\exp[-(y - l^2k)^2/2l^2] \qquad (H.25)$$

which are centered at $y = l^2k$. In a real semiconductor, the value of y is limited by the condition $0 < y < W$ so that $0 < k < W/l^2$ and there are periodic boundary conditions along x, i.e., $\psi(x,y) = \psi(x + L, y)$. The wave vector is $k = 2\pi/L$ which can be written as $k = 2\pi p/L$ where p is an integer. Therefore, the total number of states N in a Landau level is $LW/2\pi l^2$. The number of states per unit area of a full Landau level is,

$$n_B = \frac{1}{2\pi l^2} = \frac{eB}{hc}. \qquad (H.26)$$

From (H.15) we write the Hall voltage as,

$$E_y = -\frac{B}{nec} j_x . \qquad (H.27)$$

If i energy levels are fully occupied, $n = in_B$, then,

$$\rho_H = \frac{B}{in_B ec} = \frac{h}{ie^2} \qquad (H.28)$$

where i is an integer, $i = 1, 2, \cdots$. Under these circumstances ρ_{xx} becomes zero because there is no scattering and ρ_{xy} exhibits quantized plateaus in the Hall resistance. This phenomenon is called the quantum Hall effect. The fine structure constant α is a measure of interaction between matter and the electromagnetic field and it is important for the study of quantum electrodynamics. It can be expressed in terms of h/e^2 and hence is directly related to the Hall resistance as,

$$\alpha = \frac{\mu_o ce^2}{2h} , \qquad (H.29)$$

where μ_o is the permeability of vacuum and is exactly equal to $4\pi \times 10^{-7}$ Hm^{-1}. The speed of light is c =299 792 458 m/s±0.0004 ppm. Therefore,

$$\rho_H = \frac{h}{e^2 i} = \frac{25812.80}{i} \quad \text{Ohms}. \qquad (H.30)$$

According to quantum electrodynamic calculations, $\alpha^{-1} = 137.035\ 993(10)$ (±0.07 ppm). The value measured in various laboratories is $\alpha^{-1} = 137.035988$

$\pm$ 0.00002 and the internationally recommended value is $\alpha^{-1} = 137.03604$ $\pm$ 0.00011. Therefore, the measurement of the plateaus in the ρ_{xy} as a function of magnetic field are considered to be of great importance for the accurate measurement of e^2/h. The expression (H.18) shows as to how the magnetic field appears when the linear momentum of the electron is corrected for the effect of the electromagnetic field. The observation of the quantum Hall effect and the measurement of e^2/h shows that eq.(H.18) correctly predicts the effect of the electromagnetic field on the momentum of the electron. At this time it is not clear as to what is the effect of the spin of the electron on this relation. The effect of orbital angular momentum is also not determined. However, the accuracy with which e^2/h has been obtained clearly shows that the spin as well as the orbital motion of the electron does not affect the field or the effect occurs in such a way that the multiplier due to these effects is equal to unity. For $i = 1$ in (H.30) the effect of spin and orbital motion of the electron is also unity and hence not found.

H.3. Fractional Quantum Hall effect.

According to the expression (H.28), the xy resistivity is linearly dependent on the magnetic field and it is quantized in units of h/e^2 in integer multiplies, so that the conductivity becomes

$$\sigma_{xy} = i\frac{e^2}{h} \qquad (H.31)$$

where i is an integer. It was found by Tsui, Störmer and Gossard that i need not be an integer. While measuring the transverse resistivity as a function of field, a plateau was found at $i = 1/3$. Later studies found many different fractions which established that fractions of e^2/h occur and these observations gave rise to the terminology, fractional-quantized Hall effect, (FQHE).

At this time Shrivastava discovered the correct series of fractions. It was proposed that there occur energy states below the Landau levels. These states occur at a fraction of the cyclotron frequency. We consider the spin-orbit interaction so that the Landé g factor for the angular momentum $j = l \pm \frac{1}{2}$ is found to be,

$$g = (2l + 1 \pm 1)/(2l + 1) \,. \tag{H.32}$$

For $s = -\frac{1}{2}$ the g value is

$$g_- = 2l/(2l + 1) \tag{H.33}$$

and for $s = +\frac{1}{2}$ the g-value is

$$g_+ = 2(l + 1)/(2l + 1) \,. \tag{H.34}$$

The energy of the states in a magnetic field B is given by

$$\mathcal{H} = g\mu_B \vec{B}.\vec{J} \tag{H.35}$$

so that the resonance frequency is,

$$\hbar\omega_c = g\mu_B B \tag{H.36}$$

where $\mu_B = e\hbar/2mc$ is the Bohr magneton and hence

$$\omega_c = \frac{1}{2} geB/mc \tag{H.37}$$

is the cyclotron frequency. Usually for electrons in the ${}^2S_{1/2}$ state the orbital angular momentum is zero, spin 1/2 and spin degeneracy 2, the g value is equal to 2 except for a small electrodynamic correction which makes it about 2.0023, the factor $\frac{1}{2}g$ being of the order of unity is missing from the expression (H.37) for the cyclotron frequency. The occurrence of the orbital states gives rise to $\frac{1}{2}g$ different from unity. The peaks in the specific heat correspond to the energy levels as we expect from the analogy of the present problem with the Schottky anomaly. Similarly, when a parameter is varied, the Hall conductance shows the existence of levels. We find that there occur peaks in the transverse resistivity at magnetic fields determined by the resonance condition. Similarly, the Hall resistivity shows steps at the same field as that at resonance. Therefore, the steps occur at a fraction of cyclotron frequency. The fraction $\frac{1}{2}g$ is given in Table H.1. First of all we find that fractions occur in Kramers conjugate pairs. Let us now look at the Table H.1. At $l = 0, \frac{1}{2}g_- = 0$ and $\frac{1}{2}g_+ = 1$ so that only one resonance occurs at the Landau's first level. Now for $l = 1$, there is a resonance at $\frac{1}{2}g_- = \frac{1}{3}$. Indeed, it corresponds to a new resonance which can be considered to be arising from a $\frac{1}{3}$ filled Landau level. The next number in our table is $\frac{2}{3}$ for $l = 1$. The resonance at this value has also been found by Tsui et al. For $l = 2$, the fractions discovered are $\frac{2}{5}$ and $\frac{3}{5}$ both of which are seen in the experiments performed by Stormer et al. For $l = 3$ and 4 the fractions predicted are $\frac{3}{7}, \frac{4}{7}$ and $\frac{4}{9}$ and $\frac{5}{9}$ which have been obsered by Chang et al. Thus the values predicted by (H.33) and (H.34) are in accord with the experimental data. The series correctly predict precisely those fractional quantized Hall states in the Landau levels that are seen experimentally in the correct order of stability. The fractions predicted by Shrivastava are identical to those experimentally found by Clark et al and Du et al.

Table H.1: The calculated values of the rational numbers at which plateaus occur in the fractionally quantized Hall effect.

S.No.	$g_-/2 = l/(2l+1)$	$g_+/2 = (l+1)/(2l+1)$	Expt.
0	0	1	[H.1]
1	1/3	2/3	[H.3]
2	2/5	3/5	[H.5]
3	3/7	4/7	[H.6]
4	4/9	5/9	[H.6]
5	5/11	6/11	[H.7]

H.4. Chern-Simons transformation.

If ψ_e is a solution of the Schrödinger equation

$$\mathcal{H}_e\psi_e = E\psi_e \tag{H.38}$$

then ϕ is a solution of $\mathcal{H}\phi = E\phi$ with

$$\mathcal{H} = \sum_j \frac{1}{2m_b}\left[p_s + \frac{e}{c}A(r_j) - \frac{e}{c}a(r_j)\right]^2 + \sum_{i<j} v(r_i - r_j) \tag{H.39}$$

where $a(r_j)$ is the Chern-Simons vector potential,

$$a(r_i) \;=\; i\nabla_i\left[\prod_j \exp\{i\tilde{\phi}\theta(r_i - r_j)\}\right]$$

$$= \frac{\tilde{\phi}\phi_o}{2\pi} \sum_{j=1}^{N} \frac{\hat{z} \times (r_i - r_j)}{|r_i - r_j|^2} \tag{H.40}$$

and

$$\tilde{\phi} = 2m \ . \tag{H.41}$$

There is a singularity at the position of the electron,

$$\nabla \times a(r) = \tilde{\phi}\phi_o \delta(r - r_j) \, . \tag{H.42}$$

The magnetic field $b(r)$ associated with the vector potential $a(r)$ is given by,

$$b(r) = \nabla \times a(r) = \phi_o \tilde{\phi} n(r) = 2\pi \tilde{\phi} n(r) \tag{H.43}$$

where $n(r) = \sum_j \delta(r - r_j)$ is the particle density at $r = r_j$. This means that in three dimensions, $d = 3$, one can add a term $\epsilon_{ijk} A_i F_{jk}$ which is gauge and Lorentz invariant. This is the way of attaching flux tubes to the particles. The Chern-Simons transformation can be described as the exact modeling of an electron as a fermion attached to $\tilde{\phi} = 2m$ flux quantua. These fermions are called Chern-Simons or composite fermions (CF). For $\tilde{\phi} = 2m + 1$, the composite bosons emerge (CB) and for fractional values of $\tilde{\phi}$, the composite particles are called anyons. According to the Chern-Simons theory the flux tubes are independent of the angular momenta, L or S.

H.5. Particle-hole symmetry.

Yeh et al have performed the experimental work near the filling factor $\nu = \frac{1}{2}$ in support of the simplest CFs consisting of two flux quanta attached to each electron (^{2}CFs). The sequence of fractional quantum Hall effect states $\nu = p/(2p + 1)$, which is the same as Shrivastava's series given in Table H.1, converging at $\nu = \frac{1}{2}$ for $p \to \infty$ determines an effective mass. Yeh et al thought that the mass originates from the electron-electron interactions. However Yoshioka has pointed out that Yeh et al's interprestation is incorrect because the particles of equal mass did not exhibit the particle-hole symmetry. The four-flux-quanta $\tilde{\phi} = 4$ composite fermions are called ^{4}CFs. The experimental work of Yeh et al found that $m^*(\nu = 5/7)/\sqrt{B_\perp} = 0.25 \pm 0.02$ for ^{4}CF and $m^*(\nu = 8/5)/\sqrt{B_\perp} = 0.25 \pm 0.02$ for ^{2}CF. Therefore, the ^{2}CF and ^{4}CF masses are equal. Therefore, there is an effort to understand this equality of masses by Shrivastava (H.12).

The ^{4}CF quantum Hall effect occurs at the electron filling factor $\nu_p = (3p+1)/(4p+1)$ with $p = \pm 1, \pm 2, \cdots$, etc. For $p = +1$ it gives 4/5 and for $p = -1$, it gives 2/3. Yeh et al have assumed that $\nu_h = p/(4p+1)$ is the particle-hole conjugate of ν_p. For $p = 1$, we have $\nu_h = 1/5$ and for $p = -1$, we get $\nu_h = 1/3$. Therefore,

$$\nu_p + \nu_h = 1 \tag{H.44}$$

is required for the fractions to be particle-hole conjugates. For $p = 1$, we obtain $\nu_p = 4/5$ and $\nu_h = 1/5$ and the sum is unity, $4/5 + 1/5 = 1$. Similarly, for $p = -1$, $\nu_p = 2/3$ and $\nu_h = 1/3$, and the sum is again unity $2/3 + 1/3 = 1$. Hence,

$$[(3p+1)/(4p+1)] + [p/(4p+1)] = 1 \tag{H.45}$$

is correctely established for all the values of p. Here p is only a number which is not linked with any physical property of the electrons. If the spin and the charge are decoupled, ν determines the charge but not the spin. Therefore, we are free to consider the spin degrees of freedom. In the ordinary concept, the electrons and holes are created by photons. Therefore, spin of ν_p should be the same as that of the ν_h, which means that

$$\nu_{p\uparrow} + \nu_{h\uparrow} = 1$$

and

$$\nu_{p\downarrow} + \nu_{h\downarrow} = 1\ . \tag{H.46}$$

Therfore, the expression (H.44) should add up to 2. According to Yoshioka [H.11], the above expressions should be corrected as

$$\begin{aligned} \nu_p &= (3p+1)/(4p+1) \\ \nu_h &= 1 + [p/(4p+1)] \end{aligned} \tag{H.47}$$

so that for $p = 1$, $\nu_p = 4/5$ and $\nu_h = 6/5$ and $\nu_p + \upsilon_h = 2$. Similarly, for $p = -1$, $\nu_p = 2/3$ and $\nu_h = 4/3$ so that $\nu_p + \nu_h = 2$, which accounts for the spin degrees of freedom. It is obvious that there is another alternative to (H.47) such as,

$$\begin{aligned} \nu_p &= 1 + [(3p+1)/(4p+1)] \\ \nu_h &= p/(4p+1) \end{aligned} \tag{H.48}$$

which gives the correct sum

$$\nu_p + \nu_h = 2 \tag{$H.49$}$$

Actually several particle-hole conjugate states occur. Since the spin is decoupled from the values of ν_p and ν_h, which determine the charge, we have $\nu_p + \nu_h = 2$ for the conjugate states. However, it is not necessary to decouple the spin and charge. The spin-charge coupled theory gives the correct conjugate states. According to Yoshioka's expression (H.47) for $p = 1$, 4/5 and 6/5 are the particle-hole conjugate states. We shall derive these fractions without the use of spin-charge decoupling. Due to spin-charge decoupling, there is the spin degree of freedom which leads to $\nu_p + \nu_h = 2$. We show that when spin is used to determine the value of ν which gives the charge of the quasiparticles, $\nu_p + \nu_h = 1$ is reestablished. Usually at finite temperatues, the number of fermions present in a semiconductor at a wave vector k is given by

$$f_k^{(e)} = 1/[\{\exp(\epsilon_k - \epsilon_F)/k_B T\} + 1] \tag{$H.50$}$$

where ϵ_F is the Fermi energy of the electrons. Similarly, the Fermi distribution for holes is given by $f_k^{(h)}$ and the electron-hole symmetry is established by $f_k^{(e)} + f_k^{(h)} = 1$. We show that the correct theory used 1 and not 2 in the sum, $f_k^{(e)} + f_k^{(h)}$, for the prediction of correct fractions. The spin-charge coupling theory gave the charge as [H.4],

$$(1/2)g = \left(\frac{1}{2}\right)(2l + 1 \pm 1)/(2l + 1) \tag{$H.51$}$$

where ± 1 is a result of spin= $\pm\frac{1}{2}$. The above is written as

$$(1/2)g_- = l/(2l+1) \text{ and } (1/2)g_+ = (l+1)/(2l+1). \qquad (H.52)$$

The holes are produced in the theory by changing the sign of the charge so that, when electrons are negatively charged, the holes are positively charged and g_- and g_+ play the role of particle-hole conjugates. For various values of l, the values of $g_-/2$ and $g_+/2$ are given in Table H.1. Now all $g_-/2$ are quasiparticles and $g_+/2$ are quasiholes. Using formula (H.47) for $p = 1$, 4/5 and 6/5 are found to be conjugate states by Yoshioka. According to Table H.1 for $l = 2$, 2/5 and 3/5 are conjugate states when in $n = 2$ level, have apparent charges as $2\times2/5$ and $2\times3/5$ which means that 4/5 and 6/5 agree with those of Yoshioka given above. We introduce the Landau level quantum number so that the charge is multiplied by 2. The quasiparticles of charges 2/5 and 3/5 when in $n = 2$ levels, have apparent charges as $2\times2/5$ and $2\times3/5$ which means that 4/5 and 6/5 agree with those of Yoshioka given above. However, $2/5 + 3/5 = 1$ is maintained. When the same quasiparticles are in different Landau levels, they continue to have, the same charge and the particle-hole symmetry is established. The spin is used in defining the charge so that in the sum rule given by (H.49), the value of 2 need not be used and the fractions add up to 1 only. Thus our conjugate states agree with those of Yoshioka. Thus $g_-/2$ values given by Table H.1 are the particle-hole conjugates of $g_+/2$. The values 1/3, 2/5, 3/7, 4/9 agree with those seen in Fig. 2 of Eisenstein and Stpormer. The particle-hole conjugates of these states are 2/3, 3/5, 4/7 and 5/9 respectively. These predicted conjugate states are also in full agreement with those experimentally found. In all the cases, the sum of the two fractions is equal to 1 and not 2 as in Yoshioka's case.

Table H.2 shows the multiplication by Landau level number, n.

n	l (quasiparticles)	$g_-/2$ (quasiholes)	$g_+/2$
1	2	2/5	3/5
2	2	4/5	6/5

For $l \gg 1$, from the first series in (H.52), the limiting value is $(1/2)g_- = 1/2$. Similarly, for large values of l, the second series of (7) also gives 1/2. In the experimental measurements of Eisenstein and Stormer of Hall resistance as a function of magnetic field, one series which we call as the quasiparticle series, approaches the value of 1/2 starting from 1/3 and the other series, which may be called the quasihole series also approaches the value of 1/2 starting from 2/3. Thus the position of 1/2 on the side of increasing field as well as on the side of decreasing field, there are the two series of fractions, one representing quasiparticles and the other quasiholes. The entire pattern of the two series with 1/2 in the centre is repeated many times in the resistivity as a function of field. The patterns are centered at $n/2$ for n as the Landau level quantum number. The patterns with 1/2, 2/2, 3/2, 4/2, 5/2, 6/2, 7/2, 8/2 are all clearly seen in the data of Fig. 1 of Yeh et al [H.10].

The $\nu = \frac{1}{2}$ is approached by taking the $l \to \infty$ limit from both of the expressions in (H.52). This means that there is a quasiparticle of charge $-\frac{1}{2}$ as well as a quasihole of charge $+\frac{1}{2}$. Thus the excitation of $\nu = \frac{1}{2}$ seen from a naive viewpoint is electrically neutral. Actually, there is a problem of mapping the charge since the coordinates of the $+\frac{1}{2}$ charge are not the same as that of $-\frac{1}{2}$. It is clear that we may say that $\nu = \frac{1}{2}$ is an electrical dipole and hence is compressible. Therefore, for small wave vector the response function is linear in q. The excitation has a finite life time and the positively charged quasihole will be annihilated by the negatively charged quasiparticle. Therefore, the response function will have a resonant shape and hence will cease to be linear in q.

H.6. High Landau levels.

At large values of the Landau level quantum number, n, there is much less structure in the diagonal resistivity as a function of field than at low values of n. The rich structure of fractions of the fractionally quantized Hall effect, FQHE, found at small values of n is vertually absent for large values of n; only fragile and poorly understood states at Landau filling fractions $\nu = 7/3, 5/3$ and $8/3$ are seen in the best samples. In the previous pages we have seen that 4/5 and 6/5 have equal masses because they are particle-hole conjugates. We now interpret the fractions 7/3, 5/2 and 8/3 seen in the measurements carried out by Eisenstein et al. We find that 7/3 and 8/3 are particle-hole conjugates and 5/2 is the $n = 5$ state of the level at 1/2. When $l = \infty$ both the above series approach 1/2 except that one series approaches from the right hand side and the other from the left hand side exactly as observed, [H.7, H.13]. The left and the right side approaches arise from the Kramers conjugate states and the predicted approach is exactly as observed. Since the limit is involved there is a Fermi surface at 1/2. However, we can shift the Fermi surface to higher values when higher Landau levels are occupied. The fraction 1/2 becomes $n/2$ with n as the Landau level quantum number. Thus 1/2, 3/2, 5/2, 7/2, 9/2 etc become allowed. This predicted feature with an odd numerator with 2 in the denominator is also exactly as observed. For $n = 1$ we have two series one merging from the left while the other merging from right at 1/2 and the same picture is repeated for different values of n. The entire pattern of pairwise series is observed exactly as predicted. Many fractions with 2 in the denominators have been observed [H.16]. At the magnetic field of 4 to 5 Tesla only a small number of fractions are observed, the strongest ones are at 8/3, 5/2 and 7/3. Since there is a charge versus Landau level quantum number degeneracy, it is not possible to distinguish large charge from a large Landau level quantum number. In the series (H.52) for $l = 7$ two values 7/15 and 8/15 are predicted and for $l = \infty$ value is 1/2. When the same particle occurs in different levels its charge remains unchanged. We can multiply the values by $n = 5$ so that the predicted values of 1/2, 7/15 and 8/15 become 5/2, 7/3 and 8/3. These predicted values are exactly the same as those observed experimentally by Eisenstein et al [H.17]. Thus 7/3 is the particle-hole conjugate of 8/3 as seen in Table H.3. The cyclotron frequency is $\omega_c = (1/2)g_e B/mc = \nu eB/mc$. Therefore, we can replace the charge by an effecting charge $e_{eff} = (\frac{1}{2})ge = \nu e$. Alternatively we can replace the field by an effective field $B_{eff} = (\frac{1}{2})geB = \nu B$ so that the field is very much reduced. In the case of $\nu\frac{1}{3}$, the field experienced is $(1/3)B$ which is reduced from B.

Table H.3. High Landau levels

n	l	$g_-/2$	$g_+/2$	$ng_-/2$	$ng_+/2$
5	7	7/15	8/15	7/3	8/3
5	∞	1/2	1/2	5/2	5/2

The problem of Chern-Simons is that of finding the vector potential and it is clear that their correction is independent of spin and orbital effect. Our finding is that the vector potential of the electromagnetic field is to be corrected in a manner which depends on the angular momentum. The linear momentum of the electron in an electromagnetic field changes from p to $p - (e/c)A$. According to Chern-Simons it is to be changed as in eq.(H.39) where gauge invariance is maintained. Our theory of the fractional quantum Hall effect suggests that p is to be corrected as

$$p \to p - \frac{l+1 \pm 1}{2l+1}\frac{e}{c}A \qquad (H.53)$$

where $\pm$ has arisen from spin. This is quite different from that of Chern-Simons because the coefficient of the vector potential depends on l and s. However, the above expression is in full accord with a lot of experimental measurements.

The resistivity ρ_{xx} is zero when plateaus occur in ρ_{xy}. The Cooper pairs have spins $s_1 = \frac{1}{2}$, $s_2 = -\frac{1}{2}$ and hence $S = s_1 + s_2 = 0$. In the quantum Hall effect, the spin and the orbital angular momenta are combined in a more complicated manner. Therefore, FQHE is an extension of the phenomenon of superconductivity.

Suggested References

H.1 K.v. Klitzing, G. Dorda and M. Pepper, Phys. Rev. Lett. **45**, 494 (1980).

H.2 K.v. Klitzing, Rev. Mod. Phys. **58**, 519 (1986).

H.3 D.C. Tsui, H.L. Störmer and A.C. Gossard, Phys. Rev. Lett. **48**, 1559 (1982).

H.4 K.N. Shrivastava, Phys. Lett. A**113**, 435 (1986); ibid, A**115**, 459 (1986) (E).

H.5 H.L. Störmer, A. Chang, D.C. Tsui, J.C.M. Hwang, A.C. Gossard and W. Wiegmann, Phys. Rev. Lett. **50**, 1953 (1983).

H.6 A.M. Chang, P. Berglund, D.C. Tsui, H.L. Stormer and J.C.M. Hwang, Phys. Rev. Lett. **53**, 997 (1984).

H.7 R. Willett, J.P. Eisenstein, H.L. Störmer, D.C. Tsui, A.C. Gossard and J.H. English, Phys. Rev. Lett. **59**, 1776 (1987)

H.8 R.R. Du, H.L. Störmer, D.C. Tsui, L.N. Pfeiffer and K.W. West, Phys. Rev. Lett. **70**, 2944 (1993).

H.9 R.G. Clark, R.J. Nicholas, A. Vsher, C.T. Foxon and J.J. Harris, Surf. Sci. **170**, 141 (1986).

H.10 A.S. Yeh, H.L. Störmer, D.C. Tsui, L.N. Pfeiffer, K.W. Baldwin and K.W. West, Phys. Rev. Lett. **82**, 592 (1999).

H.11 D. Yoshioka, Phys. Rev. Lett. **83**, 886 (1999).

H.12 K.N. Shrivastava, Mod. Phys. Lett. B**13**, 1087 (1999).

H.13 J.P. Eisenstein and H.L. Stormer, Science **248**,1511(1990).

H.14 B.I. Halperin and A. Stern, Phys. Rev. Lett. **80**, 5457 (1998).

H.15 A. Stern, B.I. Halperin, F.v. Oppen and S.H. Simon, Phys. Rev. B**59**, 12547 (1999).

H.16 M.P. Lilly, K.B. Cooper, J.P Eisenstein, L.N. Pfeiffer and K.W. West, Phys. Rev. Lett. **83**, 824 (1999).

H.17 J.P Eisenstein, M.P. Lilly, K.B. Cooper, L.N. Pfeiffer and K.W. West, Cond-mat/9909238; 13th International Conference on the electronic properties of two dimensional systems, Ottawa, Canada, 1999.

H.18 S.S. Chern and J. Simons, Proc. Nat. Acad. Sci. (USA) **68**, 791 (1971).

A

B

C

D

E

F

G

H

I

J

K

L

M

S

T

T

V

Y

Z

सा विद्या या विमुक्तये

That **is (true) knowledge which conduces to final liberation or spiritual emancipation.**